An introduction to machinery reliability assessment

Van Nostrand Reinhold Plant Engineering Series

An introduction to machinery reliability assessment

H. P. Bloch
F. K. Geitner

VNR VAN NOSTRAND REINHOLD
New York

Library of Congress Catalog Card Number 89-5267
ISBN 0-442-23279-9

Printed in the United States of America

Van Nostrand Reinhold
115 Fifth Avenue
New York, New York 10003

Van Nostrand Reinhold International Company Limited
11 New Fetter Lane
London EC4P 4EE, England

Van Nostrand Reinhold
480 La Trobe Street
Melbourne, Victoria 3000, Australia

Nelson Canada
1120 Birchmount Road
Scarborough, Ontario
Canada M1K 5G4

16 15 14 13 12 11 10 9 8 7 6 5 4 3 2

Library of Congress Cataloging-in-Publication Data

Bloch, Heinz P., 1933–
    An introduction to machinery reliability assessment / H. P. Bloch
  F. K. Geitner
        p.   cm.   ——   (VNR plant engineering series)
    Includes index.
    ISBN 0-442-23279-9
    1. Machinery—Reliability.   I. Geitner, Fred K.   II. Title.
  III. Series.
  TJ153.B582   1990
  621.8′1—dc20   89-9097
                                                                                    CIP

Contents

TO THE GREAT TEACHER

Preface

The profitability of modern industrial and process plants is significantly influenced by the reliability and maintainability of the machines applied in their numerous manufacturing processes and support services. These machines may move, package, mold, cast, cut, modify, mix, assemble, compress, squeeze, dry, moisten, sift, condition, or otherwise manipulate the gases, liquids, and solids which move through the plant or factory at any given time. To describe all imaginable processing steps or machine types would, in itself, be an encyclopedic undertaking and any attempt to define how the reliability of each of these machine types can be assessed is not within the scope of this text.

However, large multinational petrochemical companies have for a number of years subjected such process equipment as compressors, extruders, pumps and prime movers, including gas and steam turbines, to a review process which has proven cost-effective and valuable. Specifically, many machines which were proposed to petrochemical plants during competitive bidding were closely scrutinized and compared in an attempt to assess their respective strengths and vulnerabilities, quantify the merits and risks of their respective differences, and finally to combine subjective and objective findings in a definitive recommendation. This recommendation could take the form of an unqualified approval, or perhaps a disqualification of the proposed equipment. In many cases, the assessment led to the request that the manufacturer *upgrade* his machine to make it meet the purchaser's objectives, standards, or perceptions.

This text outlines the approach which should be taken by engineers wishing to make similar reliability assessments for any given machine. It is by no means intended to be an all-encompassing "cook book" but aims, instead, at highlighting the principles which over the years have worked well for the authors. In other cases, it gives typical examples of what to look for, what to investigate with somewhat cursory screening studies, and when to go back to the equipment manufacturers with questions or an outright challenge.

We wish to acknowledge the constructive suggestions received from John W. Dufour and Dr Helmut G. Naumann, who reviewed our manuscript. Their helpful comments certainly improved this text.

An introduction to machinery reliability assessment

PART I

Probabilistic approaches

1

Scope and limitations

Everywhere we look we notice a re-emergence and a new emphasis of the quality idea. We see how workers are encouraged to collect data and convert it to information using statistical quality control (SQC) methods in order to manage their affairs [1]. Quality and reliability have three important things in common. First, one is not thinkable without the other (see Fig. 1.1). Second, both are old and basic concepts and are intuitively understood by most people. Third, all quantitative understanding of quality and reliability is founded on probability and statistics.

We are interested in the last two traits – quality and reliability – as we look at our topic. The purpose of the following sections is thus to first highlight the intuitive aspects of machinery reliability assessment. We then briefly describe when and how it is applied and finally come to a probabilistic and statistical way of thinking when dealing with matters of process machinery reliability, availability, and safety.

INTUITIVE ASPECTS OF RELIABILITY

The intuitive aspects of reliability are illustrated by the semantic confusion that surrounds the terms "availability", "maintainability", and "reliability". Yet, when someone talks about the good or bad reliability of a given machine, for instance, without adhering to the definitions in our glossary at the end of this book, we still understand intuitively what is meant. Figure 1.2 helps us to understand reliability and its relationship with other terms. It also shows that reliability and related activities cannot exist without some overall directive which we call "operating standards". Standards are basic beliefs, such as "Excellence", "To Be the Best", or simply "Quality", as illustrated in the following:

Worth Thinking About

Beginning for the end

When's the best time to finish a job?

3

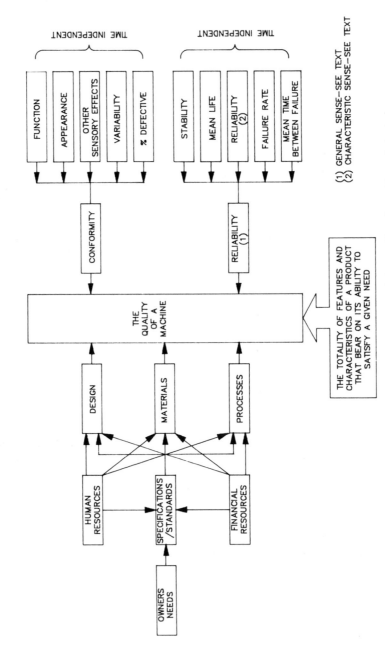

Fig. 1.1 Some of the determinants and measures of the quality of machinery (adapted from ref. 2).

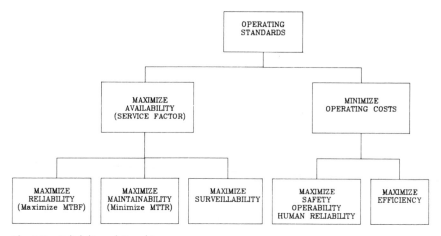

Fig. 1.2 Reliability relationships.

Take mowing your lawn. You pull in the driveway one day and see a lawn that looks like a hayfield, bushes like giant porcupines and a flower bed tangled with weeds.

Your mind flashes to an evergreen carpet, trimmed hedges and flowers lining the walk to your front door. Quickly you grab the mower, and you don't stop till the last weed is pulled.

With every type of work, quality starts right at the beginning. So the best time to start finishing a job is when you begin it.*

Operating standards must generate in us the will to pursue machinery reliability. Once these or similar intuitive standards exist one can proceed to build a framework aimed at optimizing the reliability function. Figure 1.3 illustrates such a framework in the form of a cause-and-effect diagram. The far left of the diagram refers to the philosophical or intuitive aspects of our subject, whereas the right-hand side shows the physical aspects, namely the technology and tools of machinery reliability assessment, the subject of this text.

With an organizational framework in place to control reliability we must allocate our resources to produce the best results with the least expenditure. The amount of effort that can be directed towards a particular aspect of the reliability activity is governed by the cost-effectiveness of that effort.

The effectiveness of a reliability assessment of a given machine is of no value if it does not result in a decision. The same is true of any other assessment, be it related to stress, physics, hydraulics, or whatever. If there is no intention of acting on a decision based on a machinery reliability assessment there is no reason for performing the assessment in the first place.

When should we assess machinery reliability? The answer is, throughout

---

* *Quality Promotion*, by Marlin Industrial Division, Inc., North Haven, Conn. By permission.

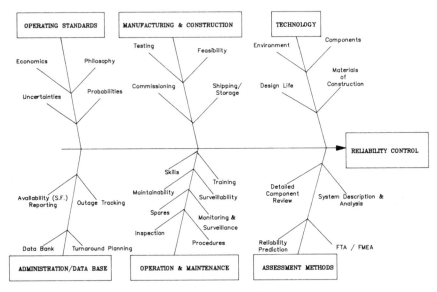

Fig. 1.3   Elements contributing to machinery reliability.

the life-cycle of the equipment, whenever we are faced with the prospects or consequences of poor machinery reliability. Machinery life-cycle phases are typically specification, procurement, testing, installation, start-up, operation, re-rating, maintenance, repair, overhaul, inspection, and replacement. However, we must understand that the alternatives available for improving reliability diminish as the machinery system passes through its life-cycle, that is, from the design stage to the operational stage.

Often the most difficult decision in the context of machinery reliability assessment relates to the extent and the location of the assessment effort. There is a large element of judgment involved, but some guidelines exist nevertheless.

The allocated effort should be applied on a priority basis to those areas which are likely to produce the greatest returns. A first approach would be to determine equipment criticality. The following list serves as a guide for this task:

- part of a continuous process;
- single train;
- unspared;
- unique or unprecedented application;
- new and unproven design;
- proven design; little in-house experience;
- major scale-up of past design or experience;

- operating at high, or dangerous
  - temperatures,
  - pressures,
  - voltage,
  - process fluids;
- accessibility;
- high-velocity/high-inertia components;
- market conditions;
- required skill level;
- availability and cost of spare parts.

More detailed considerations are shown by way of a logic diagram (Fig. 1.4). For a new piece of machinery, for instance, question one recognizes that experience is the best proof of performance. An affirmative answer can be obtained if experience is both favorable and relevant. Experience is relevant only if it comes from a similar machine, in similar service, and with similar operating parameters. Changes in operating conditions or parts and components may often seem small and insignificant to the outsider. They may, however, result in significant differences in reliability performance. A thorough and detailed comparison of seemingly similar services is therefore necessary.

Question two ("Is production affected?") differentiates between machinery essential to operation – such as equipment belonging to utilities – and those

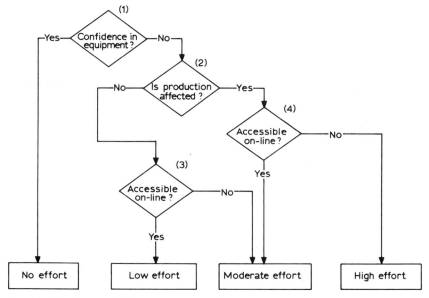

Fig. 1.4 Machinery reliability assessment effort.

whose functions can be interrupted without immediate loss of production. A negative answer indicates a requirement for high reliability. Another question would naturally be: "What does the forced outage of a machine really cost?"

The last question ("Is on-line access possible?") addresses the frequently very important parameter of machinery reliability considerations, namely maintainability. Without on-line access, no on-line repair is possible and high reliability is required.

METHODOLOGY

After the degree of reliability assessment effort at a given point in the life-cycle of a machine has been determined, we should ask "What assessment technique is likely to be most effective?" The answer depends largely on circumstances. Among the principal analytical tools available we find:

- cause-and-effect diagrams;
- reliability estimating and prediction;
- failure mode and effect analysis (FMEA);
- availability analysis;
- fault tree analysis;
- hazard analysis;
- field investigation;
- detailed design review.

Each of these techniques is a systematic method of assembling information for decision making and will be examined in more detail later. The various techniques help us to ask questions because unsatisfactory reliability experience stems more often from failure to ask questions than from the inability to answer them. These methods help to organize information in both a qualitative and quantitative form in order to be able to make decisions.

Accordingly, the method selected depends on the information needed. Table 1.1 lists the degree of effort required to apply these methods.

Cause-and-effect diagrams. Frequently, machinery reliability assessment is part of a problem-solving activity. Problem solving is defined by Kepner-Tregoe as "a logical process that includes identifying the problem, defining the problem and correcting the problem". The Kepner-Tregoe approach is a synthesis of classical problem-solving methodology [3].

We found another tool extremely useful for solving machinery reliability related problems, that is, cause-and-effect diagrams. Cause-and-effect diagrams were developed to represent the relationship between some "effect" and all the "causes" influencing it. The effect or problem – in our case the achievement of reliability – is stated on the right-hand side of the chart and

Table 1.1 Machinery reliability assessment methods and degree of effort

Reliability assessment methods	Assessment effort			
	No effort	*Low effort*	*Moderate effort*	*High effort*
Cause-and-effect diagrams	●			
Reliability estimating and predicting			●	
Failure Mode and Effect Analysis (FMEA) – qualitative		●		
FMEA – quantitative				●
Fault Tree Analysis (FTA) – qualitative			●	
FTA – quantitative				●
Field inspection		●		
Detailed reviews and audits			●	●
Weighted factor analysis			●	

the major influences or "causes" are listed to the left. Cause-and-effect diagrams are drawn to illustrate clearly the various causes and components of a machinery reliability problem. It helps to sort out and relate the causes. A well-defined cause-and-effect diagram will take on the shape of fishbones and has therefore received the name "fishbone diagram". From this well-defined list of causes we can identify and select additional ones that are most likely to influence machinery reliability. Figure 1.5 is an example of a cause-and-effect diagram used in reliability assessment work.

The cause-and-effect diagram identifies the various components of the problem and serves as a communications tool. Figure 1.5 depicts the considerations for reliability achievement of a complex lubrication system for a centrifugal process gas compressor train.

Cause-and-effect diagrams are important. They will facilitate the identification of causes that contribute to a specific machinery reliability problem.

Reliability estimating and prediction uses statistical reliability modelling from available reliability data to predict the reliability of machinery components and systems.

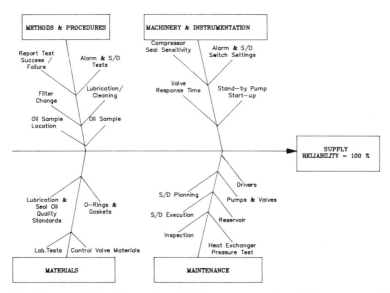

Fig. 1.5   Cause-and-effect diagram for reliability assessment of a critical process machinery lubrication system.

Failure mode and effect analysis (FMEA) is a basic screening tool. The method is an analytical technique that ensures that all possible failure modes of a machine have been addressed. FMEA allows the assessment of the probability of a failure occurrence as well as the effect of a failure. This systematic approach parallels the mental discipline that a designer or reviewer goes through in any design process [4]. With additional effort numerical values can be assigned to failure probabilities and consequences. This quantitative assessment permits relative ranking of failure risks and provides input to other analyses.

Availability analysis is an activity which also uses input data from FMEA. The extent of this activity can range from examination of only the most critical parts of a machine to complete machinery system assessment.

Fault tree analysis, unlike the preceding methods, is specific in application – it identifies the possible causes of a particular failure. It can range from a simple sketch to a complex diagram or a computer program for numerical solution. In one form or another the fault tree is useful for troubleshooting at any level from component to complete machinery system.

Hazard analysis covers HAZOP (Hazard and Operability) reviews specifically directed at machinery. This analysis defines hazards around machinery and associated risks by trying to find the answer to three questions:

- What can go wrong?
- How likely is it to go wrong?
- What are the consequences?

Field investigations help us to arrive at machinery reliability and complexity indices that can be used to determine reliability management needs.

The detailed design review backtracks all engineering assumptions for verification and proof of reliability. Beginning with Chapter 11, our readers will find examples for this approach. We will see that this approach is essentially a design audit of individual machinery components. It is an integrity or vulnerability assessment to ensure fitness for purpose under in-service conditions within the total machinery system. The analysis is performed in detail and takes into account such requirements as:

- maintainability;
- surveillability;
- component capacity;
- force and stress tolerances;
- reactive and environmental compatibility;
- time endurance;
- temperature resistance, etc.

THE MEANING OF RELIABILITY

In order to set the stage for our readers we would like to define some basic terms used in this text.

Machines are man-made and concrete *systems* consisting of a totality of orderly arranged and functionally connected elements. A system is characterized by having a boundary to its environment. The system's connection to its environment is maintained by *input* and *output* parameters. Each system can usually be subdivided into two or more subsystems. Generally, these subdivisions may be made with a varying degree of detail depending on our overall purpose. Consider, for example, the "clutch coupling" system shown in Figure 1.6. We would usually find this "system" as an assembly within a machine. However, if we want to investigate the system from a functional point of view, we could dissect it into the subsystems "elastic coupling" and "clutch". These subsystems, in turn, could be broken down into system components or individual parts.

For the purpose of reliability assessments we have found the following definitions useful.

System and mission

A system is any composite of hardware or software items that work together to perform a mission or a set of related missions. A mission is the external "goal" of a system. A function in turn is the internal "purpose" of a system

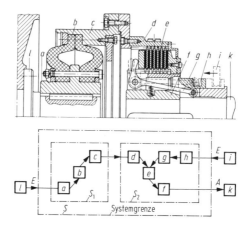

Fig. 1.6 System "coupling". *a–h* are system elements, *i–l* are connecting elements, *S* is the total system, S_1 the subsystem "elastomeric coupling", S_2 the subsystem "clutch", *E* inputs, *A* outputs (from ref. 15). (Reprinted from Dubbel, Taschenbuch fuer den Maschinenbau, edited by W. Beitz and K.H. Kuettner, p. 315, Fig. 1, by courtesy of Springer-Verlag.)

or system components needed to accomplish the mission. A complex system may be made up of two or more groupings of hardware or software items, each of which has a distinct role in performing the mission of the system. The definition of function and mission in a given case is frequently subject to personal interpretation but should be as thorough as possible. Consider, for example, the oil system of an oil-injected rotary screw compressor (Fig. 1.7). Cursory examination may lead to the definition of the system mission or function as "Supplying oil for lubrication and cooling to the compressor". A better idea would be to subdivide this "function" into at least four related but distinct subfunctions and subsystems:

Sub-functions	*Subsystems*
1. Oil admission when compressor is running	1. Oil system – oil stop valve, item 28
2. Oil cooling and temperature control	2. Oil system/water system – coolers
3. Oil filtering	3. Oil system – filters
4. Air/oil separation	4. Oil/air system – separators

This more thorough breakdown will lead to a better understanding of the system mission as well as its function. The example also reveals that there are several functions that are performed simultaneously by one system, subsystem, or their components. It stands to reason that one would want to determine primary and secondary functions in these cases and rank them according to their criticality values.

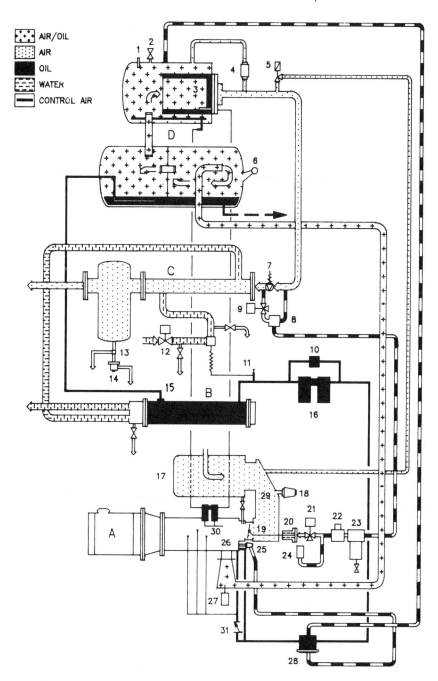

AIR/OIL
AIR
OIL
WATER
CONTROL AIR

Fig. 1.7 System diagram of a two-stage oil-flooded screw compressor (Demag).

Assembly and part

An assembly is any functional component that can be disassembled into two or more subordinate components without disrupting permanent physical bonds [5]. A simple example of a mechanical seal assembly drawing is shown in Figure 1.8. The components of an assembly may be any combination of subassemblies or parts. A part in turn is defined as any hardware item that cannot be disassembled into subordinate components without severing permanent physical bonds. We have already seen how an assembly can be investigated regarding its functional characteristics. It is important in machinery reliability assessment to consider the geometric aspects of machinery parts. We introduce the term "element" [6] to define four internal functions used in machinery assemblies. There are four types of elements:

1. Transmitting elements, such as gear tooth surfaces.
2. Constraining, confining, and containing elements, such as bearings or seals.
3. Fixing elements, such as threaded fasteners.
4. Elements that have no direct functions but which are inevitably needed to support the above functions (e.g. gear wheels or bearing supports).

The term "component" is used almost interchangeably with "assembly". However, "component" will have a somewhat more independent or stand-alone character. Machinery components, for example, are clutches, couplings, drive belts, gear boxes, or pneumatic and hydraulic systems.

Assembly hierarchy

From the foregoing it can easily be understood that machinery systems have a hierarchical structure (see Table 1.2). Assembly hierarchy describes the

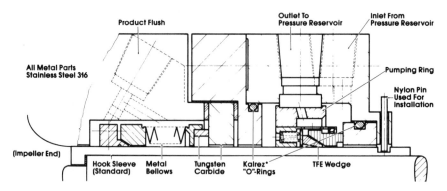

Fig. 1.8   Mechanical shaft seal assembly (EG&G Sealol).

Table 1.2   Assembly hierarchy

System level	*Example*
System	Screw compressor package
Subsystem	Compressor or driver
Assembly	Gear assembly
Part	See parts list
Element	Gear tooth, bearing, bolt

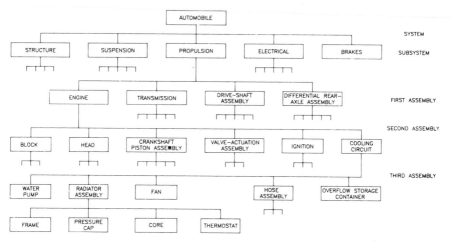

Fig. 1.9   Assembly hierarchy for an automotive engine cooling circuit (from ref. 7). (Reprinted from Moss, M.A., Designing for Minimal Maintenance Expense, p. 64, Fig. 4-1, by courtesy of Marcel Dekker, Inc.)

organization of system hardware elements into assembly levels. Assembly levels descend from the top – or system level – on the basis of functional and sometimes static relationships (see also Fig. 1.9). Thorough reliability assessments are carried out in reverse hierarchic sequence: first, we take a look at the lowest-level components; then the components of the next-highest level are assessed, and so on until the top level (the system level) has been reached.

Failure

Machinery systems are subject to failure. In its simplest form, failure can be defined as any change in a machinery part or component which causes it to be unable to perform its intended function or mission satisfactorily. A popular yardstick for measuring *failure experience* of machinery parts, assemblies, components, or systems is to determine a failure rate. Failure rate is obtained by dividing the number of failures experienced on a number of homogeneous items, also called "population", within a time period, by the population.

For example, if we had 10 injection pumps, and 3 failed during a period of 12 months, our failure rate (λ) would be:

$$\lambda = \frac{3\,\text{failures}}{10\,\text{machine-year}} = 0.3\,\frac{\text{failures}}{\text{machine-year}}$$

or

$$\lambda = \frac{0.3}{365 \times 24} = 0.000\,034 \qquad \text{or} \qquad (34 \times 10^{-6})\ \text{failures per machine-hour}$$

For reliability assessments, failures are frequently classified as either chargeable or non-chargeable. A chargeable failure, for example, would be a failure that can be attributed to a defect in design or manufacture. A non-chargeable failure would be a failure caused by exposure of the part to operational, environmental, or structural stresses beyond the limits specified for the design. Other non-chargeable failures are those attributable to operator error or improper handling or maintenance.

Other terms used in the context of machinery failure experience are "malfunction" and "fault" that should, when used, be clearly defined.

Failure mode

A failure mode is the appearance, manner, or form in which a machinery component failure manifests itself [8]. It should not be confused with the failure *cause*, as the former is the effect and the latter the cause of the failure event.

Failure modes can be defined for all levels of the system and the assembly hierarchy. For example, deterioration of the oil stop valve (Fig. 1.7, item 28) of the oil-injected compressor system could have one of the following failure modes:

1. Fail open. *Consequence:* The compressor is flooded and cannot be started.
2. Fail close. *Consequence:* The compressor will shut-down due to high discharge temperature.
3. Fail not fully open or fully closed. *Consequence:* Gradual deterioration of system performance.

The causes of these failure modes could either be common, such as dirt or foreign objects in the valve, or specific to each failure mode – a broken return spring would keep the valve open, insufficient discharge pressure would keep it closed, and so forth.

Service life

Service life designates the time-span during which a product can be expected to operate safely and meet specified performance standards, when maintained in accordance with the manufacturer's instructions and not subjected to environmental or operational stresses beyond specified limits [9]. The service life for a given machinery part represents a prediction that no less than a certain proportion of the machinery system or its components will operate successfully for the stated time period, number of cycles, or distance travelled. Service life is clearly a probabilistic term subject to a confidence limit. A good example is anti-friction bearings. Since a bearing failure generally results in the failure of the machine in which it is installed, bearing manufacturers have made a considerable effort to identify the factors that are responsible for bearing failures. A typical equation for determining ball bearing service life shows the rated life to be inversely proportional to the rotational speed of the inner ring and the third power of the applied radial load. Rated life in this case is the so-called L_{10} life, which is the number of bearing revolutions, or the number of working hours at a certain rotational speed and load, which will be reached or exceeded by 90% of all bearings.

Reliability

Reliability, finally, in general terms, is the ability of a system or components thereof to perform a required function under stated conditions for a stated period of time. It is also apparent that "reliability" is frequently used as a characteristic denoting a probability of success or success ratio [10]. This means that it may be stated that:

1. A component or piece of machinery should operate successfully for X hours on Y% of occasions on which it is required to operate; or
2. A machine should not fail more frequently than X times in Y running hours; or
3. The mean life of a population of similar components or machinery should be equal to or greater than Y hours with a standard deviation of S hours.

Maintainability

Many machinery components are designed to receive some form of attention during their life. The goal is to compensate for the effects of wear or to allow for the replacement of consumable or sacrificial elements. The ease with which this kind of work can be done is termed "maintainability". The operational and organizational function of this work is called "maintenance". Maintenance possibilities are illustrated in Fig. 1.10. It has been shown that, if maintenance on process machinery has to be performed at all, predictive maintenance is the most cost-effective mode [11–13].

Maintainability then is the ability of an item, under stated conditions of use, to be retained in, or restored to, a state in which it can perform its

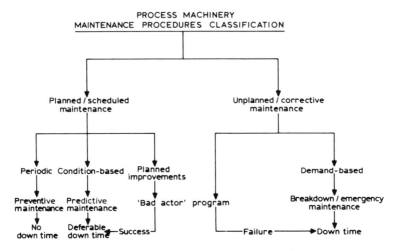

Fig. 1.10 Process machinery maintenance procedures classification.

required functions, when maintanence is performed under stated conditions and using prescribed procedures and resources [14].

Maintainability has a direct influence on the reliability of machinery systems. We will see that maintainability parameters must be considered an integral part of the machinery reliability assessment effort.

Surveillability

We have already acquainted ourselves with this term in Figure 1.2. Surveillability is closely related to maintainability and will receive the same attention within the overall reliability assessment activity. We have already stated that process machinery maintenance can be optimized by practicing condition-based or predictive maintenance. Surveillability is the key. It is defined as a quantitative parameter that includes:

- accessibility for surveillance;
- operability if required;
- ability to monitor machinery component deterioration;
- provision of indicating and annunciation devices.

Availability

Maintainability together with reliability determine the availability of a machinery system (see Fig. 1.2). Availability is influenced by the time demand made by preventive and corrective maintenance measures. Maintenance activities which are performed during planned downtimes or on-line without affecting operation do not have an impact on availability. Availability (A) is measured by:

$$A = \frac{\text{MTBF}}{\text{MTBF} + \text{MTTR}} \tag{1.1}$$

where   MTBF = mean time between failures,
        MTTR = mean time to repair or mean repair time.

REFERENCES

1. Isikawa, K., *Guide to Quality Control*. White Plains: UNIPUB, 1987, p. 1.
2. *BSI Handbook*, Vol. 22 *Quality Assurance*. London: British Standards Institution, 1983, p. 4.
3. Kepner, C. H. and Tregoe, B. B., *The Rational Manager*. New York: McGraw-Hill, 1965.
4. Ford Motor Company, *Reliability Methods, Module XIV*. Detroit: North American Automotive Operations, 1972.
5. Moss, M. A., *Designing for Minimal Maintenance Expense*. New York: Marcel Dekker, 1985, p. 13.
6. Yoshikawa, H. and Taniguchi, N., Fundamentals of mechanical reliability and its application to computer aided design. *Annals of the CIRP*, **24** (1), 1975, p. 298.
7. Moss, op. cit., p. 64.
8. Bloch, H. P. and Geitner, F. K., *Machinery Failure Analysis and Troubleshooting*. Houston: Gulf Publishing, 1983, pp. 2–3.
9. Moss, op. cit., p. 16.
10. British Standards Institution, op. cit., p. 14.
11. Finley, H. F., Maintenance management for today's high technology plants. *Hydrocarbon Processing*, Jan. 1978, pp. 101–105.
12. Bloch and Geitner, op. cit., p. 610.
13. Matteson, T. D., Overhauling our ideas about maintenance. *Mechanical Engineering*, May 1986, pp. 86–88.
14. British Standards Institution, op. cit., p. 19.
15. Dubbel, *Taschenbuch für den Maschinenbau*, edited by W. Beitz and K. H. Küttner. New York: Springer Verlag, 1981.

2

Reliability as probability of success

Probabilistic thinking is based on very old ideas which go back to De Mere [1], La Place [2], and Bayes [3]. What is probability and how does it relate to frequency, statistics and, finally, machinery reliability? The word probability has several meanings. At least three will be considered here.

One definition of probability has to do with the concept of equal likelihood. If a situation has N equally likely and mutually exclusive outcomes, and if n of these outcomes are event E, then the probability $P(E)$ of event E is:

$$P(E) = \frac{n}{N} \qquad (2.1)$$

This probability can be calculated *a priori* and without doing experiments.

The example usually given is the throw of an unbiased die, which has six equally likely outcomes – the probability of throwing a one is $1:6$. Another example is the withdrawal of a ball from a bag containing four white balls and two red ones – the probability of picking a red one is $1:3$. The concept of equal likelihood applies to the second example also, because, even though the likelihoods of picking a red ball and a white one are unequal, the likelihoods of withdrawing any individual ball are equal.

This definition of probability is often of limited usefulness in engineering because of the difficulty of defining situations with equally likely and mutually exclusive outcomes.

A second definition of probability is based on the concept of relative frequency. If an experiment is performed N times, and if event E occurs on n of these occasions, then the probability of $P(E)$ of event E is:

$$P(E) = \lim_{n \to \infty} \frac{n}{N} \qquad (2.2)$$

$P(E)$ can only be determined by experiment. This definition is frequently used in engineering. In particular it is this definition which is implied when we estimate the probability of failure from field failure data.

Thus, when we talk about the measurable results of probability experiments – such as rolling dies or counting the number of failures of a machinery component – we use the word "frequency". The discipline that deals with such measurements and their interpretation is called statistics. When we discuss a state of knowledge, a degree of confidence, which we derive from statistical experiments, we use the term "probability". The science of such states of confidence, and how they in turn change with new information, is what is meant by "probability theory".

The best definition of probability in our opinion was given by E. T. Jaynes of the University of California in 1960:

> Probability theory is an extension of logic, which describes the inductive reasoning of an idealized being who represents degrees of plausibility by real numbers. The numerical value of any probability $(A \mid B)$ will in general depend not only on A and B, but also on the entire background of other propositions that this being is taking into account. A probability assignment is 'subjective' in the sense that it describes a state of knowledge rather than any property of the 'real' world. But it is completely 'objective', in the sense that it is independent of the personality of the user: two beings faced with the same total background and knowledge must assign the same probabilities.

Later, Warren Weaver [5] defined the difference between probability theory and statistics:

> Probability theory computes the probability that 'future' (and hence presently unknown) samples out of a 'known' population turn out to have stated characteristics.
>
> Statistics looks at a 'present' and hence 'known' sample taken out of an 'unknown' population, makes estimates of what the population may be, compares the likelihood of various populations, and tells how confident you have a right to be about these estimates.
>
> Stated still more compactly, probability argues from populations to samples, and statistics argues from samples to populations.

Whenever there is an event E which may have outcomes $E_1, E_2, \ldots, E_n$, and whose probabilities of occurrence are $P_1, P_2, \ldots, P_n$, we can speak of the set of probability numbers as the "probability distribution" associated with the various ways in which the event may occur. This is a very natural and sensible terminology, for it refers to the way in which the available supply of probability (namely unity) is "distributed" over the various things that may happen.

Consider the example of the tossing of six coins. If we want to know "how many heads there are", then the probability distribution can be shown as follows:*

No. of heads	0	1	2	3	4	5	6
Probability	1/64	6/64	15/64	20/64	15/64	6/64	1/64

* See Appendix A for calculation of probability values.

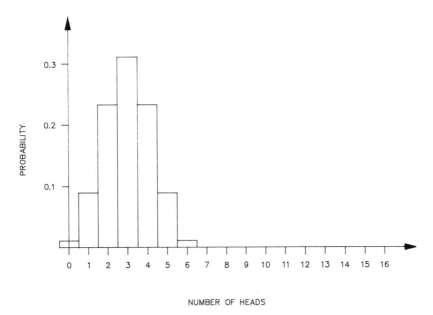

NUMBER OF HEADS

Fig. 2.1 Distribution of probabilities measured by the vertical height as well as by areas of the rectangles (the six-coin case).

These same facts could be depicted graphically (Fig. 2.1). Accordingly, we arrive at a probability curve versus "frequency" as a way of expressing our state of knowledge.

As another application of the probability-of-frequency concept, consider the reliability of a specific machine or machinery system. In order to quantify reliability, frequency type numbers are usually introduced. These numbers are mean times between two failures or MTBF, for instance, which are based on failures per trial or per operating period. Usually they are referred to in months.

Most of the time we are uncertain about what the MTBF is. "All machines and their components are not created equal", their load-cycles and operating conditions are unknown, and maintenance attention can vary from neglect to too frequent intervention. Consider for instance the MTBF of a sleeve bearing of a crane trolley wheel. At a MTBF of 18 months (30% utilization factor), early failures can be experienced after 6 months. The longest life experience may be five times the shortest life (see Fig. 2.2). Even though the data were derived from actual field experience, we cannot expect exact duplication of the failure experience in the future. Therefore, Figure 2.2 is our probabilistic model for the future of a similarly designed, operated, and maintained crane wheel.

It is important to distinguish the above idea from the concept of "frequency of frequency". Let *R* denote the historical reliability of an individual

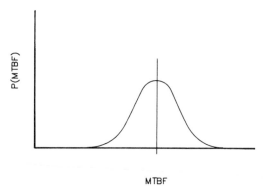

Fig. 2.2   Probability of frequency curve for a machinery component.

designated machine, selected at random from a population of similar machines. The historical reliability of a machine is defined as:

$$R = 1 - \frac{H_1}{H_1 + H} \qquad (2.3)$$

where H_1 = total time on forced outage (h),
        H = total service time (h).

We can build a frequency distribution using historical reliability for each machine showing what fraction of the population belongs to each reliability increment. If the population is large enough we can express this distribution as a continuous curve – a "frequency density" distribution, $\Phi(R)$. The units of $\Phi(R)$ are consequently frequency per unit R, or fraction of population per unit reliability.

This curve is an experimental quantity. It portrays the variability of the population, which is a measurable quantity. The value of R varies with the individual selected. It is a truly fluctuating or random variable.

Contrast this with the relationship shown in Figure 2.2, where we selected a specific machinery component and asked what its future reliability would be. That future reliability is the result of an experiment to be done. It is not a random variable: it is a definite number not known at this time. This goes to show that we must distinguish between a frequency distribution expressing the variability of a random variable and a probability distribution representing our state of knowledge about a fixed variable.

A third definition of probability is degree of belief. It is the numerical measure of the belief which a person has that an event will occur.

Often this corresponds to the relative frequency of the event. This need not always be so for several reasons. One is that the relative frequency data available to the individual may be limited or non-existent. Another is that although somebody has such data, he or she may have other information which causes doubt that the whole truth is available. There are many possible reasons for this.

Several branches of probability theory attempt to accommodate personal probability. These include ranking techniques, which give the numerical encoding of judgments on the probability ranking of items. Bayesian methods allow probabilities to be modified in the light of additional information [6].

The key idea of the latter branch of probability theory is based on Bayes' Theorem, which is further defined below.

In basic probability theory, $P(A)$ is used to represent the probability of the occurrence of event A; similarly, $P(B)$ represents the probability of event B. To represent the joint probability of A and B, we use $P(A \wedge B)$, the probability of the occurrence of both event A and event B. Finally, the conditional probability, $P(A|B)$, is defined as the probability of event A, given that B has already occurred.

From a basic axiom of probability theory, the probability of the two simultaneous events A and B, can be expressed by two products:

$$P(A \wedge B) = P(A) \times P(B|A) \tag{2.4}$$

$$P(A \wedge B) = P(B) \times P(A|B) \tag{2.5}$$

Equating the right sides of the two equations and dividing by $P(B)$, we have what is known as Bayes' Theorem:

$$P(A|B) = P(A) \times [P(B|A)/P(B)] \tag{2.6}$$

In words, it says that $P(A|B)$, the probability of A with information B already given, is the product of two factors: the probability of A prior to having information B, and the correction factor given in the brackets. Stated in general terms:

Posterior probability $\propto$ Prior probability $\times$ Likelihood

where the symbol $\propto$ means "proportional to" [7]. This relationship has been formulated as follows:

1. The A_i's are a set of mutually exclusive events for $i = 1 \cdots n$.
2. $P(A_i)$ is the prior probability of A_i before testing.
3. B is the observation event.
4. $P(B|A_i)$ is the probability of the observation, given that A_i is true.

Then

$$P(A_i|B) = \frac{P(A_i)P(B|A_i)}{\sum\limits_{i}^{n} P(A_i)P(B|A_i)} \tag{2.7}$$

where $P(A_i|B)$ is the posterior probability, or the probability of A_i now that B is known. Note that the denominator of equation 2.7 is a normalizing factor for $P(A_i|B)$ which ensures that the sum of $P(A_i|B) = 1$.

As powerful as it is simple, this theorem shows us how our probability – that is, our state of confidence with respect to A_i – rationally changes upon getting a new piece of information. It is the theorem we would use, for example, to evaluate the significance of a body of experience in the operation of a specific machine.

To illustrate the application of Bayes' Theorem let us consider some examples. If C represents the event that a certain pump is in hot oil service and G is the event that the pump has had its seals replaced during the last year, then $P(G|C)$ is the probability that the pump will have had its seals replaced some time during the last three years *given that it is actually in hot oil service*. Similarly, $P(C|G)$ is the probability that a pump did have its seals replaced within the last three years *given that the pump is in hot oil service*. Clearly there is a big difference between the events to which these two conditional probabilities refer. One could use equation 2.6 to relate such pairs of conditional probabilities.

Although there is no question as to the validity of the equation, there is some question as to its applicability. This is due to the fact that it involves a "backward" sort of reasoning – namely, reasoning from effect to cause.

Example: In a large plant, records show that 70% of the bearing vibration checks are performed by the operators and the rest by central inspection. Furthermore, the records show that the operators detect a problem 3% of the time while the entire force (operators and central inspectors) detect a problem 2.7% of the time. What is the probability that a problem bearing, checked by the entire force, was inspected by an operator? If we let A denote the event that a problem bearing is detected and B denote the event that the inspection was made by an operator, the above information can be expressed by writing $P(B) = 0.70$, $P(A|B) = 0.03$, and $P(A) = 0.027$, so that substitution into Bayes' formula yields:

$$P(B|A) = \frac{P(B) * P(A|B)}{P(A)} \tag{2.8}$$

Numerically, this is:

$$P(B|A) = \frac{(0.70)(0.03)}{(0.027)} = \frac{0.021}{0.027} = \frac{7}{9} = 78\%$$

This is the probability that the inspection was made by an operator given that a problem bearing was found [8].

According to the foregoing, our understanding of reliability here is a probability rather than merely a historical value. It is statistical rather than individual.

REFERENCES

1. Weaver, W., *Lady Luck. The Theory of Probability*. Garden City: Doubleday, 1963, pp. 45–52.

2. Weaver, op. cit., p. 251.
3. Weaver, op. cit., pp. 308–309.
4. Kaplan, S. J. and Garrick, B. J., Try probabilistic thinking to improve powerplant reliability. *Power*, March 1980, pp. 56–61.
5. Weaver, op. cit., pp. 306–307.
6. Lees, F. P., *Loss Prevention in the Process Industries*. Boston: Butterworth, 1980, vol. 1, p. 81.
7. Henley, E. J. and Kumamoto, H., *Reliability Engineering and Risk Assessment*. Englewood Cliffs: Prentice-Hall, 1981, p. 256.
8. Freund, J. E. and Williams, F. J., *Elementary Business Statistics*. Englewood Cliffs: Prentice-Hall, 2nd ed., 1972.

3

Estimating machinery reliability

The reliability of a machinery system may be mathematically described by defining distribution functions using discrete and random variables. An example of a discrete variable is the number of failures in a given time interval. Examples of continuous random variables are the time from part installation to failure or the time between successive equipment failures.

This approach has been particularly useful in the field of electronic engineering where it has been applied to the design and evaluation of electronic devices. Using reliability theory one can estimate the reliability of complex electronic systems. Calculation methods, specific to electronic systems, make use of failure probability data compiled for this purpose.

To evaluate electronic component reliability, the concept of constant failure rate is used, that is failure rates of electronic components remain constant during the useful life of the component. However, this is frequently not the case when evaluating mechanical component reliability. There are several reasons for this. It is, for example, an established fact that in many cases machinery components follow an increasing failure rate pattern. Another reason is the fact that machinery components are not well standardized. Finally, there seem to be many more failure modes experienced by machinery parts than by electronic parts. Consequently, reliability data for mechanical components and assemblies is scarce, and, when available, caution is advised. From this it follows that there is no accurate method available for absolute reliability prediction that takes the specific nature of machinery systems into account. As we will see later, it seems that only *relative* reliability predictions can be made for machinery. What is the specific nature of machinery? Figure 3.1 illustrates a machinery system by comparing it with an electric system. Consider, for example, the reliability of a tribo-mechanical system* in which wear behavior is a function of time. Three main characteristics may be determined for the loss–output wear rates of such a system [1]:

1. Self-accommodation ("running-in")

---

* A system with parts in rubbing contact.

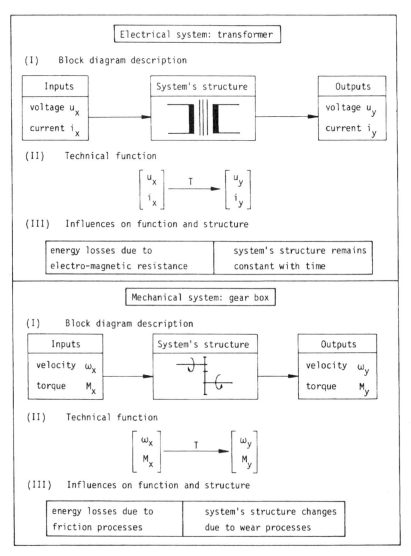

Fig. 3.1 Comparison of the characteristics of an electrical and a mechanical system (from ref. 1). (Reprinted from Czichos, H., Tribology—A Systems Approach to the Science and Technology of Friction, Lubrication, and Wear, 1978, p. 26, Fig. 3-1, by courtesy of Elsevier Science Publishers, Physical Sciences & Engineering Div.)

2. Steady-state
3. Self-acceleration ("catastrophic damage")

These three phase changes in the system behavior may follow each other in time (Fig. 3.2). Here, Z_{lim}^{M} denotes a maximum allowable level of wear loss.

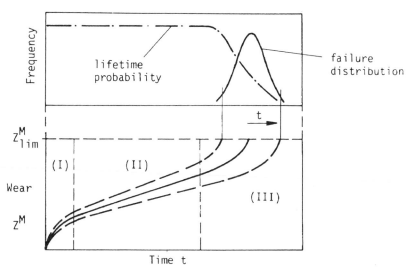

Fig. 3.2   Wear curves and failure distribution (from ref. 1).

At this level the system structure has changed in such a way that the functional input–output relationship of the system has been severely disturbed. Repeated measurements show random data variations as indicated by the dashed lines in Fig. 3.2. A distribution of the "life" of the system or a failure distribution can be derived from sample functions of the wear process.

Earlier, we familiarized ourselves with the concept of relative frequency. The reader is referred to Figure 2.2, which for convenience, is reproduced in Figure 3.3. If we wish to determine the probability of failure occurring between the times t_b and t_c, we multiply the y-axis value by the interval $(t_c - t_b)$. Figure 3.3 is also called a probability density function where the equation of the curve is denoted by $f(t)$. As an example, if $f(t) = 0.6 \exp(-0.6t)$, we obtain the curve shown in Figure 3.4, a negative exponential distribution which will be dealt with later.

Returning to Figure 3.3, the probability of a failure occurring between t_b and t_c is the area of the hatched portion of the distribution. This area is the integral between t_b and t_c of $f(t)$ or:

$$\int_{t_b}^{t_c} f(t)\, dt \tag{3.1}$$

Consequently, the probability of a failure occurring between times t_a and t_z is:

$$\int_{t_a}^{t_z} f(t)\, dt = 1 \tag{3.2}$$

We stated earlier that the failure distributions of different types of machinery systems are not the same. Even the failure distributions of identical machines may not be the same if they are subjected to different levels of

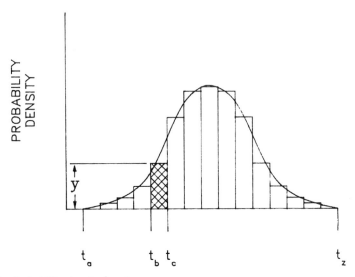

Fig. 3.3   Probability density function.

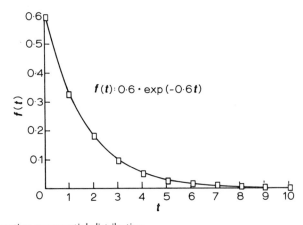

Fig. 3.4   Negative exponential distribution.

Force, Reactive Environment, Temperature, and Time (FRETT). There are a number of well-known probability density functions which have been found in practice to describe the failure characteristics of machinery (see Fig. 3.5) [2].

The cumulative distribution function. In reliability estimations we want to determine the probability of a failure occurring before some specified time *t*. This probability can be calculated by using the appropriate density function as follows:

$$\text{Probability of failure before time} = \int_{-\infty}^{t} f(t)\, dt \qquad (3.3)$$

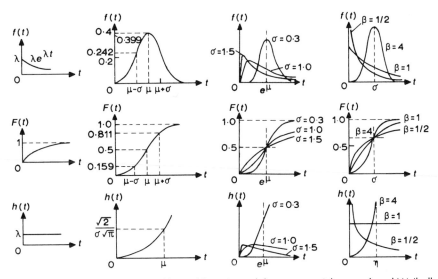

Fig. 3.5 Density, reliability, and hazard functions of the exponential, normal, and Weibull distributions.

The integral $\int_{-\infty}^{t} f(t)\,dt$ is termed $F(t)$ and is called the cumulative distribution function. One can state that as t approaches infinity, $F(t)$ approaches unity.

The reliability function. The function complementary to the cumulative distribution function is the reliability function, also called survival function. This function can be used to determine the probability that equipment will survive to a specified time t. The reliability function is denoted $R(t)$ and is defined by:

$$R(t) = \int_{t}^{\infty} f(t)\,dt \qquad (3.4)$$

and, obviously:

$$R(t) = 1 - F(t) \qquad (3.5)$$

The failure rate or hazard function. The last type of function derived from the other functions is the hazard function. It has other names in the literature, such as intensity function, force of mortality, and also failure rate in a certain context. It is denoted $h(t)$ and defined as:

$$h(t) = \frac{f(t)}{R(t)} = \frac{f(t)}{1 - F(t)} \qquad (3.6)$$

The hazard function is a conditional probability that a system will fail during the time t and dt under the condition that the system is safe until time t. Someone once had a simple explanation of the hazard function. It was made by analogy. Suppose someone takes an automobile trip of 200 miles and completes the trip in 4 h. The average travel rate was 50 mph, although the person drove faster at some times and slower at others. The rate at any given instant could have been determined by reading the speed indicated on the speedometer at that instant. The 50 mph is analogous to the failure rate and the speed of any point is analogous to the hazard rate.

The foregoing definitions rely on some rather involved mathematics. The reader is referred to Green and Bourne [3] and Henley and Kumamoto [4] for more detailed explanations. However, we believe that there is no need to burden oneself with the mathematics of failure distributions. As we will see later, there has been considerable progress in the application of computerized models and appropriate software.

Specific distribution functions. A number of distributions have been proposed for machinery failure probabilities. Their definitions in terms of density function, reliability function, and hazard rate are depicted in Figure 3.5.

The exponential distribution is the most important function due to its wide acceptance in the reliability analysis work of electronic systems. As shown in Figure 3.5, this function is defined as:

$$f(t) = \lambda \cdot \exp(-\lambda t) \qquad \text{for} \quad t \geqslant 0 \qquad (3.7)$$

where

$$\lambda = \frac{1}{\mu} \qquad (3.8)$$

The exponential distribution is an appropriate model where failure of an item is due not to deterioration as a result of wear, but rather to random events. This feature of the exponential distribution also implies a constant hazard rate. The exponential distribution has been successfully applied as a time-to-failure model for complex systems consisting of a large number of components in series, none of which individually contributes significantly to the total failure density [5]. This distribution is often used because of its universal applicability to systems that are repairable. Many kinds of electronic components follow an exponential distribution. Machinery parts behave in this mode when they succumb to brittle failure. For example, Figure 3.6 shows that Diesel engine control unit failures followed an exponential distribution.

The normal distribution. Although the normal distribution has only limited applicability to life data, it is used where failures are due to wear processes. The hazard or failure rate of this distribution cannot be expressed in a simple form.

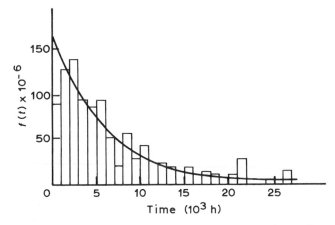

Fig. 3.6 Density function *f*(t) of the failure of diesel engine control units (from ref. 6).

The lognormal distribution is defined by:

$$f(t) = \frac{1}{t\sigma\sqrt{2\pi}} \exp \frac{-[\log(t/t_{50})]^2}{2\sigma^2} \tag{3.9}$$

where t_{50} = median = $\exp(u)$,
u = the mean of the logarithms of the times to failure,
σ = standard deviation.

The limited applicability of normal distribution to life data has been mentioned [7]. This is not the case for the *lognormal distribution* which enjoys wide acceptance in reliability work. It has been applied in machinery maintainability consideration and where failure is due to crack propagation or corrosion. Nelson and Hayashi [8] give an exhaustive account of stress–temperature related furnace tube failure phenomena modelled by the lognormal distribution.

The Weibull distribution is defined by two parameters – η, the nominal or characteristic life,* and a constant β, a non-dimensional shape parameter. A typical Weibull distribution fit for life of a ball bearing is shown in Figure 3.7.

The ability of the Weibull function to model failure distributions makes it one of the most useful distributions for analyzing failure data. If the shape parameter $\beta > 1$, an increasing $h(t)$ is indicated which is symptomatic of wear-out failures. Where $\beta = 1$, we find an exponential function, which obviously is a special case of the Weibull distribution. With $\beta = 1$, a constant hazard or failure rate is indicated.

Where $\beta = 2$, this means that $h(t)$ is linearly increasing with t. The resulting

---

* Also called scale factor.

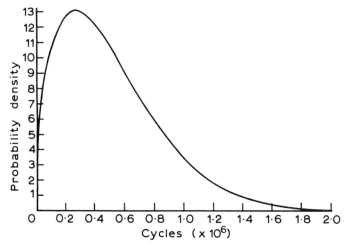

Fig. 3.7 Weibull function for a ball bearing (from ref. 9). (Reprinted from Sidall, J.N., Probabilistic Engineering design, 1983, p. 361, Fig. 11-3, by courtesy of Marcel Dekker, Inc.)

distribution is a special case of the Weibull function known as the Rayleigh distribution [5].

If $\beta < 1$, a decreasing failure rate $h(t)$ is indicated. This would be typical for machinery components where run-in or initial self-accommodation takes place. Mechanical shaft seals would be a typical example.

The mean and standard deviation of the Weibull distribution involves complex calculations. For most engineering problems where the shape factor is greater than 0.5, they can be found from:

$$\mu = \eta(0.9 + 0.1/\beta^3) \tag{3.10}$$

$$= \mu/\beta \tag{3.11}$$

In cases where the shape factor is greater than 1, the mean is nearly equal to characteristic life (η). The error involved in this assumption will generally be small compared to other errors stemming from the quality of data.

One difficulty in attempting to fit theoretical distribution to failure or "life" data arises when a part or an assembly is subject to different failure modes. Table 3.1 lists some of the basic machinery component failure modes and shows the distributions they tend to follow. There are three different possibilities in which these failure modes appear:

1. Simultaneously with some time differences. Fit corrosion, for instance, in an anti-friction (ball) bearing would appear as wear first and then as corrosion. Sidall [9] shows how to evaluate simultaneous failure mode occurrences in the context of failure distributions.
2. Failure modes occur singularly and exclusive of others. This is a somewhat theoretical assumption that we will not deal with any further.

Table 3.1 Selected basic machinery component failure modes and their statistical distributions

	Probability distribution		
Basic failure mode	*Exponential*	*Normal*	*Weibull*
1.0 Force/stress			
1.1 Deformation			●
1.2 Fracture	●		
1.3 Yielding	●		
2.0 Reactive environment			
2.1 Corrosion			●
2.2 Rusting			●
2.3 Staining	●		
3.0 Temperature/thermal			
3.1 Creep		●	
4.0 Time effects			
4.1 Fatigue			●
4.2 Erosion			●
4.3 Wear		●	

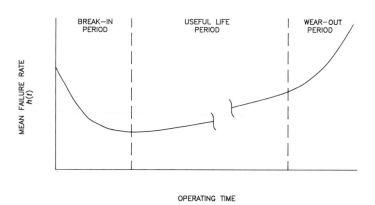

Fig. 3.8 Mean failure rate curve as a function of time.

3. A more realistic model can be created by assuming that failure modes occur consecutively in time. A commonly accepted concept is shown in Figure 3.8. In this curve, called the bath tub curve, three conditions can be distinguished: (1) early or infant mortality failures, (2) random failures, and (3) wear-out failures.

Condition 1 describes the early time period of a machinery system or part by showing a decreasing failure rate over time. It is usually assumed that this period of "infant mortality" or "burn-in" is caused by the existence of material and manufacturing flaws together with assembly errors. Parts or systems that would exclusively exhibit this behavior would fit a Weibull distribution with $\beta < 1$. Condition 2, the area of constant failure rate, is the region of normal performance. This period is termed "useful life", during which time only random failures will occur. Parts or systems that would exclusively exhibit this failure behavior would fit a Weibull distribution with $\beta = 1$ or for that matter an exponential distribution. Condition 3 ($\beta > 1$) is characterized by an increase of failure rate with time. As mentioned before, failures may be due to aging and wear-out.

It has been said that the bath tub curve concept is purely theoretical and only serves the purpose of promoting a better understanding of failure events. However, real-world examples can be cited. For example, if a large number of light bulbs or anti-friction bearings operate continuously, some fail due to defects. During the useful life, occasional random failures occur, but most survive to old age, when the failure rate rises.

ESTIMATION OF FAILURE DISTRIBUTIONS FOR MACHINERY COMPONENTS

The data required to determine failure distributions are the individual times to failure of the equipment.

The procedure is to convert the data to become representative of the cumulative failure distribution $F(t)$. This is done by plotting times to failure against $F(t)$ on a scale which corresponds to the distribution to be fitted. For the exponential distribution this would be:

$$F(t) = 1 - \exp(-\lambda t) \tag{3.12}$$

Consequently:

$$t = \frac{1}{\lambda} \ln \frac{1}{1 - F(t)} \tag{3.13}$$

A plot of $1/[1 - F(t)]$ on a log scale against time on a linear scale produces a straight line. For the Weibull distribution:

$$\ln(t) = \frac{1}{\beta} \ln \ln \frac{1}{1 - F(t)} + \ln \eta \tag{3.14}$$

For most distributions, special graph papers are available which allow direct plotting of $F(t)$ versus t (Fig. 3.9 illustrates a Weibull graph). Nelson [10] describes distributions and the fitting of life or failure data. Again, we encourage our readers to investigate the possible use of computer software

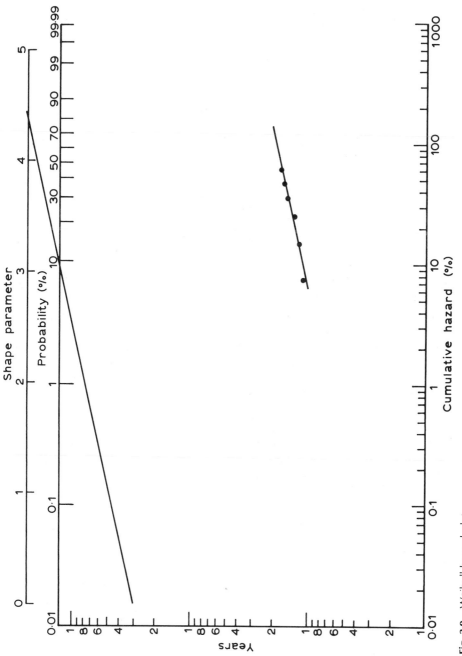

Fig. 3.9 Weibull hazard plot.

packages developed for the statistical analysis of data relating to the failures and successful performance of machinery or components. Their analysis capabilities range from simple calculations such as mean life, to the fitting of Weibull and other distribution models.*

APPLICATION OF FAILURE DISTRIBUTIONS

The application of failure distributions for reliability predictions has been described in numerous references. With the emergence of improved data bases there is a new interest in these applications. Exhaustive information covering the application of distribution functions to equipment maintenance, replacement, and reliability decisions can be obtained, for example, from Jardine [11].

Our *example* will cover a replacement decision in connection with large (>1500 hp) electric motors in a petrochemical process plant. The motors considered for replacement had served this particular plant well for 18 years, but failure experience with similar motors at the same time had raised doubt in the owner's mind as to whether or not an 18-year-old motor could still be called reliable. All motors were 4000 kVA, 3 phase, 60 cycle, pipe ventilated squirrel cage induction motors.

The failure experience of similar motors is listed in Table 3.2. Motors shown as having failed are denoted by a superscript ([a]). These motors had stopped suddenly on-line through winding failures. Mean forced outage penalties were in the neighborhood of 1600 k$ considering the availability or unavailability of motor rewind shops and materials. The cost of an emergency rewind amounted to 125 k$, whereas the cost of a preventive rewind was 100 k$ with no penalty cost for loss of production. The problem was simply to balance the cost of preventive rewinds against their benefits. In order to do this one needs to determine the optimal preventive replacement age of the motor windings to minimize the total expected cost of replacement per unit time. Obviously, one requires a probabilistic model of the motor winding life in order to make a reliability assessment.

Obtaining the Weibull function

The Weibull function was obtained by plotting the data contained in Table 3.2 on appropriate Weibull paper (Fig. 3.9).

The plotting method used has been proposed by Nelson [10] for "multiply censored" life data consisting of times to failure on failed units, and running times – called censoring times – on unfailed units. The method is known as hazard plotting. It has been used effectively to analyze field and life test data

---

* "RECLODE" program, University of Windsor, Ontario.

Table 3.2 Large motor winding failures: Failure data and hazard calculation

Motor	Rank	Years	Hazard	Cumulative hazard
C-70	20	8		
C-71A	19	8		
C-71B	18	8		
P-70A	17	8		
P-70B	16	8		
P-71	15	8		
C-25	14	10		
C-11	13	11^a	7.69	7.69
C-52	12	12^a	8.33	16.03
C-13	11	13^a	9.09	25.12
C-31	10	13		
C-53	9	15^a	11.11	36.23
C-41	8	16^a	12.58	48.73
C-91	7	17^a	14.29	63.01
C-32A	6	17		
C-32B	5	17		
C-01	4	18^b		
C-30	3	18		
C-50	2	18		
C-51	1	18		

[a] Winding failure.
[b] Preventive winding replacement.

on products consisting of electronic and mechanical parts ranging from small electric appliances to heavy industrial equipment. The hazard plotting method originally appeared in Nelson [12], which also contains more details.

Steps

1. The *n* times, or years in our case, are placed in order from the smallest to the largest as shown in Table 3.2. The times are labelled with reverse ranks, that is the first time is labelled *n*, the second labelled $n-1, \ldots,$ and the *n*th is labelled 1. The failure times are each marked by a superscript ([a]) to distinguish them from the censoring times.
2. Calculate a hazard value for each failure as $100/k$, where *k* is its reverse rank. The hazard values for the large motor winding failures are shown in Table 3.2. For example, for the winding failure after 13 years, the reverse rank is 11 and the corresponding hazard value is $100/11 = 9.1\%$.
3. Proceed to calculate the cumulative hazard value for each failure as the sum of its hazard value and the cumulative hazard value of the preceding failure. For instance, for the motor failure after 13 years of operation, the

cumulative hazard value of 25.12 is calculated by adding the hazard value of 9.1 to the cumulative hazard value of 16.03 of the preceding failure.

4. For plotting purposes, the hazard paper of a theoretical distribution of time to failure was chosen. The Weibull distribution seemed appropriate. On the vertical axis of the Weibull hazard paper, make a time-scale that includes the sample range of failure times (i.e. years).

5. Plot each failure time vertically against its corresponding cumulative value on the horizontal axis. The plot of the large motor winding failures is shown in Figure 3.9. If the plot of the sample times to failure is reasonably straight on a hazard paper, one can conclude that the underlying distribution fits the data adequately. By eye, fit a straight line through the data points (Fig. 3.9). Practical advice and more tips on making hazard plots are given by Nelson [12] and King [13].

A hazard plot provides information on:

- the percentage of items failing by a given age;
- percentiles of the distribution;
- the behavior of the failure rate of the units as a function of their age;
- distribution parameters.

In our context we are mainly interested in the distribution parameters. We already know that the Weibull distribution has an increasing or decreasing failure rate depending on whether its shape parameter has a value greater than, equal to, or less than 1. To obtain the shape parameter, β, draw a straight line parallel to the fitted line so it passes through the dot in the upper left-hand corner of the paper and through the shape parameter scale. Nautical chart parallel rulers are ideally suited for this task. Figure 3.9 shows the result. The value on the shape parameter scale is the estimate and is $\hat{\beta} = 4.3$.* A β-estimate of 4.3 suggests that the winding failure rate increases with age – that is, in a wear-out mode. It also suggests that the machines should be rewound at some age when they are too prone to failure.

In order to estimate the other parameter of the Weibull function, we enter the hazard plot on the cumulative hazard scale at 100 or 63% on the probability scale. If we move up the fitting line on Figure 3.9 and then sideways to the time-scale, we find the scale parameter, that is 18.5 years.

We can now proceed to define the Weibull distribution function that describes the large motor winding population. From Figure 3.5 we determine the basic form:

$$f(t) = \frac{\beta}{\eta} \left(\frac{t}{\eta}\right)^{\beta-1} \exp -\left(\frac{t}{\eta}\right)^{\beta} \qquad \text{for} \quad t \geqslant 0 \qquad (3.15)$$

---

* The circumflex or "hat" symbol (ˆ) means "estimated" value.

$$R(t) = \exp - \left(\frac{t}{\eta}\right)^{\beta} \tag{3.16}$$

$$h(t) = \frac{f(t)}{R(t)} \tag{3.6}$$

Applying the estimated parameters $\hat{\beta}$ and $\hat{\eta}$, Figure 3.10 was produced by using a simple Lotus® program. After having made this reliability assessment one can now proceed to work the economic decision of how to optimize motor replacement.

Construction of the replacement model

The construction of the replacement model is credited to A. K. S. Jardine [11] and A. D. S. Carter [14].

1. C_p is the cost of preventive replacement.
2. C_f is the cost of forced outage replacement.
3. $f(t)$ is the probability density function of the failure times of the motor windings.
4. The replacement strategy is to preventively replace the motors or their windings once they have reached a specified age t_p. Also, there will be replacement upon failures as necessary. This strategy is shown in Figure 3.11.
5. The goal is to determine the optimal replacement age of the motor windings to minimize the total expected replacement cost per unit time.

The equation describing the model of relating replacement age t_p to total expected replacement cost per unit time is:

$$C(t_p) = \frac{C_p \times R(t_p) + C_f \times [1 - R(t_p)]}{t_p \times R(t_p) + \int_{-\infty}^{t_p} t f(t) \, dt} \tag{3.17}$$

For the motor winding replacement case:

$$C_p = 100 \text{ k\$}$$

$$C_f = 1600 \text{ k\$} + 125 \text{ k\$} = 1725 \text{ k\$}$$

The numerical solution to the problem is presented in Table 3.3 The various columns of Table 3.3 show the values of the variables in equation 3.17 as a function of t_p. Finally, Figure 3.12 illustrates $C(t_p)$ and shows that the optimal decision would have been to preventively rewind the company's large motors after 7–8 years.

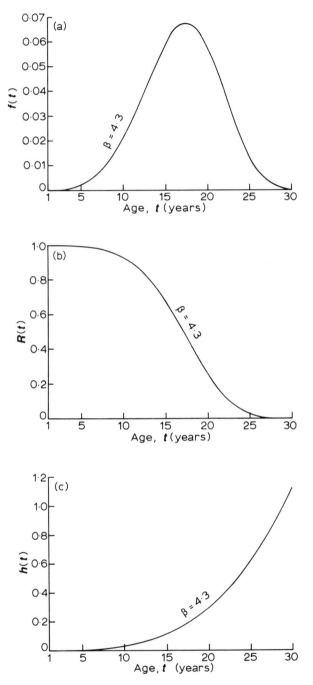

Fig. 3.10   (a) Probability density curve for large motor windings; (b) reliability curve for large motor windings; (c) hazard curve for large motor windings.

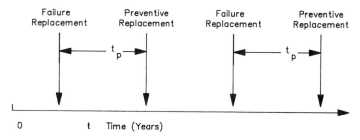

Fig. 3.11 Large motor replacement strategy.

Table 3.3 Calculation results for large motor replacement case

t_p	$R(t_p)$	$1 - R(t_p)$	$\int t\,f(t_p)\,dt$	$C(t_p)$	$\int h(t_p)\,dt$	5-year k$ incentive
0						
1	1.000	0.000	0.000	100.01	0.00	
2	1.000	·0.000	0.000	50.06	0.00	
3	1.000	0.000	0.001	33.55	0.00	
4	0.999	0.001	0.005	25.56	0.00	
5	0.996	0.004	0.016	21.18	0.00	
6	0.992	0.008	0.040	18.82	0.01	
7	0.985	0.015	0.090	17.86	0.02	
8	0.973	0.027	0.179	18.03	0.03	
9	0.956	0.044	0.328	19.22	0.05	
10	0.931	0.069	0.563	21.40	0.07	
11	0.899	0.101	0.912	21.53	0.11	
12	0.856	0.144	1.405	28.60	0.16	
13	0.803	0.197	2.070	33.58	0.22	
14	0.710	0.260	2.928	39.39	0.30	
15	0.666	0.334	3.989	45.91	0.41	
16	0.585	0.415	5.245	52.97	0.51	
17	0.499	0.501	6.665	60.35	0.70	
18	0.411	0.589	8.196	67.77	0.89	
19	0.326	0.674	9.767	74.92	1.13	
20	0.217	0.753	11.295	81.52	1.10	418
21	0.178	0.822	12.699	87.30	1.73	563
22	0.122	0.878	13.911	92.08	2.11	660
23	0.078	0.922	11.890	95.78	2.56	166
24	0.047	0.953	15.626	98.46	3.07	884
25	0.026	0.974	16.138	100.24	3.66	
26	0.013	0.987	16.465	104.33	4.33	
27	0.006	0.994	16.656	101.93	5.09	
28	0.003	0.997	16.758	102.23	5.95	
29	0.001	0.999	16.806	102.37	6.92	
30	0.000	1.000	16.827	102.12	8.00	

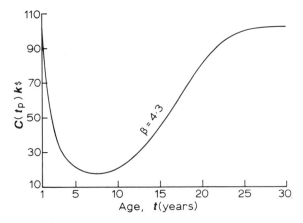

Fig. 3.12   Expected replacement cost as a function of time.

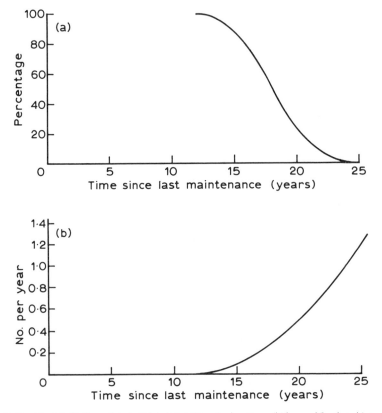

Fig. 3.13   Model of failure data in Table 3.2. (a) Survival pattern; (b) hazard (bath-tub) curve.

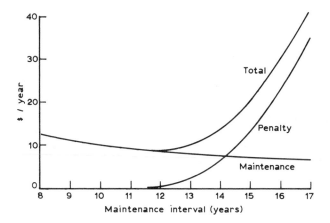

INTERVAL (YEARS)	MAINTENANCE COST PER YEAR ($)	PENALTY COST PER YEAR ($)	TOTAL COST PER YEAR ($)	
8	12.5	0	12.5	
9	11.1	0	11.1	
10	10	0	10	
11	9.09	0	9.09	
12	8.33	.27	8.6	
13	7.72	1.98	9.71	MINIMUM
14	7.25	6.12	13.3	
15	6.92	13.1	20	
16	6.71	22.9	29.6	
17	6.62	34.7	41.3	
18	6.6	47.5	54.1	

Fig. 3.14   Large motor replacement optimization calculations.

A more acceptable decision was reached by a maintenance consulting company dealing in maintenance strategy and optimization.* Using a decision-making "black box" computer program they first modelled the petrochemical plant's motor experience from the data in Table 3.2, column 3 (see Fig. 3.13). The advantage of such a tool is that no tedious calculations have to be gone through. The maintenance consultant then proceeded to utilize the information from Figure 3.13 to show how optimization of the problem is achieved by calculating "total cost per year" for every year and then obtain a minimum at year 12 (see Fig. 3.14). Column 1 shows the years under consideration, column 2 represents the annualized costs for scheduled motor rewinds, and column 3 contains the annualized forced outage associated costs or penalties. This average yearly cost of a forced outage is

* The Halcyon Systems Corporation, Wimberley, Texas 78676.

given by: $$\frac{C_p}{T} \int_0^T h(t)\, dt \qquad (3.18)$$

where T = time between two maintenance actions.

Finally, column 4 is the sum of columns 2 and 3 reflecting optimization for the year of its minimum value. The bottom of Figure 3.14 shows a graphic representation of annual costs versus time.

The petrochemical company obviously missed out on optimizing its large motor rewind strategy, given the validity of the Weibull function based model. The question arose whether or not it would now be economical, into the 19th year of their large motor operations, to plan for a preventive rewind of their three oldest motors during an upcoming shutdown. Using equation 3.18 above, annual penalties for the next 5 years* were determined as shown in column 7 of Table 3.3. These amounts were in turn claimed as credits in a discounted cash flow (DCF) analysis. They felt it was a sound decision to preventively rewind their three old motors during the shutdown.

REFERENCES

1. Czichos, H., *Tribology – A Systems Approach to the Science and Technology of Friction, Lubrication and Wear*. New York: Elsevier, 1978.
2. Yoshikawa, H. and Taniguchi, N., Fundamentals of mechanical reliability and its application to computer aided machine design. *Annals of the CIRP*, **24** (1), 1975, p. 300.
3. Green, A. E. and Bourne, A. J., *Reliability Technology*. New York: Wiley-Interscience, 1972.
4. Henley, E. J. and Kumamoto, H., *Reliability Engineering and Risk Assessment*. Englewood Cliffs: Prentice-Hall, 1981.
5. BSI BS 5760:Part 2:1981, Reliability of systems, equipments and components. Part 2. Guide to the assessment of reliability. London: British Standards Institution, 1981, pp. 9–11.
6. Fleischer, G., Probleme der Zuverlässigkeit von Maschinen. *Wiss. Z. TH Magdeburg*, **16**, 1972, p. 289.
7. British Standards Institution, op. cit., p. 12.
8. Nelson, N. W. and Hayashi, K., Reliability and economic analysis applied to mechanical equipment. *Journal of Engineering for Industry*, Feb. 1974, pp. 311–316.
9. Sidall, J. N., *Probabilistic Engineering Design*. New York: Marcel Dekker, 1983.
10. Nelson, W., *Applied Life Data Analysis*. New York: John Wiley, 1982.
11. Jardine, A. K. S., *Maintenance, Replacement and Reliability*. Bath, UK: Pitman Press, 1973.
12. Nelson, W. B., A method for statistical hazard plotting of incomplete failure data that are arbitrarily censored. Schenectady: GE R&D Center TIS Report 68-C-007, Jan. 1968.
13. King, J. R., *Probability Charts for Decision Making*. New York: Industrial Press, 1971.
14. Carter, A. D. S., *Mechanical Reliability*. New York: John Wiley, 1972.

* Average time between planned shutdowns.

4

A universal approach to predicting machinery reliability

In the preceding chapter we showed the usefulness of hazard functions in estimating machinery reliability. Frequently, it is not possible to arrive at an appropriate distribution function due to a lack of specific data and the need for complicated calculations. In many cases, and especially when comparing competing solutions to a technical problem (i.e. *relative* reliability), a constant failure rate for machinery components may be assumed and judiciously applied.

A *constant failure rate* assumption does not deviate too much from the real world for at least two reasons. First, different distribution functions for a variety of components when combined produce a random failure pattern. Second, repair at failure tends to produce a constant failure rate when the population is large. This has been demonstrated in the literature [1].

With a constant failure rate the reliability of components or systems follows the exponential distribution:

$$R(t) = \exp(-\lambda t) \qquad (4.1)$$

We have already seen that the reciprocal of failure rate is called Mean Time Between Failure (MTBF), or μ, the mean of the distribution. For example, small electric motors have typical failure rates of $\lambda = 14.3 \times 10^{-6}$ per h. What is the MTBF of the motor and what is its reliability for a 8000-h operating period?

$$\text{MTBF} = \frac{1}{\lambda} = 8 \text{ years}$$

$$\text{reliability, } R(t) = \exp(-14.3 \times 8000 \times 10^{-6})$$

$$= 0.891 \quad \text{or} \quad 89.1\%$$

47

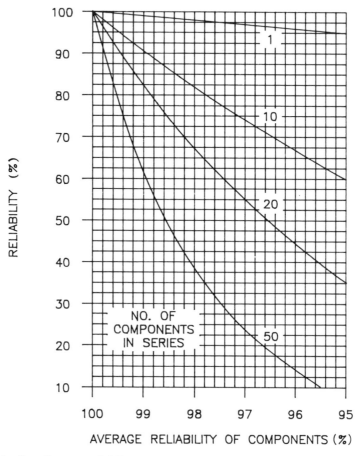

Fig. 4.1   Overall system reliability: components in series.

RELIABILITY OF PARTS IN SERIES

The reliability of parts or components in series is:

$$R_s = R_1 \times R_2 \times \cdots \times R_n \tag{4.2}$$

$$= \exp[-(\lambda_1 + \lambda_2 + \cdots + \lambda_n)t] \tag{4.3}$$

If the components have identical failure rates, then

$$R_s = \exp(-n\lambda t) \tag{4.4}$$

Usually, this approach leads to a demand for very high component reliability in any system consisting of many parts (Fig. 4.1). For instance, we see that in order to obtain an 80% reliability in a unit with 50 components in series, an average component reliability of 99.4% is required.

A. S. Carter [2] describes a simple everyday experience of automotive transport that suggests that this approach is oversimplified:

> At peak hour traffic conditions 20 to 30 vehicles may be held up at a traffic light. Each vehicle has at least 100 components in series in its transmission system, giving some 2000 components in series at each traffic light. Yet how often does the queue fail to move when the traffic lights change to green due to mechanical failure? Chaddock [3] has carried out a more scientific investigation of the supposed correlation between reliability and number of components, studying a number of weapons for which accurate data existed. He concludes there is no such correlation.
>
> The truth is that success is achieved when the weakest or least adequate individual component of a system is capable of coping with the most severe loading or environment to be encountered, that is the strength of the chain equals that of its weakest link. This has been emphasized by other researchers who at the same time recognize the fact of variability, or scatter, both in the capability or strength of the product. It has been further emphasized in the duty it will have to face, that is the load which will be imposed on it.

The author then goes on to explain this phenomenon by a model in which both load and strength are distributed and where the strength distribution, due to some form of progressive weakening, invades the load distribution causing more and more of the population to fail (see Fig. 4.2). Carter's work also shows that where a component proves inadequate in changed duty or environment, it has only to be strengthened a little to restore the failure rate to an acceptable level. This goes to show that a "management-by-exception" approach to machinery reliability assessment is justified, that is vulnerabilities, as we will see later, have to be exposed.

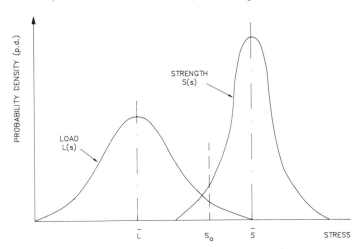

Fig. 4.2 Distribution of load and strength (from ref. 2). (Reprinted from Carter, A.D.S., Mechanical Reliability, 1972, p. 5, by courtesy of Macmillan Press Ltd.)

Concepts as shown in Figure 4.1 have nevertheless sometimes led to unjustified waste in the process industries by providing spares, for instance, that are poorly or not at all utilized. We are alluding to cases where spares are almost "automatically" furnished without prior evaluation of the alternatives. The alternatives are to procure machinery reliable enough so that spares are not required, or to weigh the risks of not furnishing spares against the incentives of providing them [4].

TWO COMPONENTS IN PARALLEL

The combined reliability of two identical components in parallel depends on the system requirement (Fig. 4.3). Two cases are possible. First, the failure of either component disables the system. Both A and B must survive. They are in series from the reliability point of view:

$$R_s = R_A \times R_B = \exp(-2\lambda t) \qquad (4.5)$$

Second, survival of one component is sufficient. Here, system reliability (R_s) is the probability that A or B or both survive:

$$R_s = R_A \times R_B - R_A R_B \qquad (4.6)$$

$$= 2 \exp(-\lambda t) - \exp(-2\lambda t) \qquad (4.7)$$

which is valid for identical components.

An example for this case can be found in petrochemical pumping services. Here the need for parallel redundancy is based on cost of lost production, cost of unscheduled versus emergency repairs, and capital cost [5]. In order to determine the need for a spare or standby pump, one would first evaluate equation 4.1. With a failure rate of $\lambda = 1.5$ per year, the resulting reliability referred to 1 year would be:

$$R_s = \exp(-1.5 \times 1) = 0.22 \qquad \text{or} \qquad 22\%$$

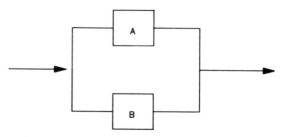

Fig. 4.3   Diagram of two components in parallel.

That is, the probability of failure (P_f) would be:

$$P_f = 1 - R = 0.78 \quad \text{or} \quad 78\%$$

Obviously, this is an unacceptable proposition.
We will now evaluate equation 4.7:

$$R_s = 2 \exp(-1.5) - \exp(-3) = 0.446 - 0.050$$
$$R_s = 0.40 \quad \text{or} \quad 40\%$$

This represents an improvement by almost a factor of 2. However, the result of equation 4.7 does not tell the whole story. Remember the definition – "Probability of survival of A or B or both". Obviously, we have to consider the fact that the system tends only to fail if the operating pump fails while the spare is out for repair. For $\lambda = 1.5$ per year, $\mu = 8$ months MTBF, $t = 5$ days repair time $(= 0.0137$ years):

$$R_s = \exp(-1.5 \times 0.0137)$$
$$= 0.98 \quad \text{or} \quad 98\%$$

or

$$P_f = 1 - 0.98 = 0.02$$

Installing a spare pump in our system reduces probability of failure of the system during one operating year from 78 to 2%.

Often the repair quality of spared machinery is unduly compromised by shortening repair times as much as possible. Obviously, this is done intuitively in order to maintain reliability of the system. Figure 4.4 explains the relationship between MTBF of a spared machinery installation, time to repair the spare, and the resulting reliability factor. It assumes a "mature" machinery population, meaning that failures occur mutually independent of each other or perhaps as the result of some random outside influence such as the result of a unit start-up or upset. We assume that no common failure causes exist, such as suction system or shared utility service problems.

Suppose we wanted to know how long a spared pump can be out for repair without endangering the process unit reliability goal which has been decreed to be 98.5%. The presently unspared pump was started and is running satisfactorily. It belongs to a population of similar pumps in similar service with an MTBF of 15 months. We move vertically from 15 on the horizontal axis and intersect the reliability line of 98.5% at a horizontal line corresponding to an allowable spare pump outage of 7 days. We conclude that there should be no need to rush the repair of the spare pump.

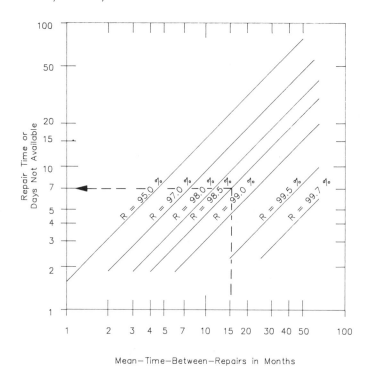

Mean−Time−Between−Repairs in Months

Fig. 4.4   Reliability versus mean-time-between-failure and repair time - spared service.

THREE IDENTICAL COMPONENTS IN PARALLEL

1. Failure of any component disables the system illustrated in Figure 4.5:

$$R_s = R_A \times R_B \times R_C = \exp(-3\lambda t) \tag{4.8}$$

2. The system can stand failure of one component. Two or more components must survive.

$$R_s = 3R^2 - 2R^3 \tag{4.9}$$

$$= 3 \exp(-2\lambda t) - 2 \exp(-3\lambda t) \tag{4.10}$$

3. The system can stand failure of any two components. One of the three must survive.

$$R_s = 1 - (1 - R)^3 \tag{4.11}$$

$$= 1 - [1 - \exp(-\lambda t)]^3 \tag{4.12}$$

$$= 3 \exp(-\lambda t) - 3 \exp(-2\lambda t) + \exp(-3\lambda t) \tag{4.13}$$

4. Consider now a system as shown in Figure 4.6, where the failure of any one component can cause a system failure. Assume that the failure of any

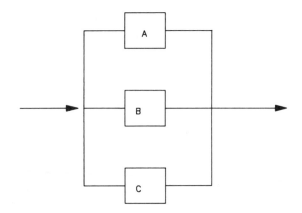

Fig. 4.5 Diagram of three components in parallel.

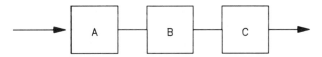

Fig. 4.6 Diagram of components in series.

one part in this series system is independent of the failure of another. The probability that the system will survive is, according to equation 4.2:

$$R_s(t) = R_A(t) \times R_B(t) \times R_C(t) \qquad (4.14)$$

For an exponential time-to-failure density of each individual part, we can write:

$$R_s(t) = \exp(-\lambda_A t) \times \exp(-\lambda_B t) \times \exp(-\lambda_C t) \qquad (4.15)$$

$$= \exp[-(\lambda_A + \lambda_B + \lambda_C) \times t] = \exp(-\lambda_s t) \qquad (4.16)$$

PREDICTION PROCEDURES

Most reliability engineering prediction procedures are based upon the above described exponential time-to-failure density. This permits simple addition of average component failure rates in order to arrive at the equipment or system failure rate from which the MTBF or reliability function may be obtained. The mathematical basis for this approach was demonstrated by equations 4.2 and 4.16. In this technique, we merely add the number of indispensable or non-redundant components of each type, multiply this by the basic average failure rate for each type of component, and add these figures to obtain the machinery unit failure rate. The MTBF is then the reciprocal of that failure rate.

As a more sophisticated approach for electronic systems MIL-HBK-217D [6] advocates the above method for predicting electronic equipment failure rates using data contained in it. The method depends on the quality of generic failure rates. These are derived from the equation:

$$\lambda_P = \lambda_b(\pi_E \times \pi_Q) \tag{4.17}$$

where λ_P = predicted failure rate,
   λ_b = base failure rate for the generic part,
   π_E = an environmental factor,
   π_Q = a quality factor.

π_Q relates to sources and specified quality, π_E to the general environment in which the part will be used. The generic part values themselves are based on a large amount of data from laboratory, development, and field sources.

A similar source exists for mechanical generic data [7]. It subdivides the data by source so that the environment is known, but does not classify by quality. With its aid, however, a parts count for a piece of mechanical equipment can be performed, taking into account the environmental effects, providing that all the required data are in the lists.

Other methods take part stress levels into account. This is done, for example, by determining ratios of operating versus design pressure, operating versus design temperature, design size versus median size, or other important parameters. Relevant stress ratios are then weighted and the result applied to a suitable distribution function from which failure rates are determined [8].

Since our purpose is to make *relative* machinery reliability assessments we feel that a judicious application of the parts count method is justified.

We have already used the concept of expressing failure rates in terms of failures per 1 million hours. This would amount to 150 years if we assume continuous around-the-clock operation. This seems like quite a long time. However, we can look at it in another way. An equivalent experience would be if a process unit with 150 different kinds of failures had one outage a year.

Expressing failure rates per million hours is convenient because many failures occur at a rate of 1 per million operating hours. Machinery component failures would lie mostly between 1 and 100 failures per 1 million hours or $1–100 \times 10^{-6}$ h. Table 4.1 illustrates how these failure rates relate subjectively to various levels of reliability.

FAILURE RATE DATA

Failure rate data is best obtained from operating experience. Table 4.2 illustrates how failure rate data for machinery components can be obtained from field statistics. Column 2 shows the actual service experience of reciprocating compressors based on a company's experience in several plants.

Table 4.1

Reliability	$\lambda \times 10^{-6}$
Extremely reliable	0.01
Highly reliable, OK in large numbers	0.01–0.1
Good reliability for moderate numbers	0.1–1.0
Average reliability, OK in small numbers	1.0–10
Very unreliable	10–100
Intolerable	>100

Source: Atomic Energy of Canada Ltd.

Table 4.2 Failure rate statistics: Reciprocating compressor

Elements	Failures (%)	Rate per 1×10^6 h
Valves	43.0	98.4
Pistons and cylinders	19.0	43.0
Lube systems	18.0	41.0
Piston rods	10.0	22.8
Packings	10.0	22.8
Total	100.0	228.0[a]

[a] Equivalent to 2 incidents per year.

Column 3, the failure rates, are obtained by first postulating two incidents per year on these particular machines at a given plant site. The failure rates are then calculated by multiplying the field data percentages by the failure rate equivalent of two incidents per year (i.e. 228×10^{-6} h).

Another important aspect of reliability prediction using failure rates is the consideration of failure modes. Failure modes have distinct failure rates and the component or part failure rate is the sum of its mode failure rate.

Failure modes are typically first a description of loss of function or malfunction and then a more detailed expansion in terms of the basic failure mode, namely the appearance of the failure (see Table 6.1). Earlier we looked at some basic failure modes in connection with failure distributions. We refer our readers to Table 3.1. Basic failure modes and the failure mechanisms associated with them play a central role in machinery failure analysis [9].

In using failure rate data for machinery reliability assessment it is a good idea to work with "worst", "best", and "expected" concepts. This reflects the fact that machinery parts and components can have different qualities. In order to make things less complicated we will calculate reliability based on two qualities – best and worst. We will then investigate if the worst case is viable. If that is the case, we need not worry because the actual quality will be closer to the expected value. If, however, our reliability based on the

Table 4.3 Failure rates for machinery components [10–13]

	$\lambda\ (10^{-6})$	
	Best	*Worst*
1.0 Transmitting elements		
1.1 Couplings		
1.1.1 Elastomeric	20.0	30.0
1.1.2 Gear	8.0	20.0
1.1.3 Disc/diaphragm	0.01	0.1
1.2 Gear sets		
1.2.1 General purpose	8.0	50.0
1.2.2 High-speed helical	0.5	15.0
1.3 Shafts		
1.3.1 Lightly stressed	0.02	0.1
1.3.2 Heavily stressed	0.1	0.5
1.3.3 Crankshafts (R.C.)	5.0	8.0
1.4 Clutches		
1.4.1 Friction	2.0	8.0
1.4.2 Magnetic	4.0	10.0
1.5 Drive belts		
1.5.1 V-belts	20.0	80.0
1.5.2 Timing belts	40.0	80.0
1.6 Springs		
1.6.1 Lightly stressed	0.01	0.1
1.6.2 Heavily stressed	0.8	2.5
2.0 Constraining, confining, containing elements		
2.1 Bearings		
2.1.1 Sleeve bearings	4.0	10.0
2.1.2 Ball bearings	5.0	50.0
2.1.3 Roller bearings	3.0	10.0
2.2 Seals		
2.2.1 O-rings	0.1	0.7
2.2.2 Oil seals	8.0	10.0
2.2.3 Mechanical seals	25.0	200.0
2.3 Valves		
2.3.1 R.C. (Recip. comp.)	50.0	150.0
2.3.2 Check valves	0.8	10.0
2.3.3 Manual valves	0.4	6.0
2.3.4 Relief valves	1.0	10.0
3.0 Fixing elements		
3.1 Threaded fasteners		
3.1.1 Bolts	0.001	0.007
3.1.2 Pins	8.0	40.0
3.1.3 Set screws	0.03	1.0
3.1.4 Rivets	0.001	0.01

Table 4.3 (cont.)

	$\lambda\ (10^{-6})$	
	Best	*Worst*
4.0 Support elements		
4.1 Casings		
4.1.1 R.C. cylinder jackets	0.01	0.1
4.1.2 R.C. cylinder liners	10.0	30.0
4.1.3 Pump casings	0.01	1.0
4.2 Vibration mounts		
4.2.1 Elastomeric	6.0	20.0
4.2.2 Wire rope coils	0.1	1.0
4.3 Motor windings		
4.3.1 Small motors <250	10.0	20.0
4.3.2 Large motors >250	5.0	10.0
5.0 Basic failure modes		
5.1 Force/stress/impact		
5.1.1 Deformation	0.01	0.1
5.1.2 Fracture	0.001	0.01
5.1.3 Binding/seizure	0.1	1.0
5.1.4 Misalignment	0.1	1.0
5.1.5 Displacement	0.01	0.1
5.1.6 Loosening (fastener)	0.1	1.0
5.2 Reactive environment		
5.2.1 Corrosion		
1. Accessible parts	0.01	0.1
2. Inaccessible parts	0.1	1.0
5.2.2 Fretting		
1. Mostly stationary	0.1	1.0
2. Exposed to dirt	1.0	10.0
5.3 Temperature effects (see under aging)		
5.4 Time effects		
5.4.1 Wear/relative motion		
1. Non-lubricated	0.1	1.0
2. Lubricated	0.01	0.1
5.4.2 Erosion		
1. Accessible parts	0.01	0.1
2. Inaccessible parts	0.1	1.0
5.4.3 Aging		
1. Lubricants	0.01	0.1
2. Rubber	0.01	0.1
3. Metals, thermally stressed	0.1	1.0
5.4.4 Contamination		
1. Accessible parts	0.01	0.1
2. Inaccessible parts	0.1	1.0
5.4.5 Fouling/plugging		
1. High-velocity areas	0.01	0.1
2. Low-velocity areas	0.1	1.0

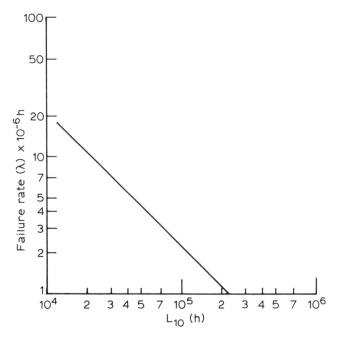

Fig. 4.7   Failure rate for anti-friction bearings versus L_{10} life (adapted from ref. 13).

best case scenario is unacceptable, we have to take corrective action by looking for improved designs.

Table 4.3 lists failure rates for machinery components and parts as well as failure modes compiled from various literature sources and the authors' experience. Figure 4.7 may be used to obtain failure rates for anti-friction bearings based on known or assumed L_{10} lives.

THE PROCEDURE

The procedure to calculate reliability based on failure rates is simple. It can be applied to predict reliability of machinery on the assembly, hierarchy and system level within the limits of underlying assumptions. Figure 4.8 shows the form used in this effort.

The first step is to list all parts essential to the successful functioning of the system under study. The second step is to determine the quantity of parts. Third, after checking Table 4.3 for failure rate information on the specific component under consideration, determine the most probable failure mode for it; typically, fractures with shafts, wear or contamination with simple oil seals, and bearing or winding failures with motors, and so forth. If more than one failure mode is expected consider the mode with the highest failure rate. Compare this with the failure rate for the part if it is available.

Non-redundant components	Quantity	Failure rate per 10^6 h									
		0.01		0.1		1.0		10.0		100.0	
		Best	Worst	Best	Worst	Best	Worst	Best	Worst	Best	Worst
1. Clutch cplg (3)											
Bearing	1			1			1				
Oil seals	2					2	2				
Coupling	1					1	1				
2. Gear Box (4)											
Bearing	2			2			2				
Oil seals	2					2	2				
Gear set	1	1			1						
3. Coupling (5)											
Elastomer	1					1	1				
4. Motor (6)											
Bearing	2			2			2				
Winding	1			1	1						
5. Gear box (7)											
Bearing	6			6			6				
Oil seals	2					2	2				
Gear set	3	3			3						
Sum best: 9.24		0.04		1.20		8.00					
Sum worst: 19.50					0.50		19.00				

Fig. 4.8 Calculating machinery failure rates.

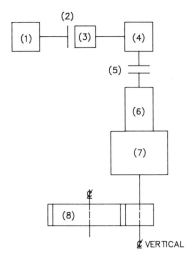

(1) AIR MOTOR

(2) COUPLING NO.1

(3) OVER–RUNNING CLUTCH
    3–REPAIRS

(4) RIGHT ANGLE GEARBOX
    2–REPAIRS

(5) COUPLING NO.2
    1–REPAIR

(6) ELECTRIC MOTOR
    3–REPAIRS

(7) TRIPLE REDUCTION GEARBOX
    3–REPAIRS

(8) DRIVE GEARS

Fig. 4.9   Rotary air pre-heater drive train.

Multiply the highest failure rate in terms of best and worst and enter the values in the appropriate columns. It stands to reason that this analysis will be as accurate as one is able to recognize the elements of a part and their corresponding failure modes.

The fourth step is to add the best and worst values separately. One has to determine now whether or not the sum of the worst values is tolerable. If the conclusion is affirmative no action is necessary. If the "worst quality" assumption is not tolerable we have to look for improvements. Usually, the individual failure rate values will be a clue! If there are some particularly high values try to substitute their "best quality" value and see how this affects the overall failure rate. If this does not satisfy our expectations we have to embark on a design change.

The example illustrated in Figure 4.8 pertains to a drive critical to the operation of a rotary furnace air pre-heater in a large process plant. The drive consisted of eight machinery components schematically shown in Figure 4.9. The reliability analysis was made in order to determine whether or not the drive was the weak element in an otherwise highly reliable and cost-effective scheme to recover waste heat.

Finally, how can we reduce component failure rates? The following questions need to be asked:

● Can the component be replaced by a known improved component?
● Has a review of the component design been done?
● Have all known weaknesses been eliminated?
● Have all uncertainties been identified?
● Are the uncertainties being eliminated by analysis or test?

- Can the design be simplified
  - by reducing the number of parts?
  - by eliminating need for high precision?
  - by requiring less maintenance skill?
- Have new features and new materials been proven by analysis or test?
- Can components tolerate abnormal conditions?

We have chosen to lump abnormal conditions under the acronym "FRETT". The letters F-R-E-T-T stand for:

F: Forces, mechanical loads, deflections and pressure
R: Reactive agents
E: Environment
T: Temperature
T: Time, exposure to long-term and short-term loads, and deflections, i.e. vibration and shock.

Whenever F-R-E-T-T are outside the as-designed or anticipated values or quantities, the part, machine, or component will be prone to fail prematurely.

THE USE OF RELIABILITY DATA FOR MACHINERY SCOPING STUDIES

Reliability data in the form of failure rates have been used in the past for scoping studies and machinery reliability assessments. A typical example is the following material which was excerpted from a technical report. It illustrates how a critical shaft sealing system for a process gas compressor was chosen by using relative reliability and availability failure rate estimates.

The engineer responsible for the selection of the sealing system prepared the following report,* which basically stated that:

A recycle gas compressor for a gas methanation project handles gas which is toxic and contains methane. The gas suction temperature is above 300 deg. C. A review of the sealing arrangement for the compressor was made with regard to the gas composition and other important operating parameters. The suitability of various buffer gases available at the plant site was considered and recommendations made for an appropriate sealing arrangement and buffer gas.

The conclusions were:

- A labyrinth sealing system with buffer gas will provide the safest and most reliable sealing system for the compressor package.

---

* Courtesy of S. A. Pradhan, formerly British Gas.

62

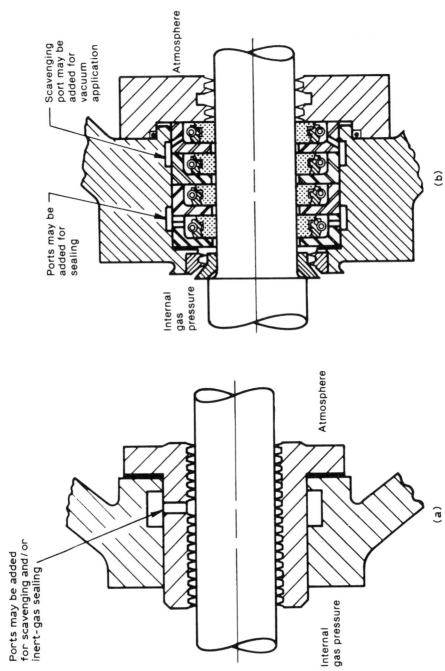

Fig. 4.10 (a) Labyrinth shaft seal; (b) restrictive bushing shaft seal (from ref. 14).

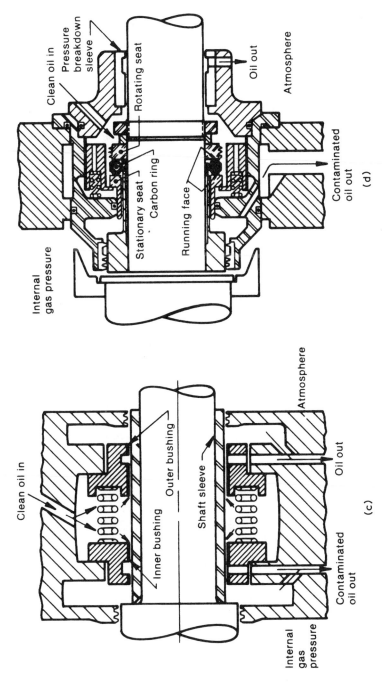

Fig. 4.10 (c) Liquid film shaft seal; (d) mechanical contact shaft seal (from ref. 14).

63

- Steam is the most suitable buffer medium. Its ready availability at the plant site makes it more attractive than nitrogen.
- Nitrogen is also a suitable buffer gas if the cost factor introduced by its use is acceptable.

The details were as follows.

The hot recycle gas compressor was to handle a mixture of gases containing a high proportion of carbon monoxide, carbon dioxide, and steam. It was designed to work in a hazardous area Zone 1 with group II B gases.* Owing to the possible toxic effects of the gas mixture and plant and personnel safety, the compressor sealing arrangement had to be carefully selected and be very reliable. Further, the sealing system had to satisfy the above requirements during all operating conditions.

The hot recycle gas compressor was to handle a mixture of hydrocarbon gas with steam. Its composition was as follows:

Gas	% by volume
CO_2	36.79
CO	0.76
H_2	3.90
CH_4	26.39
N_2	0.62
Steam	31.54

Molecular weight	$= 26.57$
Gas suction pressure	$= 340$ psia
Gas suction temperature	$= 200°C$
C_P/C_V ratio	$= 1.256$
Gas discharge temperature	$= 220°C$

In order to prevent IN and OUT leakage an industrial standard recommends four basic types of sealing designs as follows [1]:

1. Labyrinth seal with or without buffer gas (Fig. 4.10a).
2. Restrictive seal (Fig. 4.10b).
3. Liquid film seal (Fig. 4.10c).
4. Mechanical contact seal (Fig. 4.10d).

The sealing system for the compressor had to satisfy the following criteria.

1. Prevent IN and OUT leakage of gas during all operating conditions of the compressor and also during compressor shutdown.
2. Involve as few mechanical moving parts as possible for utmost reliability.

* A process plant area classification.

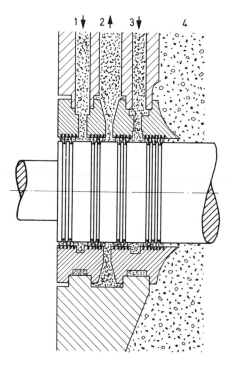

Fig. 4.11 Labyrinth shaft seal system with ejector. 1, Buffer gas supply; 2, leak-off port; 3, balance port; 4, process gas.

3. Be simple in design and easy to maintain.
4. All parts making up the sealing system had to be compatible with process gas and be able to withstand temperatures of 220°C.
5. Act as heat sink to limit temperature rise and heat transmission to bearings and other lubricated parts.
6. Be suitable for the larger compressor sets to be used in future commercial scale SNG (synthesized gas) plants.

The available seal types were evaluated as follows. First, the labyrinth seal. A simple seal that will only restrict gas flow but will not provide a positive seal. This seal type therefore does not satisfy all the seal selection criteria. Second, the labyrinth seal with buffer gas. In this design a balance chamber connects the high pressure discharge side to the suction side. With this arrangement, seals at both shaft ends operate against the same gas pressure under normal working conditions.

The seal has three chambers separated by groups of labyrinths (see Fig. 4.11). Buffer gas is admitted to the middle chamber at 0.5–1.0 bar above the reference gas pressure and expands on either side. Buffer gas leaking from the middle chamber to the inner chamber mixes there with any gas leaking through the labyrinths from inside the compressor. The resulting mixture is at slightly less than suction pressure and is vented to the atmosphere. Buffer gas leaking from the middle chamber to the outer chamber is not contaminated and can be exhausted to the atmosphere. This system provides positive sealing during standstill and shutdown if the supply of buffer gas is uninterrupted.

Reference to Table 4.4 shows that this sealing system satisfies the selection criteria.

Subsequently, the decision had to be made which buffer gas to select. The following gases were available at the plant site:

- instrument air, not suitable for safety reasons;
- gaseous nitrogen at ambient temperature;
- natural gas;
- steam at 450 psia.

It appeared as though either nitrogen or natural gas could have been used as buffer gas. Both gases were readily available at the site and could have acted as a cooling medium for the hot leakage process gas.

During compressor standstill, a supply of either nitrogen or natural gas would prevent leakage of air into, or process gas from, the compressor casing. A controlled leakage of process/buffer gas mixture to vent would also occur. It was estimated that about 100 Nm^3/h of buffer gas would leak. Similar protection was to be provided during compressor operation or if any deviations from suction or discharge conditions occurred.

At this site, nitrogen was produced in an air separation plant. The nitrogen produced was mixed with a small percentage of hydrogen and used as a circulation gas during SNG production and shutdown cycles. In addition, there were other processes where nitrogen was used. This placed a heavy demand on nitrogen which, depending on the demand, was likely to lead to fluctuations in the supply. The use of nitrogen as a buffer gas for the sealing system could thus reduce availability.

On larger compressors for commercial scale SNG plants in the future, nitrogen sealing systems were judged not to be very suitable as nitrogen may not be produced on site. Also, nitrogen costs (at 1980 prices) at the equivalent of $200/ton could impose a severe constraint on its use:

Nitrogen Costs in HCM Compressor Sealing
Price of nitrogen (1980) = $200 per ton
Molecular weight of nitrogen = 28
Estimated leakage of nitrogen from the seal system = 100 Nm^3/h

$$m = \frac{P_1 V_1}{R T_1} = \frac{1.0132 \times 10^5 \times 100}{\dfrac{8.3143 \times 10^3}{28} \times 273} = 124.98 \approx 125 \text{ kg}$$

Table 4.4 Selection criteria for process compressor seal system

Seal type	Prevent in and out leakage	Involves mechanical parts	Simple design	Compatibility with gas	Temperature resistance	Comments
Labyrinth seal	Will only restrict leakage	No	Yes	Yes	Yes	Not suitable, does not provide positive seal
Labyrinth seal with buffer gas and ejector	Yes. Positive sealing if buffer gas supply maintained	Ejector with motive gas	Yes	Yes	Yes	Fulfill all selection criteria
Restrictive seal	Will only restrict leakage	Yes. Dry rub between shaft and ring	Yes	Yes	Maximum temperature for carbon rings 200°C	Not suitable. Frequent replacement of rings
Liquid film seal	Yes	Yes. Floating rings with seal oil system	No	Yes	Seal oil temperature limit 200°C	Not suitable. Compressor pressure not high enough to warrant this system
Mechanical contact seal	Yes	Yes	No	Yes	Maximum temperature for carbon rings 200°C	Not suitable. Frequent maintenance, increased instrumentation. Over-sealing

Total consumption of nitrogen per annum (200 days) $= 125 \times 24 \times 200$
$$= 600,000 \text{ kg}$$
$$= 600 \text{ tons}$$

Total nitrogen cost $= 600 \times \$200/\text{ton}$
$$= \$120,000/\text{year}$$

Similarly, cost factors associated with the use of natural gas, together with the safety hazards associated with venting of large volumes of combustible gas to the atmosphere, made the use of natural gas undesirable as buffer gas for future SNG plant compressors.

Looking at steam as buffer gas the conclusions were:

Steam is a safe and suitable buffer gas. But in this case an ejector will have to be fitted to the inner chamber. This is necessary because during start up, steam from the middle chamber will expand to the inner chamber and mix with a cold stream of H_2/N_2 mixture [see Fig. 4.12]. This will induce some condensation of the steam and reduce the sealing. However, if the middle chamber is always under suction due to the ejector, steam will never leak into the compressor casing, except in cases where the ejector has failed to operate.

The process gas contains 31.54% of steam. A common steam generating plant can supply the process requirements as well as steam for sealing requirements.

Since the process gas requires a continual supply of steam, a failure of the steam supply will shut down the compressor plant. The steam supply to the sealing system will also be integrated. However, since process gas supply has stopped, the compressor, in this situation, could be vented to atmosphere. Another advantage of using steam sealing systems is that the experience gained on the pilot plant can be used on the future larger commercial plants. The use of steam will be more cost effective and also more reliable than nitrogen which will not necessarily be produced on site.

When investigating the suitability of the liquid film seal, it was concluded:

In this seal system in addition to a supply of buffer gas, oil is supplied between two seal rings at a pressure slightly above reference gas pressure. This system prevents process gas from escaping even at highest gas pressure. The seal oil also acts as a cooling medium for the rings.

As with the labyrinth seal system with buffer gas, liquid film seal systems only seal a stationary compressor when the buffer gas system is operating. With this system a separate seal oil supply system together with pumps, filters and oil separator is necessary. Unlike the labyrinth seal system with steam and ejector, this system is complex and has many mechanical components liable to break down and malfunction.

Reference to [Table 4.4] shows that this system satisfies most of the selection criteria. However the use of a seal oil system introduces some unreliability into the system.

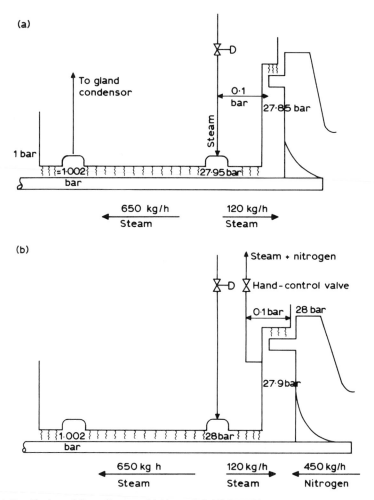

Fig. 4.12 Shaft seal flow diagram. (a) Normal operation; (b) start-up.

[Table 4.5] shows the reliability data for compressor and auxiliaries including lube and seal oil systems.* It shows that in the worst case a seal oil system introduces 1.4 failures per 2000 hours of operation. This failure rate is only exceeded by the trip system.† Moreover, the operating pressures involved in the process are not high enough to justify a liquid film seal oil system. On this basis, a liquid film sealing system is not recommended.

---

* From a national reliability data base.
† This experience has been superseded by modern petrochemical process plants, i.e. 0.1 failures/2000 h.

Table 4.5   Failure rates and expected downtime: Recycle gas compressor

Component group	Failure rates per 10^6 h $(\lambda)^a$		Expected no. of failures per 2000 h (λT)	Downtime per breakdown (d)	Total downtime (λTd)
Compressor	λ_w	301	0.6		201.6
	λ_b	24	0.048	336	16.1
	λ_e	110	0.22		73.9
Motor and gearbox	λ_w	346	0.69		41.4
	λ_b	53	0.11	60	6.6
	λ_e	128	0.26		15.6
Couplings	λ_w	158	0.32		1.28
	λ_b	20	0.04	4	0.16
	λ_e	60	0.12		0.48
Lubrication and seal oil system	λ_w	703	1.4		2.8
	λ_b	272	0.54	2	1.08
	λ_e	272	0.54		1.08
Trip system	λ_w	1144	2.3		2.3
	λ_b	1	0.002	1	0.002
	λ_e	347	0.69		0.69

^{a} λ_w = worst case values; λ_b = best case values; λ_e = expected values.

When considering the restrictive seal it was observed:

This is very similar to the labyrinth seal but here the labyrinths are replaced by dry carbon rings which provide a more tortuous path to passage of gas along the shaft. As for a simple labyrinth, the dry carbon seal is not a positive seal. Due to close tolerance the gas leakage is minimal, however wear on the rings is greater, and they may require more frequent replacement. This sealing system is thus unlikely to be suitable for hot recycle compressor sealing without special development.

Finally, the mechanical contact seal was investigated. In this system a seal element slides on a rotating sealing ring. It has a liquid sealing medium. This sealing system will provide positive sealing at standstill. As with restrictive seals the rate of wear of the carbon rings is greater than that of the harder mating material. Since the overall system has many components including separate lube and seal oil systems, the expected overall reliability is relatively low. Requirements for control and instrumentation to monitor the system are very much increased. This increases the cost and also introduces the possibility of spurious trips and alarms with unnecessary interruption of the process.

On the basis of the foregoing, Table 4.4 indicates that this arrangement is unsuitable for hot recycle gas compressor sealing.

The above example is admittedly simplistic and rather unique. We would like to point out that in some cases, this kind of selection process is greatly influenced by outside forces or interests independent of reliability considerations: environmental protection agencies, national and local code requirements, operating versus expense costs, and so forth.

REFERENCES

1. Henley, E. J. and Kumamoto, H., *Reliability Engineering and Risk Assessment*. Englewood Cliffs: Prentice-Hall, 1981, pp. 198–205.
2. Carter, A. D. S., *Mechanical Reliability*. New York: John Wiley, 1972, pp. 5–11.
3. Chaddock, D. M., The reliability of complicated machines. *Inter Services Symposium on the Reliability of Service Equipment*. London: Institute of Mechanical Engineers, 1960.
4. Simmons, P. E., The optimum provision of installed spares. *Hydrocarbon Processing*, Apr. 1982.
5. Davis, G. O., How to make the correct economic decision on spare equipment. *Chemical Engineering*, Nov. 1977.
6. MIL-HBK-217D, *Reliability Prediction of Electronic Equipment*, ANSI, 1982.
7. *Guided Weapon System Reliability Prediction Manual*, MOD (PE) Report DX/99/013-00.
8. Venton, A. O. F., *Component-based Prediction for Mechanical Reliability. Mechanical Reliability in the Process Industries*. London: MEP, 1984.
9. Bloch, H. P. and Geitner, F. K., *Machinery Failure Analysis and Troubleshooting*. Houston: Gulf Publishing, 1983, pp. 527–530.
10. Hauck, D., *A Literature Survey*. AECL Report No. CRNL-739, 1973.
11. Grothus, D., *Die Total Vorbeugende Instandhaltung*. Dorsten, Germany: Grothus Verlag, 1974.
12. Giacomelli, E., Agostini, M., and Cappelli, M., *Availability and Reliability in Reciprocating Compressors Evaluated to Modern Criteria*. Florence, Italy: Quaderni Pignone, 42/12/1986.
13. *Standard Handbook of Machine Design*. Edited by J. E. Shigley and C. R. Mischke. New York, Toronto: McGraw-Hill, 1986, pp. 27-1 to 27-17.
14. API Standard 617, *Centrifugal Compressors for General Refinery Service*, 5th edn, April 1988.
15. Stankovich, I., Modern pneumatic handling. *Bulk Solids Hardware*, **4**(1), March 1984, pp. 183–188.

5

Predicting reliability of turbomachinery*

Numerical, statistical methods, as shown in the preceding chapters, have the advantage of being able to identify areas of vulnerability before the analyst is actually confronted by the hardware and its potential problems. A similar systematic reliability evaluation of major turbomachinery components, for example, warns of potential problems in future or existing installations.

The major turbomachinery train shown in Figure 5.1 features a steam turbine driving a low-pressure (LP) and a high-pressure (HP) turbocompressor.

A procedure and a set of curves can be developed to coordinate the major factors influencing reliability of this type of equipment. These factors are type of machine, unit size, speed, pressures and temperatures, coupling effects, number of start–stop cycles, starting cycle time, characteristics of supports, foundation, piping, and the effects of operating practices and maintenance provisions.

Reliability factors were established to improve the accuracy of equipment evaluations, and to make sure a maximum number of remedies can be considered quickly. Reliability factor curves presented in the following pages are based on personal experience and extensive use of references. Such curves can never be highly accurate, and they can never cover all possible types of installation. Common sense must be used in their application. The curves are given more to outline a systematic procedure than to provide numbers ready for use.

INTERPRETATION OF RELIABILITY FACTORS (RF)

- RF = 2.0 or above: Excellent probability of trouble-free operation. Breakdown rate is estimated about half that of normal.

---

* Adapted from ref. 13, with the kind permission of the author, John S. Sohre, Turbomachinery Consultant, Ware, Massachusetts 01082.

Fig. 5.1 Two-casing oxygen compressor with main and intermediate gearing. *Source:* Mannesmann-Demag, Duisburg, West Germany.

- RF = 1.0: Average installation with normal probability of failures and breakdowns.
- RF = 0.5: Probability of problems is about twice that of normal.
- RF = 0.1: Probability of problems is about 10 times normal. In other words, chances of trouble-free operation will be very poor, and time between breakdowns will be short. Usually, basic changes will be required to correct the situation.

The interpretation of reliability factors should be valid for individual components as well as for the overall installation. Table 5.1 gives a random example to illustrate the procedure. This assumed example indicates a good overall plant design but poor installation and facilities. Unusual trouble is not likely to occur. But if it should happen, significant improvement could be obtained quickly by correcting the piping and foundation rather than by looking into couplings, bearings, or other basic equipment details. Unusual troubles could just as easily develop the other way around. It depends on where the weak spots of an installation are located. This same installation could become marginal if it were started and stopped every day or if it ran at higher speeds or if it were to be quick-started, etc.

The example shown would be a relatively easy project to work on. It would be far more difficult to come up with a solution where all the factors run close to 1.0. When all factors are 1.0, a random breakdown is probably involved. Troubleshooting would then require a well-planned and coordinated effort. Many different symptoms would have to be analyzed before an improvement could be made.

Table 5.1   Example of overall reliability determination

Equipment: 10,000 rpm, turbine driven compressors
From the design curves (Fig. 5.2)

Type of equipment:	Turbine	RF =	1.0
	Compressor		1.2
Equipment size:	Turbine, 5-ft bearing span		1.0
	Compressor, 1st body 3-ft bearing span		1.5
	Compressor, 2nd body 4-ft bearing span		1.2
Number of bearings in train:	6		0.9
Start-up time:	40 min		1.0
Maximum pressures:	Turbine: 600 psi		1.6
	No. 1 compressor: 75 psi		1.9
	No. 2 compressor: 300 psi		1.8
Maximum temperature:	Turbine: 750°F		0.8
	Compressor No. 1: 150°F		2.0
	Compressor No. 2: 250°F		1.9
Coupling:	Gear, curved teeth		1.0
Casing support:	Turbine, flexplates, centerline		1.1
	Compressor No. 1, non-centerline		0.8
	Compressor No. 2, centerline, sliding		1.0
Starting frequency:	One per year		1.8
Multiplied subtotal, design features			51.3

Installation (from Figs 5.3 and 5.4)

Piping strains:	Turbine: 150% NEMA and API	0.7
	No. 1 compressor: 100%	1.00
	No. 2 compressor: 100%	1.00
Pipe supports	Turbine: springs	1.00
	Compressor: rack and rod	0.7
Expansion joints:	One poorly restrained joint on compressor	0.6
Foundation:		
Rigidity:	Mat and slab weak	0.7
Vibration:	Non-resonant (weight 1 × unit weight)	0.8
Vibration isolation:	Not isolated, significant non-resonant transmission	0.8
Installation, multiplied subtotal		0.132

Operation:		
Operators:	Average	1.0
Maintenance personnel:	Good	1.5
Maintenance facilities	Poor	0.5
Operation, multiplied subtotal		0.75

Total overall reliability of installations: RF = 51.4(0.132)(0.75) = 5.1

FACTORS INFLUENCING RELIABILITY

Type of equipment (*Fig. 5.1*)

Electric motor driven units are highly reliable at low speeds. As speeds and gear ratios increase, double increaser gear trains become necessary. As units become more and more sophisticated, the reliability drops off rather sharply. Reliability drop is especially a problem with long equipment trains.

Because of the extremely short starting time with motor driven high-speed trains, it is practically impossible to supervise the unit during start-up. The short starting time may result in a very high damage level if trouble occurs. Frequent starting aggravates the situation.

Synchronous motor drives introduce the additional risk of torsional failure when passing through slip-frequency resonance. Heavy torsional shock and vibration can be the result of relatively minor malfunctions during the synchronizing cycle. Other shock and vibration problems are caused by short-circuit and phase faults. Particularly susceptible are long trains exposed to frequent starting.

Steam turbines require a considerable amount of auxiliary equipment such as boilers, piping, or condensers. Auxiliaries affect overall realiability, but reliability is very high once the unit is running. High speeds present no more problems than would be expected with a centrifugal compressor, allowing for some effects of temperature, pressure, and auxiliaries.

Equipment size (*Fig. 5.2*)

The faster a machine runs, the smaller it must be. Otherwise problems of stress, and especially vibration problems, will develop. Critical speeds and other rotor instabilities become a dominating factor at high speeds. Situations can arise where it becomes impossible to pass through a critical phase without risking destruction of the machine. Such conditions are mainly affected by bearing span. Therefore, the reliability curves are plotted for various spans.

For three-bearing machines, the stability of the rotor improves considerably and longer spans can be used. For a given reliability, the longest span between adjacent bearings can be increased by 10–50%. Span depends on speed and design features. An increase in span can more than double the shaft length without loss in reliability. Obviously, there must be provisions to hold the three bearings lined up nearly perfect under all normal and abnormal conditions. Otherwise, the reliability can go down rather than up. Alignment is the major problem with three-bearing machines.

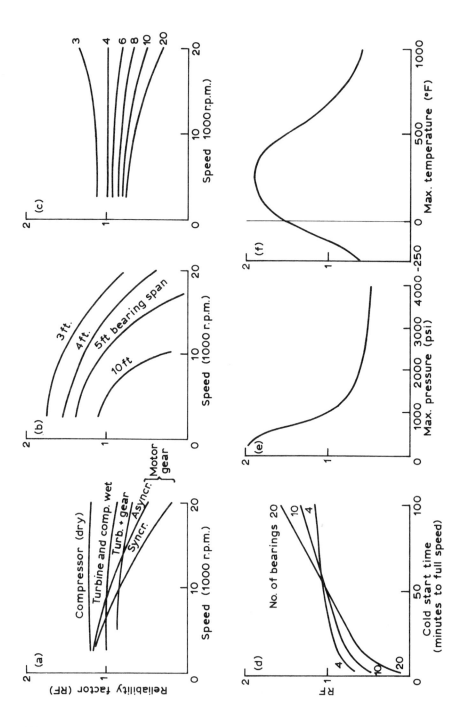

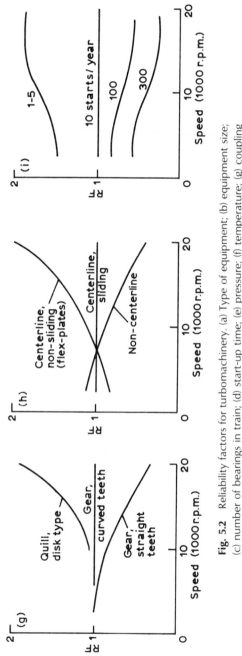

Fig. 5.2 Reliability factors for turbomachinery. (a) Type of equipment; (b) equipment size; (c) number of bearings in train; (d) start-up time; (e) pressure; (f) temperature; (g) coupling type; (h) type of casing support; (i) starting frequency.

Number of bearings in the train

The reliability of a long train with many couplings will be less than that of a simple, short unit. Reliability can be expressed by the number of bearings in the train. If there are gears in the train, multiply the reliabilities of the low-speed section with the one for the high-speed section to get an overall reliability.

Start-up time

Quick starts (motor drive) are more likely to damage long trains than short trains. There is no time to supervise the long unit during the few seconds it takes to come to full speed. A curve has been included to show the effect of start-up time. Much depends on supervisory instrumentation, protective devices, starting shock severity, temperature, pressures, surging, switch gear operation, etc. Individual estimates must be made to include these factors.

Pressure

The pressure factors are reflected in the reliability curves. Pressures shown are maximum pressures on the unit. The advantages and disadvantages of high-pressure machines are:

Advantages	*Disadvantages*
Compact design	Small machines, sensitive to pipe strain
Small distortions	Heavy pipe walls
Small piping	Small supports, close together
Small internals	High impact loads on internals
	High thrust load and thrust load variations
	Long seals
	Thick casing walls
	Tight clearances

Temperature

Temperature is the main offender where reliability is concerned. Most compressors, gears, and motors are only exposed to moderate temperatures, as compared to turbines. Lower temperatures compensate for some other shortcomings of motors such as short starting cycle, torsional vibrations, electrical problems, etc.

Temperature can cause distortion of the casing, foot, and foundations as well as misalignment and problems with pipe expansion. Temperature inflicts restrictions on materials. Seals are only one example. Material restrictions affect the entire design philosophy, as well as the efficiency and life-expectancy of the unit.

Coupling types

Gear couplings are often considered standard for large, high-speed equipment. At high speeds straight teeth can contribute to certain rotor instabilities. Curved or barrelled teeth often give smaller exciting forces on the rotor. Much depends on the design and coupling quality.

Someone once said: "You can never waste money buying the best coupling you can get." This statement is especially true for large, fast machines and long trains. The problem is not so much that the coupling breaks down – although this may also happen – but that the coupling excites the rotor system into vibrations and instabilities which can be very violent. This excitation is caused by the interaction of periodic tooth friction forces with the rotor–stator damping system. Other problems, such as those caused by misalignment or rotor critical speeds, may also be emphasized or de-emphasized by variations of coupling design and quality.

Well-designed quill shafts or flexible disk-type couplings are much lighter. They do not generate the instabilities caused by looseness, friction, and lubricant contamination or lubricant breakdown, which are inherent in gear couplings to a greater or lesser degree. This makes flexible, dry-disk couplings more reliable, especially at high speeds.

Improper installation or poor maintenance can easily cause failures. Damage to highly stressed quills and membranes is one reason why these couplings are sometimes not used. Couplings of this type have been used successfully in aircraft engines, where a very high level of maintenance control is standard.

Casing support

A casing support structure is sometimes suspected of causing trouble in an area where it cannot do much harm or even where it is the best type to apply. Each type has its advantages and disadvantages. A rugged sliding foot support is often best for large, low-speed machines such as large turbines, gears, motors, generators, and compressors. Centerline supports become necessary at higher speeds and temperatures. If supports are of the sliding type, they may bind or lift under pipe forces and thermal distortion. Also, sliding supports can cause serious vibration. Flexible plates avoid these problems and are especially advantageous for relatively small machines running at high and very high speeds. Small, slow machines are often simply bolted down, because thermal expansions are small and can be absorbed with little distortion. A bolted machine has the advantage of ruggedness and insensitivity to piping strain.

Piping strain (*Fig. 5.3*)

The effect of piping strain on a machine reduces reliability by:

- causing misalignment and subsequent vibration;
- causing case distortion and subsequent vibration, rubs, case leakage, and possible cracking;
- causing foundation or base deflection, which may result in misalignment, case distortions, and subsequent vibrations or rubs.

Excessive piping strain may be the result of:

- Thermal expansion and contraction of the pipe, boiler, and machine. This indicates faulty pipe design. Expansion joints or loops may have to be installed.

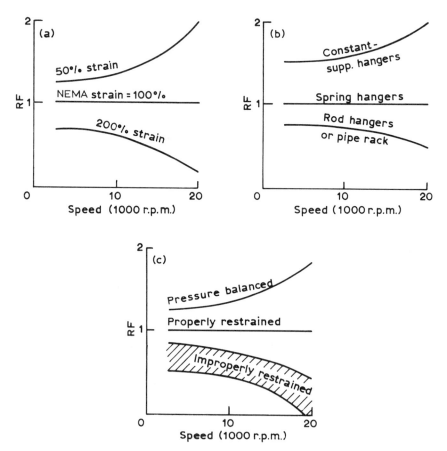

Fig. 5.3 Effect of piping on reliability. (a) Pipe strain; (b) pipe supports; (c) expansion joints.

● Improper pipe support. Frequent problems arise from indiscriminate use of rod hangers – instead of spring hangers – anchors, and other non-elastic restraints and supports. For correction, disconnect the piping at both ends and support it on spring hangers, except where anchors or restraints are required by the pipe design.

Improper pipe installation is very frequently a source of trouble and hard to find once the pipe is installed. Usually, piping is not properly lined up at the flanges. If flanges are not parallel when lined up, very large moments and forces can occur in the casing and at the case supports. To identify strains caused by flange misalignment, mount dial indicators at the coupling and supports, disconnect the pipes, and observe the movements. These installation strains are superimposed on thermal expansion strains.

Cold spring is one source of piping strains. The cold spring usually encountered is provided by cutting the pipe short by about half the anticipated thermal expansion. The pipe ends are pulled together and welded. Cold spring strain is practically unpredictable, especially the resultant moments, and quite often the equipment suffers from it.

Expansion joints

Expansion joints are often useful in low-pressure lines, but they are not anywhere nearly as flexible as many engineers believe. If expansion joints are not lined up properly, or indiscriminately exposed to shear or torsion, the strains on the machinery can cause serious problems. One must also consider the thrust caused by an unstrained expansion joint. It is equal to the cross-sectional area at the largest bellows diameter multiplied by the internal pressure, in psig. Tie rods often used on expansion joints to absorb the thrust are only effective and harmless when the joint is used in shear. If tie rods are used on a joint which is meant to move in tension-compression, they bypass the joint and make it virtually useless, because pipe forces are then transmitted through the rods. Or, if restrained only in one direction, the rods may become loose. Then, the pressure thrust will act on the machine again, as if the rods were not there.

Settling foundations of machinery, boilers or condensers, can cause serious pipe strain. Often involved are metal expansion joints between equipment and condensers (or coolers) and large, low-pressure piping with little flexibility. Concrete shrinkage and creep also belong in this category. A 10-ft column shrinks 0.06 in. during the first 6 years. Creep during the first 2 years is three to four times the original static deflection.

Size and speed

The faster a machine runs, the more sensitive it will be to pipe strain.

- A high-speed machine is smaller than a low-speed machine.
- Piping is normally sized for flow, regardless of speed. It is therefore often large compared to the casing size and support strength. The result is more severe distortion and misalignment for the faster machine. Tolerance of a machine for distortion and misalignment decreases as speed increases.
- As speed increases, the tendency for a rotor to become unstable also increases. Instability is caused by oil whirl and certain friction-induced and load-induced whirls. Therefore, a given displacement which is harmless at low speed can cause instability at high speeds.
- Bearing clearances are small (smaller journals) for high-speed machines. Thus, bearings are less tolerant of distortion and displacement.

The above factors are reflected in the curves showing the effect of piping strain on reliability (Fig. 5.3). The piping is assumed to be in accordance with API and NEMA standards. That is, allowable strains are a function of casing weight and size and therefore allowable forces and moments are smaller for fast running machines. The curves show the effect of excessive strain. This includes all strain regardless of the source such as installation, support settling, or expansion joints.

Pipe supports

Pipe supports are shown separately because their effect is pronounced during start-up, shutdown, and load changes. Also, pipe supports are a significant factor in long-term reliability due to settling, jamming of springs and slides, or plain aging effects.

It is unrealistic to base allowable pipe reactions on pipe size only, disregarding the size, mass, and speed of the equipment. Such a design allows the same pipe strain no matter whether the machine is large or small.

Foundation (*Fig. 5.4*)

The foundation is one of the most influential factors where overall reliability of a unit is concerned. A foundation must:

1. Maintain alignment under all normal and abnormal conditions. The conditions include soil settling, thermal distortion, piping forces, vacuum pull, or pressure forces in expansion joints. A heavy and rigid mat, the portion resting on the soil, is a key to good alignment. Other aids to alignment are: equal deflections of all columns under load, as well as mass, continuity, symmetry, and rigidity of the top slab on which the unit rests. The way the foundation is supported on the soil, as well as soil characteristics and soil resonances, deserve special attention.
2. Minimize vibration. The foundation must be as heavy as possible and

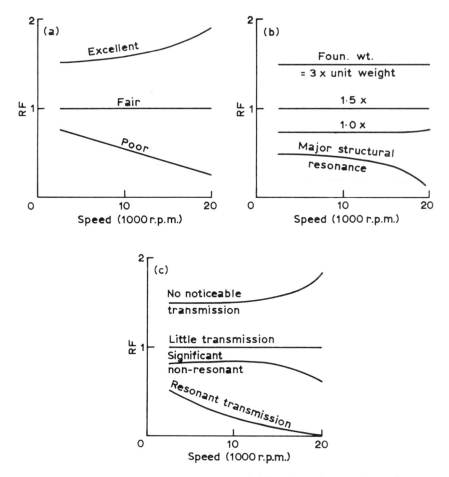

Fig. 5.4 Effect of foundation characteristics on reliability. (a) Rigidity to maintain alignment; (b) vibration characteristics; (c) isolation from surroundings.

non-resonant. If a foundation is resonant, it does not matter much whether it is a light structure or a heavy one; reliability will be greatly reduced in either instance.

3. Isolate the unit from external vibrations. For larger or more critical units, one should provide an air gap filled with mastic sealer all around the slab and mat. Vibration transmission may be from the unit to the surroundings or vice versa, and it may be aggravated by resonance at transmission frequencies. Piping, stairways, and ducts may also transmit vibration, which should be prevented by proper isolation. Ground water transmission is often serious. Reliability is reduced when units, especially large ones, are mounted on baseplates which are then mounted on top of the foundation. Baseplates introduce an additional member in the system

which increases deflections and vibrations. Usually, deflections and resonant frequencies become unpredictable and have a way of showing up at the wrong place and at the wrong time. Besides, the base usually interferes with proper foundation design. Therefore, cost savings of a unit mounted on a steel frame base should be evaluated against reduced reliability.

Operation (*Fig. 5.5*)

The larger and/or faster the unit, the more influence operators will have upon reliability. One must use one's own judgment in rating an operating

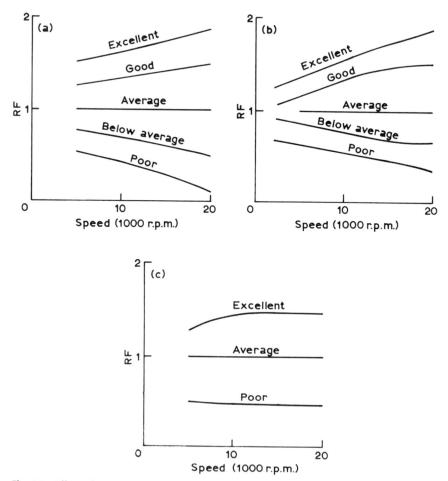

Fig. 5.5 Effect of operations and maintenance on reliability. (a) Operating personnel; (b) maintenance personnel; (c) maintenance facilities.

crew. The main factors include training, intelligence, cooperation, but especially organization and leadership.

Operating personnel as a factor in reliability may not seem important at first. One is usually stuck with a given crew when a troublesome job comes up. But evaluation of operating personnel will tell us how reliable or foolproof we must make a unit if it is to be operated successfully by such a crew.

Maintenance personnel

Maintenance crew evaluation is essentially the same as for an operating crew. The same factors must be considered together with their economic effects and calculated risks. To illustrate the effect of maintenance, consider as an example the internal inspection of a high-speed compressor which may improve the chances of successful operation. With a good crew one would make the inspection; with a poor crew one would rather take a calculated risk of a failure up to a certain level of severity. One can reason that the machine is likely to be worse off rather than better, after a poor maintenance crew inspects it. Evidently, such a unit will be considerably less reliable, whether the crew is put to work on it or not.

Maintenance facilities include working conditions with the process unit as well as shops, tools, availability of spare parts and, last but not least, availability of instruction books, drawings, and technical data.

Ruggedness of turbomachinery

Many people feel that a more massive construction provides higher reliability. Others question this point, believing that one can build a very light machine with the same reliability as a design weighing many times as much. Aircraft engines are usually referred to, and one can hardly argue the point that they are highly reliable. The question seems to be mainly whether or not the necessary sophistication went into a lightweight machine, to make up for the obvious advantages of mass and rigidity. This can only be decided by looking at the respective designs.

However, this item seems to receive increasing attention and it was suggested by a specialist that the massiveness metal content of a machine can be expressed in some comparative form, to allow evaluation. To do this, the weight not contributing to ruggedness must be disregarded.

Length is perhaps the most critical dimension of high-speed machines. A short machine is more reliable than a long one. If we calculate the weight per inch of turbine, this should tell us a good deal about its construction. For example, a turbine for the same speed, efficiency, and conditions can be built with, say, six or nine stages, depending on the thermodynamics and hardware sophistication. The shorter machine will be more compact and will have a greater average weight per inch of length, although the total weight

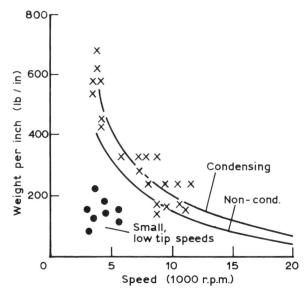

Fig. 5.6　Weight per inch of turbine (overall length).

may actually be less than that of the longer unit. Design sturdiness experienced by parameters such as wall thickness or mass will also be reflected in this number. Speed has an effect because as speeds go up machines get smaller in diameter.

Weight per inch has been plotted for several units in Figure 5.6. This figure covers several turbine generators from 3000 to 22,000 kW and several high-speed compressor drives in the 2000 to 15,000 hp range, both condensing and non-condensing. Five machine types of different manufacture are included. Most machines are of average to heavy design. Therefore, the curves indicate fairly heavy construction.

Weight used in the weight per inch ratio is overall weight of the installed turbine, including control valves, trip and throttle valve, but not baseplates, oil tanks, and the like. Length used is overall body length. But the length does not include small protrusions such as protruding shaft ends, valves, or flanges.

To get a somewhat better feel of the actual metal content of a turbine, the equivalent solid diameter (steel) has been plotted in Figure 5.7. Equivalent solid diameter is the diameter if the whole turbine were compacted until no air remained inside and then made into a round bar of the same length and weight as the turbine. Equivalent solid diameters are surprisingly large. Another way of looking at this is to multiply the weight per inch with the rated speed as a parameter of diameter. This gives us a factor which is independent of speed (since weight per inch appears to vary as a linear function of speed). We can call this factor a Mass Concentration Factor

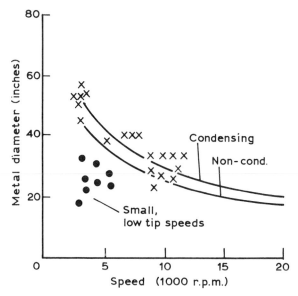

Fig. 5.7 Equivalent solid metal diameter.

(MCF). Then:

$$MCF = \frac{\text{Total turbine weight (lb)}}{\text{Total turbine length (in)}} \times (\text{rated speed, rpm})$$

For machines of comparable ruggedness of construction this factor remains surprisingly constant for a wide variety of designs and speeds. The upper curves (condensing turbines) are plotted for an MCF of 2.0×10^6. Non-condensing turbines appear to be about 10–20% lighter, with an MCF of 1.6–1.8×10^6.

Examples

A condensing turbine compressor driver with a 6000-hp rating operates at 8000 rpm with $W = 256,000$ lb and $L = 76$ in. MCF $= (26,000/76)(8000) = 2.74 \times 10^6$. This is an unusually heavy machine.

Another condensing turbine compressor drive has a 5000-hp rating and operates at 8300 rpm with $W = 20,000$ lb and $L = 108$ in. MCF $= (20,000/108)(8300) = 1.54 \times 10^6$. This is a long, relatively lightly constructed machine.

The group of light machines in the low-speed area represents small units with many stages, which are often used for this type of service.

We should use the description above only as a guide to help assemble meaningful data which can then be interpreted to suit individual preferences and requirements. Compressors and other machinery equipment can be evaluated in a similar manner. Reliability curves can then be plotted to include these factors in the overall evaluation of a unit.

BIBLIOGRAPHY

1. Sohre, J. S., *Transient Torsional Criticals of Synchronous Motor Driven, High Speed Compressor Units*. ASME paper 65-FE-22.
2. Gunter, E. J., Jr., *Dynamic Stability of Rotor-Bearing Systems*. NASA SP-113.4.5. Washington, D.C.: US Government Printing Office.
3. *American Standard Code for Pressure Piping*. ASME, 345 East 47th Street, New York, New York 10017.
4. *NEMA Approved Standards, Piping for Turbine-Generator Units, and for Mechanical-Driven Steam Turbines*. NEMA, 155 East 44th Street, New York, New York 10017.
5. *API Standards 615 (Turbines) and 617 (Compressors)*. American Petroleum Institute, 50 West 50th Street, New York, New York 10017.
6. Sohre, J. S., *Foundations for High-Speed Machinery*. ASME paper 62-WA-250.
7. Sohre, J. S., *Operating Problems with High-Speed Turbomachinery, Causes and Correction*, 3rd revision. Originally presented at ASME Petroleum Mechanical Conference, 1968.
8. Pollard, E. I., *Torsional Response of Systems*. ASME paper 66-WA/Pwr-5.
9. Naughton, D. A., *Preventable Accidents to Turbines and Speed Increasing Gear Sets*. Hartford, Connecticut: The Hartford Steam Boiler Inspection and Insurance Co.
10. Huppman, H., *Das Hydrodynamische Gleitlager im Grossmaschinenbau*. Referat #11, Allianz Versicherungs AG "Der Maschinenschaden", 8 München 22, Postfach 220, Germany.
11. Bahr, H. C., *Recent Improvements in Load Capacity of Large Steam Turbines Thrust Bearings*. ASME paper 59-A-139.
12. Schmitt-Thomas, K. G., *Das Zusammenwirken von Korrosion und mechanischer Beanspruchung an metallischen Werkstoffen bei der Auslösung von Schäden*. Allianz Versicherungs AG.
13. Sohre, J. S., *Turbomachinery Analysis and Protection*. Proceedings of the 1st, Turbomachinery Symposium, Texas A&M University, College Station, Texas, 1972.

6

Failure mode and effect analysis*

Failure Mode and Effect Analysis (FMEA) is a name given to a group of activities which are performed to ensure that all that could potentially go wrong with a product has been recognized and that actions are taken to prevent things from going wrong.

In the 1960s, the moon flight program engineers, faced with staggering consequences of malfunctioning space vehicles, devised a method of forecasting the problems that could occur with every component. They were thinking beyond the normal design considerations, all the way to the most bizarre situations one could devise. They did this in long and concentrated brainstorming sessions. The result of this approach contributed to the success of the moon landing in 1969.

With the decline of the space program in the early 1970s, many NASA engineers found jobs in other industries and brought failure forecasting with them. The technique became known eventually as the FMEA. In 1972, NAAO, a Quality Assurance organization, developed the original reliability training program which included a module on the execution of FMEA.

Although good engineers have always performed an FMEA type of analysis on their designs, most of their efforts were documented only in the form of their final parts and assembly drawings. Repetition of past mistakes, however, was possible, because people were assigned to other tasks, left the company, etc. With liability insurance besieging, for instance, the automotive industry in the 1970s, FMEA became a natural tool to lower the occurrence of failures. Since that time, the discipline has been spreading among the multibillion dollar companies. In turn, these large companies have been pressing their suppliers to adopt FMEA to improve the reliability of their products.

THE PROCESS

Fully implemented, the FMEA process is applied to each new product and to any major change of an existing product. The FMEA documents become

---

* Based on a paper presented by S. R. Jakuba, S. R. Jakuba FMEA Consultants, at the 1987 Spring National Design Engineering Conference, ASME. By permission of the author.

an integral part of the product design documentation and are as such continuously updated.

The FMEA process is used to ensure that all problems that could possibly occur in the design, procurement, and servicing of a product have been considered, documented, and analyzed.

In this regard, the Product Engineering organization has the responsibility for product performance criteria establishment, and product design and development, which includes consideration of manufacturability, serviceability, and user's potential misuse. The Manufacturing organization has the responsibility for the fabrication or purchase of a product to an engineering drawing and specification. The Marketing, Sales and Service organization has the responsibility for technical support of the product after sale.

Accordingly, there are three independent FMEA documents dealing with the three different aspects of the process. The Design FMEA lists and evaluates the failures which could be experienced with a product and the effects these failures could have in the hands of an end user. The Manufacturing FMEA lists and evaluates the variables that could influence the quality of a particular process. The Service FMEA evaluates service tools and manuals to ensure that they cannot be misused, or misrepresented. In the following, we deal with the Design FMEA procedure only.

DESIGN FMEA

The Design FMEA identifies areas that may require further consideration of design and/or test. It captures and implements design inputs, some of which might otherwise not be made, and, if made, might get lost. They include inputs from other departments such as Manufacturing, Sales, Purchasing, Service, Reliability, and Quality Assurance. Combining the different viewpoints and experience not only improves the design of a machinery product but also improves the acceptance of it throughout the company and in the field.

The Design FMEA, further referred to as FMEA, is initiated after a conceptual design has been finalized. It should be substantially completed before the production hardware is made, to ensure that the Production Release documentation includes the FMEA inputs and that the potential benefit of the FMEA information is fully utilized. Subsequent changes to a product should also be incorporated. FMEA documentation should be periodically updated to record the changes and their impact on reliability and risk.

The FMEA is initiated by the engineer responsible for release of the product for manufacturing after he has determined that the product as designed will perform to the performance specification, can be made and assembled to print, and is serviceable and "foolproof".

The objective in performing the Design FMEA is to:

- find whether the performance specification is proper and complete;
- find if and how the design could be inadequate both to the design intention as well as for reasons of overload, contamination, weather extremes, manufacturing variations, serviceability, customer misuse, or negligence, etc.;
- evaluate the consequences of a marginal product reaching a customer;
- quantify risk;
- identify the need for corrective actions and to assign priorities for their execution;
- implement and follow agreed upon actions.

DEFINITIONS AND FMEA FORMS

The following definitions apply:

Failure mode	The manner in which a part or system fails to meet the design intent
Effect of failure	The experience the owner encounters as a result of a failure mode
Cause of failure	An indication of a design weakness
Cause prevention	The in-place and scheduled design verifications and quality assurance inspections
Severity ranking	A subjective evaluation of the consequence of a failure mode on the end user
Occurrence ranking	A subjective estimate of the likelihood that if a defective part is installed it will cause the failure mode with its particular effect
Detection ranking	A subjective estimate of the probability that a cause of a potential failure will be detected and corrected before reaching the end user
Risk Priority Number (RPN)	The product of severity, occurrence and detection rankings

Failure mode, effect of failure and cause of failure serve to document all that could fail, how the failure would be perceived if it happened, and what could cause it. The cause prevention serves to document all existing and firmly scheduled measures intended to assure that the cause of a failure has been eliminated. Severity ranking, occurrence ranking and detection ranking

Table 6.1 Failure mode examples

	Failure Mode and Effect Analysis (FMEA)									Sheet No. ___ of ___	
System:		Component:									
System status:		Component status:				Operating conditions:		Documentation:			
1	2	3	4	5	6	7	8 9 10 11				12
No.	Part/ function	Failure mode	Basic failure mode/ possible cause	Failure detection (surveillability)	Available counter- measures	Failure effects	Occurrence Severity Detection RPN				Failure assessment and recommended action

Table 6.1 Sheet 2 of 2

Failure Mode and Effect Analysis (FMEA)								
1	2	3	4	5	6	7	8 9 10 11	12
No.	Part/ function	Failure mode	Basic failure mode/ possible cause	Failure detection (surveillability)	Available counter- measures	Failure effects	Occurrence / Severity / Detection / RPN	Failure assessment and recommended action

93

provide a numerical means of stating a subjective estimate of the respective parameters. Typically, on a scale of 1–10, the rankings represent a number which reflects how severe the effect of a failure is, how likely the failure is to happen, and how unlikely the cause of failure is to pass undetected.

The Risk Priority Number (RPN) is the number resulting from the multiplication of the three rankings. RPN allows prioritization of the actions that need to be performed to lessen the risk.

Table 6.1 illustrates a typical FMEA form. There is no one FMEA form that suits all companies and all applications. The first four of the above categories are, however, almost always present. Also, the form always contains a space for information needed to identify the product, such as drawing number, product application(s), where it is made, and its function. There should also be a space for the listing of corrective actions. The corrective actions are recommended by the FMEA participants, and regardless whether they will be pursued or not, they should all be recorded on the form.

Finally, there should be a space for the name of a person responsible for the implementation of a corrective action.

PROCEDURE

The Design FMEA procedure is an integral part of a product development. The engineer responsible for the product should make entries on the FMEA forms, listing his thoughts and reasoning concurrently with performing the other design and test activities. It is important that information written on the forms is concise, clear, and systematically arranged, because people unfamiliar with both the product and FMEA will later read and evaluate the entries. If the entries are vague or incomplete, the potential of the FMEA effort will not be realized; not only will the time of several people be wasted but also potentially dangerous problems may be overlooked.

Experience indicates that it is more cost-effective not to perform an FMEA at all than to produce a vague, half-hearted one. A certain writing and organizational talent is needed to produce the FMEA document. Not every engineer has the talent, and not every engineer is willing to devote the time needed for researching all the information, and write and rewrite it until it conveys the relevant message in just three or four words.

When production drawings are available the engineer contacts the person responsible for the FMEA activities. Together they select the people who should review the FMEA drafts, and amend and rank the entries. The selection of the reviewers is done on the basis of their qualification both with respect to their knowledge of the product and their ability to contribute to the FMEA process. The selected participants are briefed on the product and on the duties expected of them in the FMEA process.

After they have had a chance to study the FMEA documents, gather information related to their involvement with the product, and amend and rank the entries, an FMEA meeting is called. The meeting may last several days, so its timing must be planned. During the meeting all the entries on the form are reviewed, recommended actions confirmed, and priorities assigned.

When managerial approval of the recommended actions is obtained, the actions are given deadlines. The control of the completion is assured by entries on the FMEA form, usually on a separate, shorter form, which lists the approved actions only.

It can generally be said that the training of the FMEA participants, and the effort involved in performing FMEA is substantial. The benefits of an FMEA are reliability enhancement and cost avoidance, not a measurable saving in the bottom line. Therefore, to be carried through in an effective way, the FMEA activities require the unconditional commitment of the management and a dedicated leadership.

The FMEA technique provides the means of presenting one's thoughts in a methodical way. The objective is to document all potential flaws of a product, evaluate the risks associated with each, and prevent the occurrence of high risks.

The benefits of the FMEA process extend clearly beyond the design aspects of a machinery product. It enables the designer and owner to gain a deeper knowledge of the product. Further, it increases the awareness of the product features by all involved parties and it provides a basis for an assessment of reliability, maintainability, and safety of similar or newly designed products.

EXAMPLES

In the preceding paragraphs we have seen that a FMEA produces the following results:

- It identifies potential and known failure modes.
- It identifies the causes and the effects of each failure mode.
- It prioritizes identified failure modes according to frequency of occurrence, severity, and defect formation.
- It allows to plan for problem follow-up and corrective action.

An effective FMEA depends on certain key steps. (We refer our readers to completed examples on pp. 103 and 105). The essential steps are as follows:

1. *Describe the anticipated failure mode.* The analyst must ask the question: "How could this part, system or process fail? Could it break, deform, wear, corrode, bind, leak, short, open, etc.?" Table 6.2 and the

Table 6.2   Failure mode: Basic

2.1 Part/element level	2.2 Assembly level
2.1.1 Force/stress/impact	2.2.1 Force/stress/impact
1. Deformation	1. Binding
2. Fracture	2. Seizure
3. Yielding	3. Misalignment
4. Insulation rupture	4. Displacement
2.1.2 Reactive environment	5. Loosening
1. Corrosion	2.2.2 Reactive environment
2. Rusting	1. Fretting
3. Staining	2. Fit corrosion
4. Cold embrittlement	2.2.3 Temperature
5. Corrosion fatigue	1. Thermal growth/contraction
6. Swelling	2. Thermal misalignment
7. Softening	2.2.4 Time
2.1.3 Thermal	1. Cycle life attainment
1. Creep	2. Relative wear
2. Cold embrittlement	3. Aging
3. Insulation breakthrough	4. Degradation
4. Overheating	5. Fouling/contamination
2.1.4 Time	6. Plugging
1. Fatigue	
2. Erosion	
3. Wear	
4. Degradation	

following list of failure mode functions may serve as a guide:

1.1 Fails to open – complete or partial
1.2 Fails to remain – in position
1.3 Fails to close – complete or partial
1.4 Fails open
1.5 Fails close
1.6 Internal leakage
1.7 External leakage
1.8 Fails out of tolerance
1.9 Erroneous output
1.10 Reduced output
1.11 Loss of output
      – thrust
      – indication
      – partial
      – false
1.12 Erroneous indication
1.13 Excessive flow

1.14 Restricted flow
1.15 Fails to stop
1.16 Fails to start
1.17 Fails to switch
1.18 Premature operation
1.19 Delayed operation
1.20 Erratic operation
1.21 Instability
1.22 Intermittent operation
1.23 Inadvertent operation
1.24 Rupture
1.25 Excessive vibration

The investigator is trying to anticipate how the design being considered could possibly fail. At this point, he should not make the judgment as to whether or not it *will* fail, but concentrate on how it *could* fail.

2. *Describe the effect of the failure.* The analyst must describe the effect of the failure in terms of owner reaction. In other words "What does the operator experience as a result of the failure mode described?" For example, in considering the failure mode of a diaphragm coupling in a high-speed turbine driven process compressor application (Fig. 6.1a), the analyst would have to determine how this would affect the operation. Would there be a sudden acceleration of the turbine and would its overspeed protection device properly respond by activating the steam shut-off valve? Is there a need for a redundant emergency drive for safe run-down?

3. *Describe the cause of the failure.* The analyst will now anticipate the cause of the failure. Would temporary overload cause the coupling diaphragm failure? Would environmental conditions cause a problem? In short, the analyst investigates what conditions could bring about the failure mode. He concentrates on "FRETT", the possible effects of excessive Force, a Reactive Environment, abnormal Temperature and excessive Time.

4. *Estimate the frequency of occurrence of failure.* The analyst must estimate the probability that the given failure mode will occur. He assesses the likelihood of occurrence, based on his knowledge of the system, using an evaluation scale of 1–10. A 1 would indicate a low probability of occurrence, whereas a 10 would indicate a near certainty of occurrence.

5. *Estimate the severity of the failure.* In estimating the severity of the failure, the investigator weighs the consequence of the failure. A 1 here would indicate a minor nuisance, whereas a 10 would indicate a severe consequence such as "turbine run-away" or "stuck at wide open governor valve".

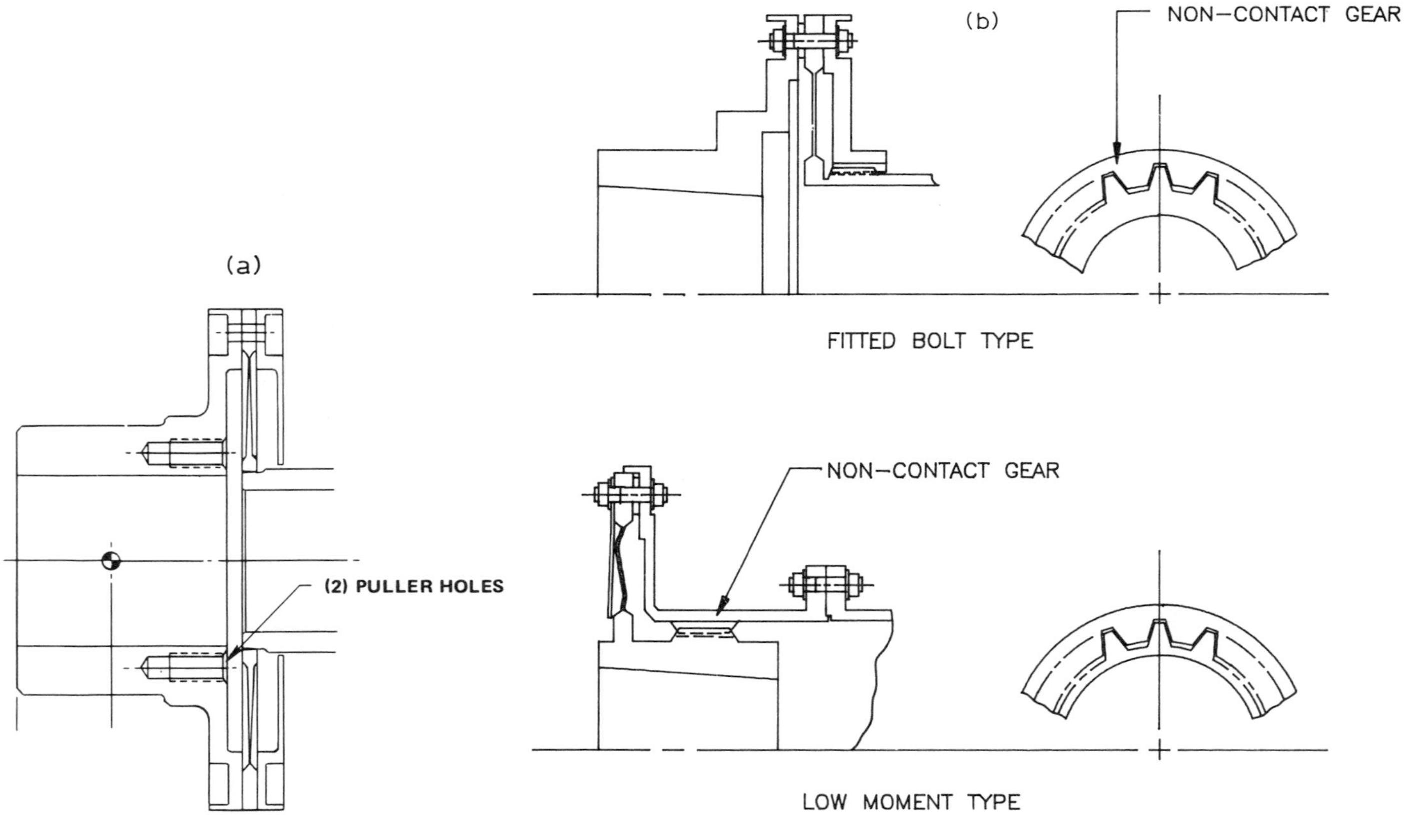

Fig. 6.1 (a) Diaphragm coupling (Koppers, Bendix similar); (b) diaphragm coupling with emergency back-up gear drive (Bendix, Koppers similar).

6. *Estimate failure detection.* The investigator will now proceed to estimate the probability that a potential failure will be detected before it can have any consequences. He will again use a 1–10 evaluation scale. A 1 would signal a very high probability that a failure would be detected before serious consequences would arise. A 10 would indicate a very low probability that the failure would be detected and consequences therefore would be appreciable. For instance, a failure of the above described diaphragm coupling might be assigned a detection probability of 10 because it would happen suddenly, without any detection possibilities. Similarly, a diaphragm coupling with an emergency run-down feature (Fig. 6.1b) would be assigned a detection probability of 4, because upon diaphragm failure there would be a detectable noise to allow initiation of contingency measures. Finally, the failure of the auxiliary resetting lever of the steam turbine overspeed trip system (Fig. 6.2) might be assigned a detection number of 1 for obvious reasons.

7. *Calculate the risk priority number.* The RPN obviously provides a relative priority of the anticipated failure mode. A high number indicates a serious failure mode. Using the risk priority numbers, a critical items summary can be developed to highlight the top priority areas that will require action.

8. *Recommended corrective action.* It is vital that the analyst takes sound corrective actions, or sees that others do the same. The follow-up aspect of the exercise is clearly critical to the success of this analytical tool. Responsible parties and timing for completion should be determined for all corrective actions.

THE FMEA FORM

The FMEA form (Table 6.1) may be used for machinery parts or assembly and systems failure mode and effect analysis. In order to complete the form the analyst needs the following information:

- system specifications;
- description of function, flow sheets, and drawings;
- description of operating conditions.

This information is entered into the appropriate rows. Components and their failure modes are numbered for identification (see column 1). Part, system or process *function* are entered into column 2; *failure mode* into column 3; failure mechanisms and possible *causes* into column 4. *Failure mechanisms* in this context are more detailed explanations of the failure mode

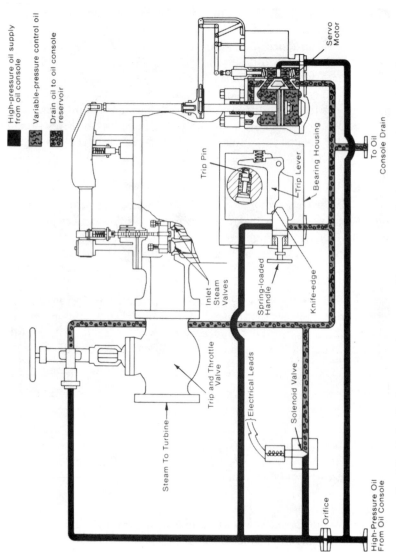

Fig. 6.2 Hydraulically actuated steam turbine overspeed trip system. *Source:* United Technologies Elliott, Jeannette, PA.

High-pressure oil supply from oil console

Variable-pressure control oil

Drain oil to oil console reservoir

Servo Motor

Trip Pin

Trip Lever

Bearing Housing

To Oil Console Drain

Inlet Steam Valves

Spring-loaded Handle

Knife-edge

Steam To Turbine

Trip and Throttle Valve

Electrical Leads

Solenoid Valve

Orifice

High-Pressure Oil From Oil Console

in terms of expanding on the mechanical, physical, or chemical mechanisms leading to the anticipated failure mode.

Failure causes should be listed as far as they are assignable to each failure mode. It would be well to assure that the list is all-inclusive so that remedial action can be directed at all pertinent causes. Examples of causes are as follows:

1.0 Design stage
    1.1 Wrong material selection, i.e. brittle when cold
    1.2 Wrong design assumptions, i.e. design temperature too low
    1.3 Design error
2.0 Materials, manufacturing, test and shipping
    2.1 Material flaw, i.e. inadequate plating thickness
    2.2 Improper fabrication, i.e. inferior welding quality
    2.3 Improper assembly, i.e. insufficient torque (fastener)
    2.4 Inadequate testing, i.e. not tested at operating conditions
    2.5 Improper preparation for shipment, i.e. part allowed to rust
    2.6 Physical damage, i.e. damaged in transit
    2.7 Insufficient protection, i.e. part or assembly dirty
3.0 Installation, commissioning and operation
    3.1 Improper foundations, i.e. foundation sagging
    3.2 Inadequate piping support, i.e. piping deflects machinery
    3.3 Wrong final assembly, i.e. built-in misalignment
    3.4 Improper start-up, i.e. shaft bow in steam turbines
    3.5 Inadequate maintenance, i.e. build-up of dirt
    3.6 Improper operation, i.e. no lubrication
4.0 Basic failure modes (FRETT)
    4.1 Failure due to high forces, stresses and impact, i.e. broken stem on gate valve
    4.2 Failure due to reactive environment, i.e. corroded casing on pump
    4.3 Failure due to thermal problems, i.e. thermal rise causes misalignment
    4.4 Time-dependent failures, i.e. aging causes O-ring leak

Column 5 shows the possibilities of *failure detection*, such as for instance automatic annunciation, inspections, and functional tests. Column 5 provides information on *surveillability*. Column 6 may contain information about appropriate countermeasures already available by design. These would be all measures and features contributing to limiting or avoiding the consequences of an anticipated failure mode. Examples are spare devices, redundancy designs, switch-over features, and devices which will limit consequential damage. When entering *failure effects* into column 7 we assume that the countermeasures listed in column 6 are effective. Effects of failure should be expressed in terms of operator reaction. The following will serve

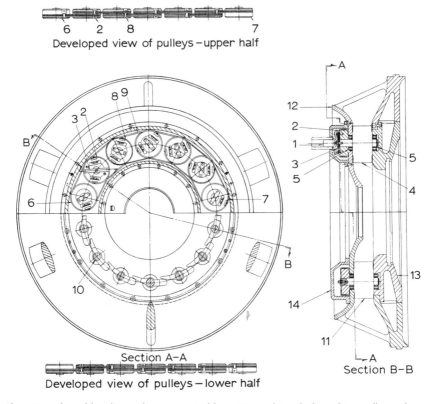

6   2   8   7
Developed view of pulleys – upper half

Section A–A
Developed view of pulleys – lower half

Section B–B

Fig. 6.3   Adjustable inlet guide vane assembly. 1, Upper drive shaft; 2, drive pulley; 3, key; 4, guide vane; 5, ball bearing; 6, end pulley; 7, end pulley; 8, pulley; 9, $\frac{1}{8}''$ diameter aircraft type cable; 10, lower drive shaft; 11, guide vane; 12, intake wall; 13, inboard support; 14, cover.

as a guide:

1. No effect.
2. Loss of redundancy, i.e. failure of one of dual shaft seals.
3. Functional degradation, i.e. excessive operating effort.
4. Loss of function, i.e. pump does not deliver.
5. Liquid/fumes/gas leakage/release, i.e. failing joint gasket.
6. Excessive noise/vibration, i.e. internal rub due to thermal expansion.
7. Violation of rules and safety standards, i.e. blocked safety valve.
8. Fails to indicate.
9. Fails to alarm.
10. Fails to trip.
11. Fails to start.
12. Fails to stop.

Table 6.3 FMEA example: Adjustable inlet guide vane assembly

Failure Mode and Effect Analysis (FMEA)

System: Variable Inlet Guide Vane Assembly

Component: As per Bill of Material or Assembly Drawing

System status: Operating

Component status: Controlling Compressor Inlet Flow

Operating conditions: Ambient Pressure 60–100°F Oily and dusty

Documentation: Comp. Outline Drawing, Inlet GV Assembly Drawing, Operation and Maintenance Manual

1	2	3	4	5	6	7	8	9	10	11	12
No.	Part/function	Failure mode	Basic failure mode/possible cause	Failure detection (surveillability)	Available counter-measures	Failure effects	Occurrence	Severity	Detection	RPN	Failure assessment and recommended action
1	Upper Drive Shaft	Fracture	Jamming	When making adjustments	Manual intervention	Reduced compressor output	2	3	6	36	Test periodically. i.e. Preventive maintenance
2	Drive Pulley	Loosening	Key sheared	Visible erratic movement	Manual intervention	Reduced compressor output	1	3	6	18	Test/exercise periodically
3	Shaft Key	Shearing/Fracture	Improper fitting procedure	See above	As above	As above	1	3	6	18	See above
4	Brake	Breakage									
5		Failure									
9		Relax									

Column 8 evaluates the probability of *occurrence* on a scale of 1–10. For example, 10 would indicate an extremely probable occurrence, whereas 1 would signify a very improbable occurrence.

Column 9 estimates the *severity* or consequence of the failure on a 1–10 scale.

The number assigned to *detection* in column 10 is based on the probability that the anticipated failure mode will be detected before it becomes a problem. Again, 10 indicates a low probability that the failure would be detected before consequences occur. A 1 means high detection probability. Column 5 will help in the evaluation of column 10.

Column 11 contains the *risk priority number* (RPN) and is calculated by multiplying the numbers in columns 8–10, inclusive. The RPN number is an indicator of relative priority.

Finally, column 12 contains the anticipated failure assessment together with a brief description of the corrective actions recommended. Under remarks one would find the persons or departments responsible for corrective actions as well as their status in terms of progress and timing.

Evaluation

The evaluation of the effects of component failures may be done according to different criteria. Examples of assessment criteria are:

- The maintenance case, i.e. the failure effect does not lead to system failure.
- System failure.

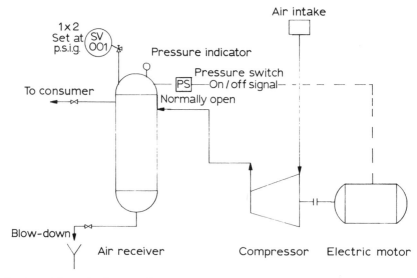

Fig. 6.4 Schematic diagram of an air compressor system.

Table 6.4 FMEA example: Air compression system component [1, 2]

No.	Part/ function	Failure mode	Basic failure mode/ possible cause	Failure detection (surveillability)	Available counter-measures	Failure effects	Occurrence	Severity	Detection	RPN	Failure assessment and recommended action
1	2	3	4	5	6	7	8	9	10	11	12
2.1	Staying closed	Leakage	Spring fatiguing	Compressor kicking off/on more often. Field inspection will detect increased noise level	Compressor keeps up supply of air	Compressor makes up for pressure loss	5	2	3	30	Preventive maintenance case, shutdown of system for repair
2.2	Staying closed	Fails open	Spring fracture	Pressure indication, field inspection	None	Rapid loss of pressure	6	7	1	42	System outage
2.3	Opening at $110 < P < 120$ psi	Fails closed	Corrosion, dirt, wrong setting	None, by valve inspection and test	None	No immediate effects, loss of safety function at over-pressuring	2	8	10	160	* Intolerable system condition * Enforce safety valve inspection and test program * Provide indicators that safety valve was activated

Failure Mode and Effect Analysis (FMEA)

System: Air supply system

System status: Undisturbed design conditions, pressurized

Component status: Design conditions (closed)

Component: Safety valve

Operating conditions: Room temperature: 50 100 F
Relative humidity: <80%
Dust-free atmosphere

Documentation: Drawings
System specifications

- Inadmissible system status, i.e. the failure effect results in a system status which violates safety rules.
- Danger status, i.e. the system's risk potential is being liberated.

Examples

Examples are presented in the form of an assembly or part FMEA covering the adjustable inlet guide vane assembly of a critical process gas compressor (Fig. 6.3). Table 6.3 shows the completed FMEA analysis. The function of this assembly is described in more detail in Appendix A3. Another example is an analysis of a compressed air system (Fig. 6.4 and Table 6.4).

REFERENCES

1. Browning, R. L., Analyzing industrial risks. *Chemical Engineering*, Oct. 20, 1969, pp. 100–114.
2. DIN 25448, *Ausfalleffektanalyse*. Berlin: Beuth Verlag, 1980.

7

Fault tree analysis

Fault Tree Analysis (FTA) is a deductive method in which a hazardous end result is postulated and the possible events, faults, and occurrences which might lead to that end event are determined. FTA also overlaps Sneak Circuit Analysis (SCA) because the FTA is concerned with all possible faults, including component failures as well as operator errors.

SCA is used to troubleshoot and improve hydraulic, electronic, shutdown instrumentation and other control interfaces around process machinery [1].

FTA is a "top-down" analysis that is basically deductive in nature. The analyst identifies failure paths by use of a fault tree drawing. A fault tree is a graphical representation of a thought process. It is constructed from events and logical operators. An event is either a component failure or system operation. The events and their graphical representation are given in Table 7.1.

A fault tree commences by selecting a top event. This event is the undesired event or ultimate disaster. From there, the analyst endeavors to find the immediate events that can, in some logical combination, cause the top event. These lower events are examined, in turn, for causes and the process is repeated to levels of greater detail. Ideally, the lowest level events will be all basic events and represented by a circle.

Fault trees provide a method for determining the logical causes of a given event. It illustrates all of the ways an undesired event can occur. It helps determine the critical components and the need for other analytical efforts. Numerical computations indicating the probability of occurrence for the top event and intermediate events can be obtained. The major drawback of the fault tree is that *there is no way to ensure that all causes have been evaluated consistently*: large fault trees are difficult to understand. On the system level, they do not resemble the system flowsheet. Complex logic is frequently involved.

FTA is performed on the system configuration, determined by the analyst. Determining the configuration of a system is generally central to all analyses.

Although this concept, like the FMEA earlier, has been intuitively used by engineers for a long time, its systematic and formal application in reliability analysis is relatively recent. Events which could cause the top event are

Table 7.1 Fault tree analysis symbols

Logic gates

"OR" – Gate denotes the situation whereby the output event will exist if any one input event is present

"AND" – Gate denotes the situation whereby all the input events are required to produce the output event

Fault events

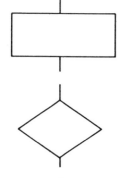

RECTANGLE denotes an event, usually a malfunction, which results from the combination of fault events through the logic gates

DIAMOND denotes a fault event of which the causes have not been developed

CIRCLE denotes a basic fault event. This category includes component failures whose frequency and failure mode are derived or known

generated and connected by logic operators AND, OR, and EOR. The AND gate provides a TRUE output if and only if all the inputs are TRUE. The OR gate provides a TRUE output if and only if one or more inputs are TRUE. The EOR, exclusive OR, gate provides a TRUE output if and only if one but not more than one input is TRUE [2, 3]. The analysis proceeds by generating events in a successive manner until the events need not be developed further. Those events are called primary events. The fault tree itself is the logic structure relating the top event to the primary events.

The linking of events according to logical rules is shown in Figure 7.1. FTA may be applied at any level from component part to full system. General

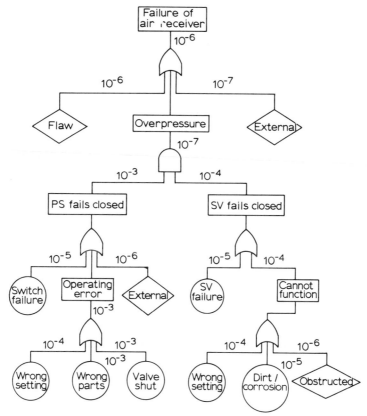

Fig. 7.1 Fault tree (adapted from ref. 7).

applications of FTA are:

- reliability assessment of machinery parts (see the compressor rotor example on p. 114);
- reliability assessment of simple subsystems;
- probability assessment of specific failure events in complex systems;
- critical failures identified by FMEA.

PROCEDURE

Although the fault tree can become complex, each part is simple. (For reasons of simplicity the reader is referred to the compressed air system in Fig. 6.4 which is used to illustrate the procedure.) Figure 7.1 depicts the fault tree. Further details may be obtained from ref. 1. The following eight steps should be followed:

1. *Define the undesired event.* "Failure of Air Receiver" has been chosen. The system can fail in other ways as shown in the preceding FMEA. A tree can be drawn for each defined event.
2. *Identify the possible prime causes of failure.* These are shown on the fault tree as "defect", "overpressure", and "external events".
3. *Identify conditions which could contribute to the prime causes.* Both "defect" and "external events" are shown as "undeveloped events". The reader should imagine that these events could of course be developed downwards to pinpoint design deficiencies for instance, or aging effects such as fatigue and corrosion. Further, external events could be shown as earthquakes or fire.

   All three conditions could result in the undesired or top event. They are therefore connected through an OR gate.
4. *Repeat step (3) at the next lower level.* Only the "overpressure" condition remains. Two causes are shown – the overpressure cut-out switch does not open when required and the relieve valve fails to blow. Both conditions are required to produce the undesired event. They are therefore connected by AND gates.
5. *Continue to the required level.* Four events were left undeveloped in Figure 7.1. They reflect the designers judgment that their probabilities were small enough to be ignored.

   We are now left with five basic inputs to which probabilities may be assigned. In a critical case, each of these could be developed further to eliminate inherent weaknesses or to reduce the likelihood of human error.
6. *Determine whether or not quantitative analysis is required.* The usefulness of the fault tree technique can be enhanced by the use of quantitative data. In this way not only can the fault paths be identified, but their probability of occurrence may be established [4].

   The decision to employ quantitative analysis should be made on the basis of experiences, system complexity and severity of consequences. One should ask the following questions:

   - What is the severity of the undesired event?
   - Are quantitative data in terms of failure rates available, meaningful, and relevant?
   - Does the fault tree contain many AND gates, expressing degrees of redundancy?
   - Does a particular branch of the fault tree appear marginal?
   - Do we want to commit resources to this tedious task?

7. *Allocate a probability value to each event.* A failure rate can be assigned to each input event. The probability of the undesired event, or top event, can then be calculated.

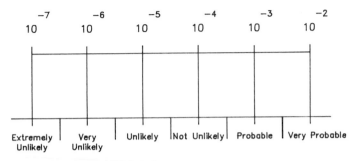

Fig. 7.2   Relative probabilities. *Source:* Atomic Energy of Canada Ltd.

Failure rate data can come from experience, test data, published data as shown in Table 4.3, or engineering judgment. The latter is applied as a first approach. Here the analyst makes use of arbitrary relative probabilities which can be selected from Figure 7.2.

We proceed to assign probabilities from Figure 7.2 to each event in Figure 7.1.

It would be well to consider maintainability and surveillability factors at this time. For instance, a high probability has been assigned to "valve shut" as the valve could be left closed after it has been maintained. As during an FMEA, we should ask ourselves the following questions in this context:

- What is the general maintenance and operational environment?
- Can an item be overlooked; can it be maintained?
- Is the part unusual or non-standard?
- Can status be easily ascertained, i.e. an open or closed gate valve?
- Is it difficult to assemble?
- Can it be installed incorrectly?
- What is the in-service failure or deterioration mode, i.e. how is it influenced by FRETT:
  - force
  - reactive environment
  - temperature
  - time

8. *Connect the input values.* The rules for each type of gate as shown in Table 7.1 might not be rigorously correct. Errors, however, are negligible for inputs smaller than 10^{-1}. The reader is referred to the list of references at the end of this chapter for further information on the subject. Two rules have to be observed:

- The output of an AND gate is the product of the input.
- The output of an OR gate is the sum of the inputs.

Because powers of ten are used we only have to add or subtract. Low probabilities do not have to be accumulated. For example:

The AND gate leading to "overpressure" in Figure 7.1:

$$10^{-3} \times 10^{-4} = 10^{-7}$$

The OR gate leading to "PS fails to Open":

$$10^{-4} + 10^{-3} + 10^{-6} \approx 10^{-3}$$

If only a few data points are available for a quantitative FTA, one might want to resort to a method proposed in ref. 6. In order to arrive at a failure prediction at the unit level, this approach combines subjective weighting of part failures and failure modes with objective data for a small number of failures or failure modes. To do this the complete tree is developed down to a part or part failure mode level. At each gate subjective probability estimates are made by using service engineers with relevant experience.

With a fully weighted tree it is possible to take one piece of hard data relating to one part failure or part failure mode and use the subjective weightings to propagate this upwards through the tree to arrive at an equipment failure rate estimate.

EXAMPLES

Three examples are shown. Example 1 demonstrates how fault trees have been used to explain failure events logically. Figure 7.4 is an expansion of one of the events in Figure 7.3. It illustrates the events leading to mechanical bearing failure. The example conveys that fault trees can be used effectively without necessarily assigning probabilities or failure rates.

Example 2 deals with an investigation made in connection with a rotor (Fig. 7.5) for a process gas compressor owned by a major petrochemical company. The effort is depicted in Figure 7.6 and had to be undertaken in order to justify the replacement of a spare rotor that had been damaged during repair and overhaul.

Example 3 explains the failure events leading to the no-flow or low-flow initiated mechanical failure of a multistage deepwell pump (Fig. 7.7). The fault tree is shown in Figure 7.8.

ASSESSMENT AND EVALUATION

Assessment of the FTA results should lead to an action plan. Several questions should therefore be asked:

• Is the overall reliability acceptable? In our last example the relative

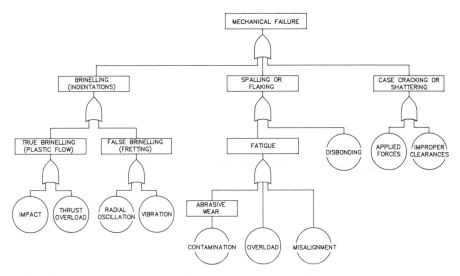

Fig. 7.3   Fault tree of mechanical bearing failure (from ref. 6).

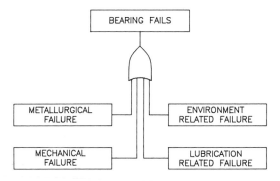

Fig. 7.4   Types of bearing failure shown as intermediate events on a fault tree (from ref. 6).

probability of the undesired event is 0.00001 per h, or "unlikely".

- What inputs are subject to large uncertainty? In our examples, there are no appreciable uncertainties.
- Is there substantial redundancy? The distribution of AND gates is an indication of the degree of redundancy.
- Could loss of redundancy go undetected? Essentially, we will ask the same questions covered in our FMEA procedures.
- Do we need help? If the risk seems high and it cannot be lowered without a major system modification, it would be a good idea to summon help from a specialist.

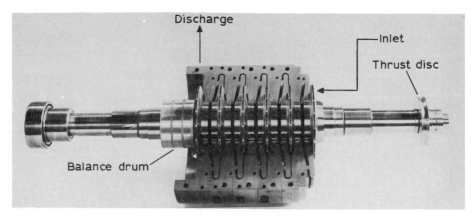

Fig. 7.5   Centrifugal compressor rotor.

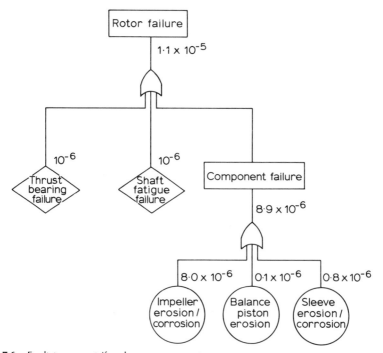

Fig. 7.6   Fault tree: centrifugal compressor rotor.

REVIEW

FTA has advantages and disadvantages. Listing the advantages motivates purpose and application of the method:

- The fault tree can serve in all phases of the machinery life-cycle because

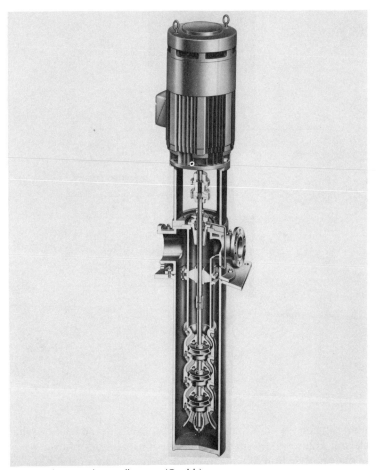

Fig. 7.7   Multistage deepwell pump (Goulds).

it can help to determine possible causes of undesirable events.

● FTA may be used to evaluate competing designs by revealing qualitative and quantitative event interdependencies.

The disadvantages of FTA can be explained by its design principle. It attempts to build a mathematical model of a complex physical condition by logical linking of events. If all peripheral, environmental, and operating conditions are not defined, then the method depends on the judgment of the analyst.

● One chief disadvantage is that there is no effective formal control against overlooking of events or the neglect of operating or environmental conditions. The best preventive measure would be to have several analysts make independent analyses.

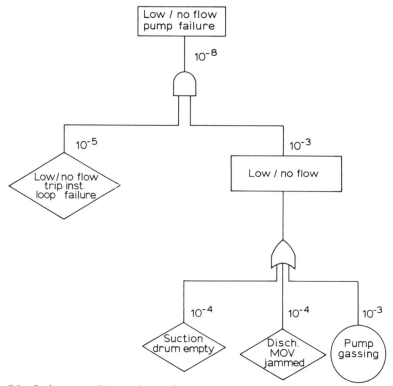

Fig. 7.8   Fault tree: multistage deepwell pump operation.

- One main difficulty with a quantitative FTA exists in the lack of reliable and relevant failure rate data as well as the probabilities of events.
- Finally, the construction of fault trees can demand a lot of effort and may become expensive.

REFERENCES

1. Bloch, H. P. and Geitner, F. K., *Machinery Failure Analysis and Troubleshooting.* Houston: Gulf Publishing, 1983, pp. 560–579.
2. Dhillon, B. S., *On Reliability and Maintainability Engineering.* Advance paper, University of Ottawa, 1980.
3. Dhillon, B. S., Bibliography of literature on fault trees. *Microelectron. Reliab.* **17**, 1978, pp. 501–503.
4. Scerbo, F. A. and Pritchard, J. J., *Fault Tree Analysis: A Technique for Product Safety Evaluation.* New York: ASME publication 75-SAF-3.
5. Innes, C. L. and Hammond, T., Predicting mechanical design reliability using weighted fault trees. *Failure Prevention and Reliability Conference*, Chicago, Sept. 1977, pp. 213–228.

6. Strauss, B. M., Fault tree analysis of bearing failures. *Lubrication Engineering*, Nov. 1984, pp. 674–679.

7. AECL-4607, *Reliability and Maintainability Manual*, edited by J. G. Melvin and R. B. Maxwell. Atomic Energy of Canada Ltd, 1974.

8

Risk and hazard assessment

The general objective of a hazard and risk assessment is the identification of machinery features which could threaten the safety of personnel, property, or the environment. Hazard and risk assessment methods are evaluated in Table 8.1. *Hazard* is defined as the source of harm, and *risk* is the possibility of experiencing this harm. For example, the hazard around a pumping service for toxic material could be the failure of the shaft seal. Two designs are suggested to prevent a leakage of the toxic material. Design A uses multiple mechanical seals (Fig. 8.1), whereas design B calls for a single mechanical seal (Fig. 8.1). Both designs may fail during operation of the pump. However, the probability of a toxic release for the first design is much less than for the second design. Consequently, for the same hazard level, design A poses less risk to the plant operators and the public than design B.

From this, risk can be defined [2] as the answer to three questions:

- What can go wrong?
- How likely is it to go wrong?
- What are the consequences?

The answer to the first question is a series of accident or incident scenarios. The answer to the second question is the probability of any given scenario. The answer to the third question lies in arriving at a measure of the extent of damage. This can be, as in our example, the number of people affected by the toxic release, the extent of damage to the environment, or the amount of business losses.

ASSESSING RISK

The methodology for assessing risks from rare events has gone through more than two decades of development. It started in the defense industry and is practiced in the nuclear industry. Today, many different industries, such as

118

Table 8.1   Risk and hazard assessment methods [1]

Method	*Characteristic*	*Advantages*	*Disadvantages*
Preliminary hazards analysis	Defines the system hazards and identifies elements for FMEA and fault tree analysis; overlaps with FMEA and criticality analysis	A required first step	None
Failure mode and effect analysis (FMEA)	Examines all failure modes of every component. Hardware oriented	Easily understood. Well accepted, standardized approach, noncontroversial, nonmathematical	Examines nondangerous failures. Time consuming. Often combinations of failures not considered
Criticality analysis	Identifies and ranks components for system upgrades. May be part of FMEA	Well-standardized technique. Easy to apply and understand. Nonmathematical	Follows FMEA. Frequently does not take into account human factors, common cause failures, system interactions
Fault tree analysis	Starts with "top event" and finds the combination of failures which cause it	Well accepted technique. Very good for finding failure relationships. Fault oriented; we look for ways system can fail	Large fault trees are difficult to understand, bear no resemblance to system flowsheet, and are not mathematically unique. Complex logic is involved
Event tree analysis	Starts with initiating events and examines alternative event sequences	Can identify (gross) effect sequences, and alternative consequence of failure	Fails in case of parallel sequences. Not suitable for detailed analysis

(continued)

Table 8.1   (cont.)

Method	Characteristic	Advantages	Disadvantages
Cause-consequence analysis	Starts at a critical event and works forward using consequence tree; backwards using fault tree	Extremely flexible. All-encompassing. Well documented. Sequential paths clearly shown	Cause-consequence diagrams can become too large very quickly. They have many of the disadvantages of fault trees
Hazards and operability studies (HAZOP)	An extended FMEA which includes cause and effect of changes in major plant variables	Suitable for large plants	Technique is not well standardized

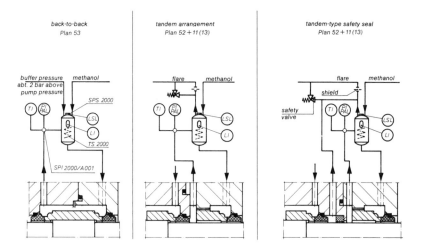

Fig. 8.1   Cross-section of multiple mechanical shaft seals for a chemical process pump (courtesy Burgmann Seals America).

the hydrocarbon processing industry (HPI), are using and modifying the basic methods to match their needs.

An example is the quantitative risk analysis procedure developed by a major HPI company [3]. The primary goal was to determine investment levels as a consequence of safety considerations. The company first analyzed their past experience with various types of process machinery and equipment.

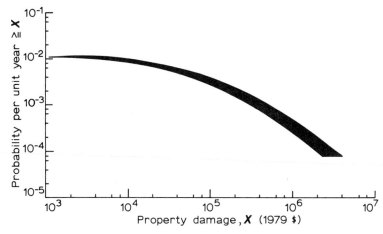

Fig. 8.2   Fires/explosions, 1961–75, in hydrocarbon processing plants (from ref. 2).

It then used the statistical data to establish a computerized data bank. The data bank allowed them to prepare a curve similar to that in Figure 8.2 for each type of equipment. Actual graphs might also show upper and lower 95% confidence limits as dashed lines above and below the main curve. Property damage is on the y-axis, and the probability of occurrence per unit-year for a dollar loss equal or greater than a given value is on the x-axis.

The curves can be used in two ways. First, they reflect historical experience and make estimates of probabilities possible. As an example, the probability of a $500,000 or greater loss for the particular equipment represented by Figure 8.2 is about one incident per 1000 unit-years of service. These curves, then, are tools for quantitative risk assessment. They are primarily used for screening purposes and are one factor in a decision-making process. Other factors are, for instance, public relations aspects, government regulations, personnel exposures, and so forth.

A second way of applying the graphs is as part of a risk analysis process. It is really a reversal of the procedure mentioned above. After the potential losses for equipment and business interruption are estimated, the probability of their occurrence is determined by finding the point of intercept of the loss curve with the y-axis. Multiplying the total loss by the probability results in the annual loss costs. Here is where the risk is quantified in dollars.

ASSESSING HAZARDS

An important first step in evaluating the risk associated with a particular machinery system is to establish the source of hazard. Often it will not be necessary to employ the whole range of methods shown in Table 8.1. A first

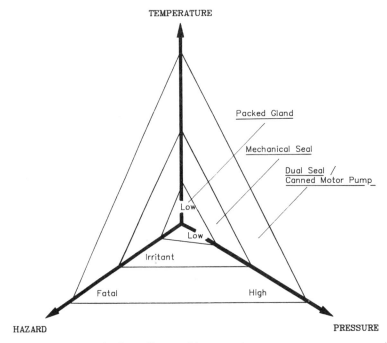

Fig. 8.3 Simple technical rule as illustrated by operating parameters versus pump shaft seal selection.

approach could be experience-based observation of the facts. Other simple methods are checking existing designs against basic technical rules (see Fig. 8.3), internal company standards for new equipment, local and national codes, or industry standards such as issued by ANSI, API, and so forth. The next step would perhaps be a preliminary hazard analysis.

PRELIMINARY HAZARD ANALYSIS (PHA)

This is the first systematic analysis of the machinery system and is designed to identify gross system hazards as the basis for more rigorous and detailed analysis later.

It can be stated that PHA is an examination of the generic hazards known to be associated with a system at its conceptual phase of development. The purpose of this analysis is to:

1. Identify hazards.
2. Determine the effects of the hazards.
3. Establish initial safety requirements.
4. Determine areas to monitor for safety problems.

5. Initiate the planning of a safety program.
6. Establish safety scheduling priority.
7. Identify areas for testing.
8. Identify the need for additional analyses.

The PHA determines the recognized and anticipated design safety pitfalls and provides the method by which these pitfalls may be avoided. When this analysis is undertaken, there is little information on design details and less on procedures.

The PHA is usually a top-level review for safety problems. In most instances, the following basic steps are undertaken for a PHA:

1. Review problems known through past experience on similar machines or systems to determine whether they could also be present in the equipment under consideration.
2. Review the functional and basic performance requirements, including the environments in which operations will take place.
3. Determine the primary hazards that could cause injury, damage, loss of function, or loss of material.
4. Determine the contributory and initiating hazards that could cause or contribute to the primary hazards listed.
5. Review possible means of eliminating or controlling the hazards, compatible with functional requirements.
6. Analyze the best methods of restricting damage in case there is a hazard due to loss of control.
7. Indicate who is to take corrective action, and the actions that each will undertake.

Three basic approaches that can be used to ensure that all hazards are being covered are the columnar form, top level fault tree, and narrative description. These methods will not in themselves find hazards. They will orient the analyst so that a thorough coverage of all aspects of the system will be performed.

The columnar form is the simplest method to implement. The chief advantage is that it is easy to review. The form has a heading that patterns questions in the mind of the analyst. The headings must at least incorporate the following terms or descriptions: Hazard, Cause, Effect, Hazard Category, Corrective or Preventive Measures.

The hazard is the generic area or condition that may influence system safety. The following is a partial list of hazards (the analyst can usually think of many more):

- acceleration,
- contamination,
- corrosion,

- chemical dissociation,
- electrical,
- explosion,
- fire,
- heat and temperature,
- leakage,
- moisture,
- oxidation,
- pressure,
- radiation,
- chemical replacement,
- shock (mechanical),
- stress concentrations,
- stress reversals,
- structural damage or failure,
- toxicity,
- vibration and noise,
- weather and environment.

The cause column of a report is used to explain when the system is exposed to the hazard. It is here that the results of system generation are considered. Project phasing must also be considered, as well as an estimate of the percentage of system operation time that the hazard will be in effect.

An effect column is system-centered. It details the action of the hazard on system operation. In this column the possibility of causing injury or death, however remote, must be stated.

The hazard category is a numerical measure of how important the hazard is. The number of categories should be kept small, usually four or less, so that attention may be placed where it will do the most good.

The corrective or preventive measures column is almost self-explanatory. Here, methods of abating the hazard are given.

A top level fault tree follows the method of FTA with generic events. Although this method helps define causes and effects, it does not follow that the system is checked hazard by hazard. Since the fault tree is event-oriented, it helps analyze undesired events, but does not determine that a particular event is a hazardous condition, element, or potential accident.

The narrative approach is less rigorous, and usually less complete, than the top level fault tree and narrative approaches. Narrative writing style is a lengthy and time-consuming task. This approach is less susceptible to systematic method or technique and, therefore, the results usually have serious gaps or incomplete areas. The hazardous conditions and potential accidents are generally identified from experience, and then are explained in great depth and detail, more on the order of a final report than an analysis.

Once a PHA has been gone through, a more thorough hazard assessment may be made using the techniques of a hazard and operability study.

HAZARD AND OPERABILITY (HAZOP) STUDY

A HAZOP study is usually a systematic technique for identifying hazards or operability problems throughout an entire facility [4]. In our context, HAZOP studies have been successfully applied around major compressor installations. Here the technique provides opportunities to think of all possible ways in which hazards or operating problems might occur. In order to reduce the chance that something might be missed, this is done in a systematic way, each pipeline and each sort of hazard being considered in turn.

A pipeline for our purposes here is one that joins two pieces of equipment. Our example is a high pressure gas supply system (Fig. 8.4) consisting of two reciprocating compressors (Fig. 8.5), necessary motorized valves for remote operation and interconnecting piping. The compressors are designed to move process gas from a common source of supply to either a customer or a high-pressure storage facility. The objective of a HAZOP review was specifically to find out whether or not the compressor piping could be simplified by eliminating some of the existing safety valves and check valves without compromising the design, safety, and operability of the system.

Table 8.2 is a summary of the HAZOP investigation. The review items

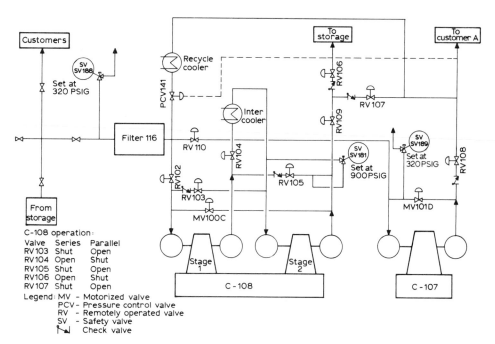

Fig. 8.4 Simplified flow plan: gas compression system.

Table 8.2   Summary of risks and suggested actions

Item no.	Equipment no.	Hazard	Possible cause	Probable severity
1	C-108 Remove check valve at RV-103	Backflow from C-108 interstage to suction piping	Failure of RV-103 (i.e. valve open but limit switches show closed)	
⋮ 5	C-107 Check valve at RV108	Backflow from Customer A to C-107 suction	During S/U of C-107 all the block valves including the bypass MV101D will be open. Only the check valve at RV108 is keeping Customer A from depressuring back	

Existing indication and/or protection	*Comments*	*Suggested actions*	*Follow up by*
– SV188 protects suction piping at 320 psi – SV181 protects C-108 2nd stage from high P. – Customer line will go high on pressure – indication at unit – at 270 psi dump valve to storage will open. – During start-up of a machine, the operator will be in the area and he will: a) Check visually the position of valves before start-up b) Hear SV release if it occurs	Valve failure of RV-103 only a concern during start-up Possibility of a RV valve opening unassisted is very remote. Single stage bypass has a check valve, however, overall bypass has *no* check. (Check if the bypass would be going in the wrong direction)	Remove check valve at RV-103	(Name)
SV189 protects C-107 suction piping. Unit set to dump to storage at 270 psi	Check valve at RV108 is required for operability of C-107. This valve should be the most reliable type of check valve available	Review alternate check valve types	(Name)

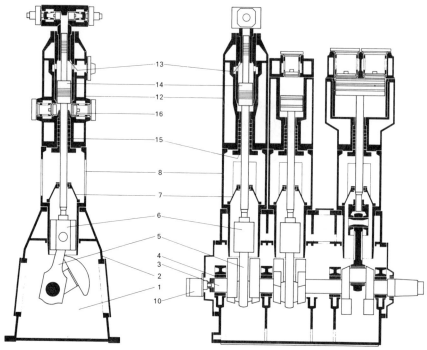

Fig. 8.5 Major reciprocating compressor (Sulzer). 1, Crankcase; 2, frame; 3, crankshaft; 4, bearing; 5, connecting rod; 6, crosshead; 7, cover; 8, distance piece; 9, purge chamber; 10, lubricating group for crankcase; 12, cylinder; 13, cylinder liner; 14, piston; 15, piston rod packing; 16, valves; 17, capacity control.

are usually dealt with by applying a series of HAZOP guide words [5]:

- None
- More of
- Less of
- Part of
- More than
- Other than

"None" in our example means no forward flow or, in effect, reverse flow when there should be forward flow. The questions to ask now are:

- Could there be reverse flow?
- If so, how could it happen?
- What are the consequences of reverse flow?
- Are the consequences hazardous or do they just prevent efficient operation?

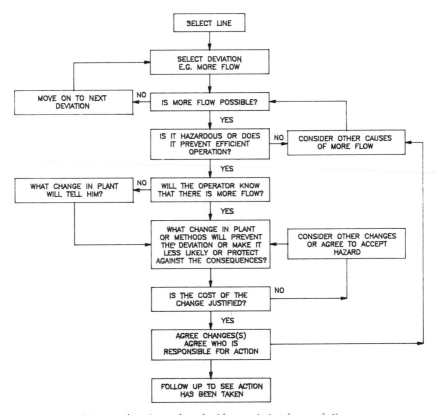

Fig. 8.6   HAZOP procedure (reproduced with permission from ref. 6).

- If so, can we prevent reverse flow by changing the design or the operating procedures?
- If so, does the size of the hazard, i.e. severity of consequences multiplied by the probability of occurrence, justify the additional expense?

Since our objective is the removal of redundant check valves, "none" or "reverse flow" were the only guide words used.

When looking at the lines containing safety valves we would of course invoke additional guide words such as "more of" (i.e. more pressure, meaning failure to open) or "less of" (i.e. less pressure, indicating failure to close or reseat). The meanings of the guide words are summarized below and Figure 8.6 reflects the entire HAZOP process.

HAZOP guide word	*Deviations*
None	No forward flow when there should be, i.e. no flow or reverse flow
More of	More of any relevant physical property than there should be, e.g. higher flow (rate or total quantity), higher temperature, higher pressure, higher viscosity, etc.
Less of	Less of any relevant physical property than there should be, e.g. lower flow (rate or total quantity), lower temperature, lower pressure, etc.
Part of	Composition of system different from what it should be, e.g. change in ratio of components, component missing, etc.
More than	More components present in the system than there should be, e.g. extra phase present (vapor, solid), impurities (air, water, acids, corrosion products), etc.
Other than	What else can happen apart from normal operation, e.g. start-up, shutdown, uprating, low running, alternative operation mode, failure of plant services, maintenance, catalyst change, etc.

HAZOP studies are now being conducted on a routine basis in many companies for all new units and major modifications. Some opinions exist in the United States that it is no longer a question of "if" the government – through the Occupational Health and Safety Administration (OSHA) – will require HAZOP reviews, but "when".

There is no doubt in the authors' opinion that the result of a HAZOP review frequently has a significant impact on the cost of a project. Since large process machinery trains are considered major subsystems, they too can affect that cost.

REFERENCES

1. Henley, E. J., *Designing for Reliability and Safety*. Gas Processors Association, annual meeting, San Antonio, Texas, March 10–12, 1986.
2. Kazarians, M. and Boykin, R. F., Assessing risk in the CP plant. *Chemical Processing*, Oct. 1986, pp. 78–82.
3. Young, R. S., *Risk Analysis Applied to Refinery Safety Expenditures*. API Committee on Safety and Fire Protection, fall meeting, Denver, Colo., Sept. 23, 1986.
4. Lawley, H. G., *Operability Studies and Hazard Analysis*. In "Loss Prevention" series, Vol. 12. New York: AIChE, 1974.
5. Kletz, T. A., Eliminating potential process hazards. *Chemical Engineering*, April 1, 1985, pp. 48–68.
6. Kletz, T. A., *HAZOP & HAZAN: Notes on the Identification and Assessment of Hazards*. Rugby, UK: ICE, 1983.

9

Machinery system availability analysis

Earlier, we defined availability of a system as *the fraction of time it is able to function*. System availability is clearly the consequence of subsystem availabilities which in turn are a function of assembly and part level availabilities.

Operating time can frequently be greater than available time. Available time therefore sets a limit on production.

The objectives of an availability analysis are:

1. To estimate system availability for comparison with a target value.
2. To identify low availability components, assemblies or parts for improvement.

Availability analysis is an extension of Failure Mode and Effect Analysis (FMEA). One specific effect, namely unavailability, is estimated. System availability analysis should therefore be performed after a FMEA has been completed.

THE APPROACH

System availability assessment is a tool that can be applied during all life-cycle phases of a machinery system. Its results can be used by management as availability control. It can identify deficiencies in certain areas, or compare alternative designs.

The work steps are as follows. The system outage time caused by each level of the hierarchy is estimated from failure and maintenance data and recorded on a work sheet. Three types of downtime are identified: forced maintenance, predictive-scheduled maintenance, and planned (turnaround) maintenance. The sum of all types of downtime is used to compute machinery system availability.

The availability (A) of a component is expressed as:

$$A = \frac{\text{total time} - \text{maintenance time}}{\text{total time}} \qquad (9.1)$$

$$= 1 - \frac{\text{maintenance time}}{\text{total time}}$$

$$= 1 - \text{unavailability } (UA)$$

The unavailability of a unit is a function of the time required for two modes of maintenance: breakdown-corrective and predictive-preventive. Unavailability resulting from breakdown maintenance, or forced unavailability is:

$$UA_F = \frac{\lambda \times t \times T_R}{T} \qquad (9.2)$$

where $t =$ operating time in hours,
$\lambda =$ failure rate per hour,
$T_R =$ average repair time in hours,
$T =$ total time in hours.

With high availability, t becomes approximately:

$$t = T$$

The error in this approximation is small compared to errors in λ and T_R. Similarly, preventive maintenance-related unavailability can be approximated to be:

$$UA_P = \frac{T_M}{T} \qquad (9.3)$$

where $T_M =$ average preventive maintenance time in h/year,
$T = 8760$ h/year.

Total unavailability of an item under scrutiny then is:

$$UA = \lambda T_R + \frac{T_M}{T} \qquad (9.4)$$

MACHINERY SYSTEM UNAVAILABILITY

In order to be able to estimate a machinery system's unavailability, we require two pieces of information:

1. The system downtime required to perform a particular job.
2. The job priority determined by business impact.

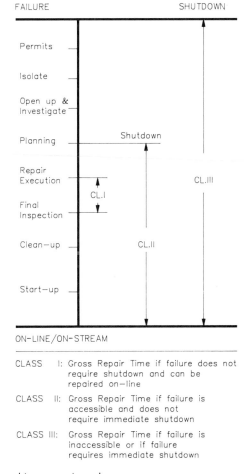

Fig. 9.1 Process machinery repair cycle.

If a machinery system must be shut down for maintenance, repair or overhaul (MRO) of a component, the resulting downtime will naturally be greater than the actual component MRO time. Therefore, the gross repair time (GRT)* is greater than T_R. This relationship is shown in Figure 9.1 for a typical repair cycle. GRT equals T_R only if an on-line repair is possible (Class I). The other extreme, Class III, may be many times greater than T_R. In order to arrive at a numerical value for GRT the length of each step in the repair cycle has to be estimated, that is obtain permit and access, identify the failure, and so forth.

---

* Often referred to as stream-to-stream time.

PRIORITIES

Machinery system unavailabilities are classified by priorities. Priorities are determined based on problem seriousness, urgency, and growth. Three types of unavailabilities can usually be identified.

Forced unavailability (UA_F)

It is caused by work that, if it would not get done, would result in high business losses, that is high seriousness, urgency, and defect growth would be the case. Usually, this is work that has to be performed in response to failures that disable the system and call for immediate repair. System forced unavailability is the sum of component forced unavailability.

$$UA_F = \sum (\lambda \times \text{GRT}) \tag{9.5}$$

Maintenance unavailability (UA_M)

Maintenance unavailability is caused by work that can be deferred for some time but usually not to a planned production unit shutdown. Operators can minimize maintenance unavailability in two ways. First, by doing the work within a suitable shutdown window (i.e. a period of time when the system is available for maintenance due to other reasons). Second, by performing several maintenance operations simultaneously.

It becomes apparent then that the design review can contribute to minimizing the total work load. Consequently, even though the sum of the component unavailabilities is not a true estimate, it is a direct estimate of the work load and a relative measure of unavailability. We already know that low availability indicates poor reliability or maintainability.

Since jobs can be done concurrently, maintenance unavailability need not include all steps in Figure 9.1. Average repair time (T_R) and maintenance time (T_M) are used rather than GRT:

$$UA_M = \sum (\text{preventive maintenance} + \text{deferrable repair}) \tag{9.6}$$

or

$$UA_M = \sum \left(\frac{T_M}{T} + \lambda T_R \right) \tag{9.7}$$

Planned unavailability (UA_P)

Planned unavailability is generated by work that can be deferred from one operating period to another. This is usually referred to as scheduled maintenance, repair, overhaul, and inspection (MRO & I). For all

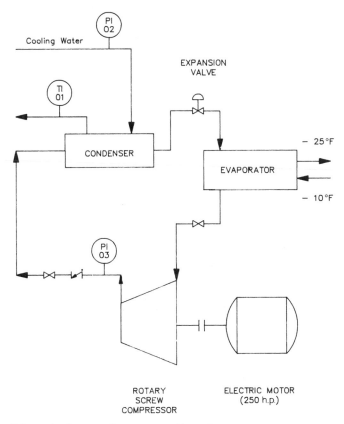

Fig. 9.2 Schematic diagram of a process refrigeration system.

components:

$$UA_P = \sum \text{(planned MRO \& I)} \tag{9.8}$$

or

$$UA_P = \sum \frac{T_M}{T} \tag{9.9}$$

Table 9.1 shows the actual procedure by way of an example. The table represents the availability assessment of three machinery subsystems as part of a process refrigeration system (Fig. 9.2).

Eight failure cases are identified for the machinery portion of the commercial refrigeration package. Two failure cases are shown for each of the two process pumps which are in unspared service. The following points may be of interest:

1. Most individual components listed in Table 9.1 have failures which demand immediate shutdown and consequently result in forced

Table 9.1 Availability analysis: Process machinery

(A)	(B)	(C)	(D)	(E)	(F)	(G)	(H)		
					Reliability maintenance		Unavailability		
Component	Failure	Rate, λ $\times E-6$ (h)	Repair time (h)	Gross repair time (h)	Predictive (h/year)	Planned (h/year)	Forced[a]	Predictive[b]	Planned[c]
1.0 Refrigeration package	1.1 Comp. seal (forced)	30		30			8.E−4	0	0
	1.2 Comp. brg. (forced & RM)[d]	9	40	72			6.E−4	4.E−4	0
	1.3 Comp. rotors (RM)[d]	25	60			10	0	2.E−7	0.0011
	1.4 Motor (forced)	8		110			8.E−4	0	0
	1.5 I&E components (forced)	7		5			3.E−5	0	0
	1.6 I&E components (RM)[d]	10	3		7		0	8.E−4	0
	1.7 Oil pump (forced)	30		10			3.E−4	0	0
	1.8 Oil pump motor (forced)	12		16			2.E−4	0	0
2.0 Glycol pump	2.1 Pump (forced & RM)[d]	30	30	40	10		0.0011	0.0011	0
	2.2 Motor (forced)	8		16			1.E−4	0	0
3.0 HC pump	3.1 Pump (forced & RM)[d]	50	10	10		10	5.E−4	5.E−8	0.0011
	3.2 Motor	16		16			2.E−4	0	0
					Total:		0.0046	0.0019	0.0023
					Unavailability:		0.0088		
					Reliability:		99.54%		
					Availability:		99.12%		

[a] (C) × (E) × E−6 × 8000/8760.
[b] (C) × (D) × E−6 × 8000/8760 + (F)/8760.
[c] (G)/8760.
[d] RM = reliability maintenance (i.e. predictive/preventive maintenance).

unavailability. These failures are caused by technical life attainments, predicated by L_{10} bearing life, for example. There are three categories of these types of failures:

- Predictable but not predicted;
- Predicted but not acted upon;
- Not predictable.

2. Failure 1.3 (Table 9.1), internals of the rotary screw compressor, includes both maintenance and planned unavailability. A planned overhaul or exchange of the compressor element every 5 years results in an average unavailability of 10 h per year. In addition, random failures are expected here at a rate of 25×10^{-6} h or one every 5 years based on an 8000-h operating year. These random failures, initiated by operational accidents, are assumed to develop gradually. They would therefore respond to predictive-preventive maintenance (PM) measures resulting in a 60-h maintenance unavailability.

3. Failure 1.6 (Table 9.1) causes unavailability due to preventive maintenance actions. They have two components: preventive adjusting of malfunctioning devices and planned instrument or electrical checks. Forced outages due to instrument malfunctions are not expected.

The resulting reliability and availability values can be compared with process unit target availability. If the discrepancy is excessive, major sources of unavailability should be identified and addressed.

10

Practical field reliability assessment

In the absence of reliability data we have to resort to on-site inspection, engineering judgment and experience in order to arrive at a reasonably consistent and comparable reliability assessment. In a large plant or an organization having many plants, it would be desirable to have a numerical value established to facilitate comparisons of similar equipment and assist in the planning and budgeting of equipment maintenance, engineering manpower support, improvements, or replacements.

In the following examples we show how a machinery index or complexity numbers may be established by actual field observation in order to assess machinery reliability management needs.

THE RELIABILITY INDEX NUMBER*

This is a relative number arrived at to represent the reliability of a particular piece of equipment and to relate it to other similar pieces. This index number can be determined for each piece of critical equipment in a process plant. It also is possible to combine these pieces and express an aggregate Reliability Index Number for the system. There would be little value in doing so, however, unless there were other like systems to be compared with it.

Because it is a relative number, we must be consistent in determining the index number for each type of equipment. Some ground rules must be established to guide craftsmen or specialists in judging the factors involved. The optimum condition would be to have one individual in a plant responsible for determining the Reliability Index Number for one type or class of equipment. The next best condition is to have one person responsible for determining the Reliability Index Number for a class of equipment and provide time for personal communication of the guide rules or guidelines to those making the inspection of that equipment.

For those who believe that perfect should always be 100%, our Reliability Index can always have a maximum value of 100. Because of the inherent

---

* Courtesy of The General Electric Company, Schenectady, N.Y.

differences in designs, to use a base of 100 may require the use of guide rules (rulers) having varying graduations. Nevertheless, this may be the simplest index to apply, providing we can rely upon the use of good judgment by qualified personnel.

DETERMINING THE RELIABILITY INDEX NUMBER

There are five basic factors that must be considered in determining the reliability of an electric motor, for instance. A perfect Reliability Index Number of 100 would be made up of:

Visual inspection	40
Tests and measurements	30
Age	10
Environment	10
Duty cycle	10
Total	100

Visual inspection

When it is made by a qualified technician, visual inspection is the most important factor in determining the reliability of critical equipment. The technician must know what to look for and how to evaluate what he sees. Critical equipment seldom fails during normal operation without giving some warning. We attempt to detect and interpret this warning before a failure occurs. The frequency of thorough visual inspections must be based upon operating experience, the recommendations of equipment manufacturers, and some consideration of the age factor. The technician should have two opportunities to view the equipment: first, in operation under load; second, when partially or completely dismantled. Also, he should have the report of the last visual inspection. A suitable checklist and report form must be used, as this enables us to determine what attention is required and to prepare a cost estimate.

Guide rules must be set up for use by the technician in evaluating the best estimate of condition versus the maximum weighted value allotted (40 in our example). If these are to be kept as simple as possible, they must be made quite broad, such as for a gear box for instance:

Power input path	10
Power conversion path	10
Power transmission path	10
Frame, housings, and base	5
Sensing, indicating, and control	5
Total (max.)	40

Using checklist below, rate each item as follows:

2 = Acceptable
1 = Keep under observation
0 = Requires immediate attention

Stator
(a) . . . . . . Insulation condition
(b) . . . . . . Winding tightness
(c) . . . . . . Cleanliness
(d) . . . . . . Lamination condition
(e) . . . . . . Condition of leads
(f) . . . . . . Air gap
(g) . . . . . . Winding temperature

Rotor
(h) . . . . . . Winding tightness
(i) . . . . . . Cleanliness
(j) . . . . . . Laminations/poles
(k) . . . . . . Bearings
(l) . . . . . . Shaft-spider-coupling
(m) . . . . . . Vibration
(n) . . . . . . Lubrication

Comments (describe condition of all items rated 0): . . . .
. . . .

Rating = sum of items [ ] (28 max.)

Fig. 10.1 Visual checklist: AC motors.

The above guide rules facilitate the use of the overall Reliability Index of 100 but require the inspector to be more flexible in applying his judgment (refer to Fig. 10.1 as applied to a motor). Regardless of the pattern of the guide rules used, we must always apply the same guide rules to similar equipment if our data are to have real significance.

Tests and measurements

These are next in importance in establishing reliability. Some may question the weighted value of visual inspection versus that of tests and measurements. If you can't make good visual inspections of equipment or if you don't have qualified personnel to make them, change the values. However, if you do lower the value of visual inspection for either of these reasons, the overall accuracy of your reliability estimates will be lowered. It might be better to

Electrical tests

A   Insulation Resistance – Megger

	Stator	Rotor
Megohm reading (1.0 min)	. . . . . . megohm	. . . . . .
Rated machine (kv + 1)	. . . . . . (kv + 1)	. . . . . .
Megohms (kv + 1)	. . . . . .	. . . . . .
Megohms (kv + 1)	*Stator rating*	*Rotor rating*
Over 10	5	5
2–10	4	4
1–2	3	3
1.0	2	2
Less than 1.0	0–1	0–1

(A) Rating = ... (Stator) + ... (Rotor) = ... (10 max.)

(B) Rating = ... (Stator) + ... (Rotor) = ... (10 max.)
(C) High-voltage D-C

If no discharge or rapid rise exists, rate 10; otherwise, rate 0–5

(C) Rating = ... (max.)

Total rating ☐ (30 max.)

Fig. 10.2   Tests and measurements: large motors and generators.

hire such qualified personnel from equipment manufacturers or service contractors on a contract basis and strive for accuracy in your ratings.

Again, we must establish guide rules to help achieve uniformity in the ratings of similar equipment. In doing so, make the ratings to be applied to each subfactor as simple as possible, such as Good = 3, Fair = 2, Poor = 1, Requires Immediate Attention = 0 (see Fig. 10.2 for ground rules that have been used for large electric motors and generators typically found in large industrial plants).

It must be pointed out that in very large motors or generators of high-voltage ratings, it is often desirable to add the A-C high-potential test, even though it is a go or no-go test. However, this should be applied only after one or two of the other tests listed have been applied or when it is necessary to establish the suitability for service of the insulation system. Likewise, a turbine-driven generator should be given other tests or measurements such as oil pressure (lubrication), bearing loading, vibration, alignment, clearance of bearings, clearance of wheels, etc. Such large machinery is usually considered individually and no attempt is made to include it here. These tests or measurements are mentioned only as examples to suggest guide rules or subfactors that might be applicable to some types

Age of insulation

Stator . . . . . .   Rotor . . . . . .
(Record age of insulation)

Age (years)	Rating
0– 2	6
2–12	10
13–15	6
16–20	4
Over 20	0–3

Note: If stator age differs from rotor age, rate older component

Rating [   ] (10 max.)

Fig. 10.3   Age guide rules: motors and generators.

of equipment. Obviously, many of our tests and measurements can be made during or at the same time as the visual inspection.

Age of equipment

Age has a definite bearing on equipment reliability, and not just because it may be very old. Most equipment has a statistical life-expectancy curve as shown in Figure 4.8. When equipment is new, it has a higher likelihood of trouble than will be the case after it has operated for 1–2 years. This is caused by manufacturing defects, design inadequacies, shipping damage, or application unknowns. As it becomes old and worn, it requires closer attention to maintenance, unless major rebuilding or upgrading has been performed, which may tend to re-establish the curve. For our use, let us pick a component of the equipment that may be most affected by age, such as the insulation system of a motor or generator, and apply a simple rating formula such as shown in Figure 10.3.

Environment and duty cycle

These are important factors but we rate them at only 10 points each (see Figs 10.4 and 10.5 for applications to motors and generators, as contrasted to the much higher values for visual inspection and tests and measurements). This is because the undesirable effects of the difficult environment and duty cycles are more important than the causes *per se*, and these effects are considered under visual inspection and tests and measurements.

Notice that under environment (Fig. 10.4) we allow for built-in features of the motor or generator that enable it to cope more effectively with the

Environment

Describe . . . .

Environment	*Open or DP	TE
(a) Warm, dry	10	10
(b) Hot (above 40°C)	7	8
(c) Corrosive gas/vapor	5	8
(d) Moisture	0–4	3–7
(e) Abrasive dust	3–5	8
(f) Conductive dust	2–4	7

* Add 3 points to (c–f) for sealed insulation systems

Rating ☐ (10 max.)

Fig. 10.4 Environment guide rules: motors and generators.

Duty cycle

Select one condition from each of the five below:

	Condition	Rate
(a)	Load	
	Smooth	1–2
	Uneven	0–1
(b)	Load	
	100% NP	1–2
	<100% NP	0–1
(c)	Duty	
	Short-time	1–2
	Continuous	1
(d)	Duty	
	Non-reverse	2
	Plug or reverse	0–1
(e)	Starts	
	Few (1/h)	1–2
	Frequent (1/h)	0–1

Rating ☐ (10 max.)

Fig. 10.5 Duty cycle guide rules: motors and generators.

problem. Again, under duty cycle (Fig. 10.5), we have favored the short-time rated motor, which may never reach name-plate maximum operating temperature, and the motor which is not plugged or reversed. However, we have penalized the motor that is plugged or reversed, or that is started and stopped frequently.

(A) *Motor data*

Unit:	Process number: . . . .
Duty:	Location: . . . .
Asset:	
Manufacturer and type:	
Installation date:	Design: . . . .
Horsepower/frame:	Class of insulation: . . . .
Synchronous speed:	Full load speed: . . . .
Serial number:	Volts/cycles/phase: . . . .
Full load amperes:	Locked motor amperes: . . . .
Enclosure:	
Temperature rise:	Maximum ambient/cooling: . . . . . .
Service factor:	

(B) *Reliability factors*

I.	Visual inspection:	. . . . . . of 40
II.	Tests and measurements:	. . . . . . of 30
III.	Age:	. . . . . . of 10
IV.	Environment:	. . . . . . of 10
V.	Duty cycle:	. . . . . . of 10
	Reliability index:	. . . . . . of 100
	Date index recorded:	. . . .

Fig. 10.6 Maintenance reliability evaluation: large motors.

It is obvious that the last three factors – age, environment, and duty cycle – can be rated with a minimum of effort.

When all five factors have been evaluated and totalled, a single Reliability Index Number results. The reliability rating report form used to establish the rating would consist of the five factors shown in Figures 10.1–10.5. These can be incorporated on a single page with appropriate headings for equipment nomenclature, location, productivity rating, and maximum Reliability Index value (see Fig. 10.6). The Reliability Index Number is not a magic number, above which all similar equipment will not fail in service and below which it will fail. One such number will have but little value; when compared with other numbers established by the same method for similar equipment, however, it can be very valuable.

Please note that our examples state that 0 = Requires Immediate Attention. If our Reliability Index report on a piece of equipment contains one or more 0 items, we must examine these before proceeding further. If they indicate that minor, or even routine, maintenance is required, it may be best to accomplish it right away and then correct the ratings accordingly. The resulting Reliability Index Numbers will be more accurate and of more value to us in planning and budgeting for equipment maintenance, rebuilding, or replacement.

EQUIPMENT REPLACEMENT AND REBUILDING

Our Reliability Index Numbers will be significant when compared with similar equipment within the same productivity rating or classification. From this comparison, we can establish and assign priorities for equipment maintenance, rebuilding and upgrading, or replacement. We can expend our maintenance effort where it is most needed. By referring to the reliability rating report forms, we can determine the action that is required to maintain operation at the normal level. An estimate of the cost of such maintenance can be established, based upon our past experience or quotations from equipment builders or maintenance contractors.

If we have used the variable base Reliability Index Numbers, we can convert them into percentages to be used as a guide in evaluating priority ratings to be assigned to different kinds of equipment within productivity ratings or classifications. An overall average Reliability Index Number can also be established for a process or an operation.

If we calculate the anticipated Reliability Index Numbers that will result from the indicated maintenance actions, we can advise management of the existing level and the anticipated level that the execution of our maintenance

Table 10.1 Rotating equipment complexity assessment

1. Complexity by train configuration	
Gas turbine-driven compressors – highest complexity	5
Steam turbine-driven compressors (turbo and/or recips)	4
Reciprocating compressors	3
Motor-gear compressors	2
GT and ST generators	2
Motor-driven compressors (direct drive)	1
2. Complexity by size, special-purpose equipment	
Over 15,000 hp	4
5001–15,000 hp	3
501–5000 hp	2
1–500 hp	1
3. Complexity by size, pumps (rating given to entire pump population in a given plant)	
Predominantly large pumps – least complex, over 100 hp	10
Predominantly medium size, 25–100 hp	20
Predominantly small pumps, less than 25 hp	30
4. Complexity by age	
Over 10 years	3
5–10 years	2
Less than 5 years	1
5. Complexity by counting driven casings	
This is a straightforward summation of casings, exclusive of drivers	

Table 10.2 Turbomachinery complexity ranking for several North American petrochemical plants

Plant	Type	Age	Size	Casings	Pumps	Total
AMOS	2 × 3	3 × 2	1 × 3	2 × 2		
	2 × 4	1 × 3	3 × 2	2 × 3		52
	14	9	9	10	10	
BRIT	4 × 2	2 × 3	1 × 4	3 × 4		
	3 × 3	3 × 1	4 × 3	3 × 3		
	1 × 1	3 × 2	3 × 2	2 × 2		100
	18	15	22	25	20	
CHET		2 × 3	1 × 4	3 × 4		
	2 × 4	1 × 2	2 × 3	1 × 2		
	3 × 3	2 × 1	2 × 1	1 × 1		84
	17	10	12	15	30	
DORA				7 × 2		
		4 × 3	2 × 3	2 × 4		
	8 × 3	3 × 2	3 × 4	1 × 3		
	3 × 2	4 × 1	6 × 2	1 × 1		128
	30	22	30	26	20	
ERIC	4 × 5	2 × 2	4 × 3			
	3 × 4	2 × 3	1 × 2	7 × 4		
	1 × 3	4 × 1	3 × 1	1 × 3		107
	35	14	17	31	10	
FRAN	2 × 4	3 × 3	2 × 4	1 × 5		
	2 × 3	1 × 2	2 × 2	2 × 4		
	1 × 2	1 × 1	1 × 1	2 × 2		88
	16	12	13	17	30	
GREG	4 × 4			1 × 4		
	3 × 3	3 × 3	4 × 4	1 × 3		
	2 × 1	3 × 2	2 × 3	6 × 2		
	1 × 2	4 × 1	4 × 2	2 × 1		119
	29	19	30	21	20	
HANK			1 × 4	2 × 4		
	3 × 4	2 × 3	1 × 3	1 × 3		
	2 × 3	1 × 2	3 × 2	2 × 2		
	1 × 2	3 × 1	1 × 1	1 × 1		71
	20	11	14	16	10	
IRIS	4 × 2	4 × 3	5 × 4	3 × 4		
	7 × 1	7 × 2	6 × 2	8 × 2		131
	15	26	32	28	30	

Note: Study includes ST-driven recips, but not motor-driven recips.

budget will accomplish. This can be done by individual pieces of equipment, productivity ratings, or processes.

THE MACHINERY COMPLEXITY NUMBER

Any assessment of machinery reliability would logically give consideration also to machinery complexity. This is sometimes recognized in engineering manpower studies for major petrochemical plants which would obviously need more personnel for large, multi-casing or old machinery trains than for smaller, less complex, or perhaps new machinery trains.

In the mid- to late 1980s, the authors investigated the merits of categorizing rotating machinery complexity on the basis of machine type, train configuration, size, and age. Five numerical gradations were proposed (Table 10.1).

Using the numbering system described in Table 10.1, a given plant would be in a position to assess its rotating equipment complexity and to make a somewhat more objective judgment than pure guessing in attempting to compare various plants.

If, for example, a plant were comprised of three major machinery trains aged 4, 7, and 16 years, the age complexity numbers would be $1 \times 1 = 1$, $1 \times 2 = 2$, and $1 \times 3 = 3$, for a total of 6 complexity points. Let us assume one of the trains to be a gas turbine-compressor–gear-compressor type, rated at 25,000 hp. Its "type complexity" would rank a 5, its casing complexity number would be 3. The remaining trains would likewise be ranked on the basis of these complexity numbers.

The relative complexity ranking for nine North American petrochemical plants is shown in Table 10.2. The purpose of this 1986 exercise was to divide the complexity numbers of various plants (e.g. CHET's 84, or DORA's 128) by the number of machinery support engineers entrusted with maintenance and surveillance. It was theorized that the resulting number might show some plants to have excessive and other plants to have insufficient staffing.

In summary, let us be sure of one thing: Our only purpose here is to acquaint the reader with yet another method of identifying potentially vulnerable machinery. As is the case with so many numerical ranking methods, our complexity assessment approach will be helpful only if it is tempered by good judgment and solid experience of the implementor.

PART II

Audits and reviews

11

The design review process

In his efforts to select the most appropriate machine for a given service or duty, the reviewing engineer must either subjectively or objectively evaluate a host of parameters. These include, of course, such traditional considerations as available capital, anticipated life, efficiency, location, environment, etc. In addition, the evaluation must attempt to measure the reliability, maintainability, and surveillability of critically important machine components or subsystems.

Whenever this assessment is attempted, the evaluating engineer must keep in mind that parts, components, assemblies, systems, or machines fail because of the action of one or more of the four principal agents of failure: Force, Reactive Environment, Time, and Temperature. Using the acronym FRETT as an easy reminder we should continually test for potential impact or susceptibility under all foreseeable operating or service conditions.

Although the following chapters can only highlight a few typical machinery components and subject these to somewhat closer scrutiny, it is hoped that our emphasis on FRETT and the interaction of reliability, maintainability, and surveillability will become evident.

According to our definition of elements critical to the functioning of machinery (Chapter 1), we have organized our detailed screening, selecting, review, and assessment examples in the ensuing chapters as follows:

- Transmitting elements: Chapters 12–16
- Constraining, confining, and containing elements: Chapter 17
- Fixing elements: Chapter 18
- Elements supporting machinery functions: Chapter 19

A review or audit initially has no effect on reliability or maintainability. The sole purpose of a review activity is to expose vulnerabilities that show where improvement is necessary so that action can be taken. Actions to improve a machinery system must affect:

- reliability or failure frequency;
- maintainability or maintenance time;

- availability or machinery capability;
- surveillability or operability.

We usually have four general types of options available to us to accomplish the above:

1. Reduce number of components as fewer components generally result in fewer failures and less maintenance.
2. Reduce component failure rate (λ) by providing improved components.
3. Reduce the effects of failures by alleviating their consequences where the failure mode cannot be eliminated by design action.
4. Reduce maintenance time by making sure that where maintenance is necessary it should first be possible and then perhaps easily accomplished.

THE CHECKLIST APPROACH

We propose an approach as shown in Fig. 11.1, the vulnerability checklist structure. In reviewing a machinery system we will resort to first general review checklists and then to checklists covering the features of a specific machinery type. After having done this the reviewer will use a checklist which covers the individual components by elements, as shown in Reviews 0.0.3.0. 1.1.1.0, 1.1.2.0, 1.1.3.0, and 2.1.1.0 to 6.1.3.0, which follow.

0.0.3.0 General review – surveillability
    0.0.3.1 Design stage
        0.0.3.1.1 Has troubleshooting and fault identification been considered
            3.1.2 If redundancy features have been provided can main, respectively spare, be monitored for failure
            3.1.3 If applicable, can wear progress of parts be monitored
            3.1.4 Is there advance warning of failure
            3.1.5 Can deterioration be detected and predicted
            3.1.6 Is there accessibility for on-line inspection
            3.1.7 Is there a risk of damage when performing on-line inspection
            3.1.8 Can protective devices be tested without causing a shutdown or an emergency situation
            3.1.9 Can important performance indicators be easily observed
            3.1.10 Surveillability assurance
                – Is it simple
                – Is it practical, can it be done on a continuous basis

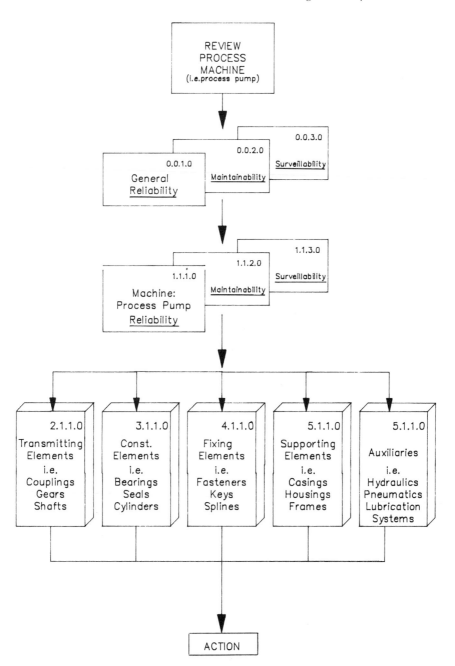

Fig. 11.1 Machinery vulnerability checklist structure.

　　　0.0.3.2 Materials, manufacturing, test and shipping
　　　　　0.0.3.2.1　Have indicators of trouble, or weak spots, been considered
　　　0.0.3.3 Installation, commissioning and operation
　　　　　0.0.3.3.1　Can predictive maintenance techniques be employed
　　　　　3.3.2　Is there open communication between the maintenance and the operating departments
1.1.1.0 Process pump review – reliability
　　　1.1.1.1 Design stage
　　　　　1.1.1.1.1 Suction performance
　　　　　　　– Is NPSH margin adequate
　　　　　　　– Has minimum continuous stable flow been considered
　　　　　　　– Has suction piping design been optimized
　　　　　1.1.2 Discharge performance
　　　　　　　– Have fluid properties (viscosity, density) been considered
　　　　　　　– Effect of curve shape considered
　　　　　1.1.3 Operating range
　　　　　　　– Has system been reviewed
　　　　　　　– Fail shut valves
　　　　　　　– . . .
　　　　　1.1.4 Pressure and temperature
　　　　　　　– . . .
　　　　　　　– . . .
　　　　　　　– . . .
　　　　　1.1.5 Casing design . . .
　　　　　1.1.6 Impeller design . . .
　　　　　1.1.7 Shaft design . . .
　　　　　1.1.8 Bearing design . . .
　　　　　1.1.9 Driver design . . .
　　　1.1.1.2 Materials, manufacturing, test and shipping
　　　　　1.1.1.2.1 Casting quality . . .
　　　　　1.2.2 Can mechanical test run be performed . . .
　　　　　1.2.3 Has protection against shipping damage on large pumps been considered . . .
　　　1.1.1.3 Installation, commissioning and operations
　　　　　1.1.1.3.1 Has water as test fluid been considered . . .
1.1.2.0 Process pump review – maintainability
　　　1.1.2.1 Design stage
　　　　　. . .
　　　1.1.2.2 Materials, manufacturing, test and shipping
　　　　　. . .
　　　1.1.2.3 Installation, commissioning and operation
　　　　　. . .

1.1.3.0 Process pump review – surveillability
    1.1.3.1 Design stage
      . . .

    1.1.3.2 Materials, manufacturing, test and shipping
      . . .

    1.1.3.3 Installation, commissioning and operation
      . . .

2.1.1.0 Coupling review – reliability
    2.1.1.1 Design stage
      . . .

    2.1.1.2 Materials, manufacturing, test and shipping
      . . .

    2.1.1.3 Installation, commissioning and operation

2.1.2.0 Coupling review – maintainability
  . . .
2.1.3.0 Coupling review – surveillability
  . . .
2.2.1.0 Gear review – reliability
  . . .
2.2.2.0 Gear review – maintainability
  . . .
2.2.3.0 Gear review – surveillability
  . . .
2.3.1.0 Shaft review – reliability
  . . .
2.3.2.0 Shaft review – maintainability
  . . .
2.3.3.0 Shaft review – surveillability
  . . .
3.1.1.0 Bearing review – reliability
  . . .
3.1.2.0 Bearing review – maintainability
  . . .
3.1.3.0 Bearing review – surveillability
  . . .
3.2.1.0 Shaft seal review – reliability
  . . .
3.2.2.0 Shaft seal review – maintainability
  . . .
3.2.3.0 Shaft seal review – surveillability
  . . .
4.1.1.0 Threaded fastener review – reliability
  . . .

4.1.2.0 Threaded fastener review – maintainability
         . . .
4.1.3.0 Threaded fastener review – surveillability
         . . .
4.2.1.0 Keys and splines review – reliability
         . . .
4.2.2.0 Keys and splines review – maintainability
         . . .
4.2.3.0 Keys and splines review – surveillability
         . . .
5.1.1.0 Casing review – reliability
         . . .
5.1.2.0 Casing review – maintainability
         . . .
5.1.3.0 Casing review – surveillability
         . . .
6.1.1.0 Auxiliaries and lube system review – reliability
         . . .
6.1.2.0 Auxiliaries and lube system review – maintainability
         . . .
6.1.3.0 Auxiliaries and lube system review – surveillability

Reviews 0.0.1.0 and 0.0.2.0 show reliability checklists that cover general and common features of most process machinery designs. We cannot even begin to develop exhaustive and all-encompassing coverage but would like to look at these checklists more as examples and guides. The same is true for Reviews 0.0.1.0 and 0.0.2.0, which show in a somewhat sketchy way what specific structure would be embodied in checklists aimed at the machine and component level.

0.0.1.0 General review – reliability
         0.0.1.1 Design stage
                  0.0.1.1.1 Is the design an extrapolation of past experience
                       1.1.2 Are component numbers minimized
                       1.1.3 Have safety risks been considered
                       1.1.4 Have environmental protection rules been considered
                       1.1.5 Has FRETT been considered:
                              – Have mechanical overload protection devices been
                                provided
                              – If wear encountered, what form will it have and what
                                protection or resistance has been provided
                              – Have sacrificial surfaces been considered
                              – Have material galling characteristics been observed
                              – Have temperature effects been considered
                              – Has the effect of long-term/short-term exposure to
                                FRETT been considered

1.1.6 Has redundancy been considered for critical components; is there manual or automatic back-up

1.1.7 Has the principle of self-closure by pressure balance (covers, seals, etc.) been employed

1.1.8 Can accidental ingestion of liquids or solids be a problem

1.1.9 Has possible dynamic interaction between drive and driven equipment been considered (torsional vibration)

1.1.10 Have support structures been isolated from damaging influences, vibration from other machines, for instance

1.1.11 Can attachments such as piping and auxiliaries exert excessive stresses

1.1.12 Have stress risers been avoided

0.0.1.2 Materials, manufacturing, test and shipping

0.0.1.2.1 Have possible material defects been considered, casting flaws, etc., for example

1.2.2 Has manufacturer's facility been inspected for quality control and organizational integrity

1.2.3 Can the machine be tested prior to final acceptance

1.2.4 Can adequate protection against shipping or storage damage be provided

0.0.1.3 Installation, commissioning and operation

0.0.1.3.1 Can machine be installed without damaging it

1.3.2 Is pre-commissioning cleaning of process fluid (liquid, gas) lines and lubrication systems possible

1.3.3 Can the machine be tested after installation

1.3.4 Has the possibility of operator error been considered

1.3.5 Have special lifting devices for installation been considered

1.3.6 Have maintenance clearances been reviewed by maintenance personnel

0.0.2.0 General review – maintainability

0.0.2.1 Design stage

0.0.2.1.1 Have on-line maintenance possibilities been maximized

2.1.2 Is there a risk of damage to machine when performing MRO (maintenance, repair, overhaul)

2.1.3 Can machine be opened up for MRO without disturbing attachments such as piping, auxiliaries, or other machines

2.1.4 Is there a risk of damage of other machine parts when performing MRO

2.1.5 Has interchangeability of parts been maximized

2.1.6 Is there a risk of damage through installation error of similar parts

  2.1.7 Are components provided that allow quick maintenance
  – Few fasteners
  – Easy assembly
  – Simple adjustments
  – Replaceable quick-change modules
  2.1.8 Can protective instruments and performance indicators be tested and changed on the run
 0.0.2.2 Materials, manufacturing, test and shipping
 0.0.2.2.1 Has long-term storage maintenance been planned
 0.0.2.3 Installation, commissioning and operations
 0.0.2.3.1 Are special skills required for MRO
  2.3.2 Are special tools, instruments, or fixtures required for MRO
  2.3.3 Have lifting devices been provided
  2.3.4 Have spare parts supply logistics been considered
  2.3.5 Has isolation of machine from process for MRO been considered
  2.3.6 Are there shaft alignment tools and fixtures available
  2.3.7 Is there an active plan to minimize MTTR
  – Illumination
  – Tools
  – Protective clothing and apparatus
  – Quick isolation, draining

Of interest to the reader should be the machinery systems completeness and reliability appraisal forms used by the more advanced multinational petrochemical companies. In the procurement, specification, evaluation, and execution of such major machines as centrifugal and reciprocating compressors, steam turbines, lubrication systems for complex machinery, etc., the engineering staff would utilize a review format similar to that shown in Appendix C. Four different forms are used for each of the four project phases. These phases are (1) the Piping and Instrument Diagram and Specification (P & ID) review; (2) the Pre-Order Review with Vendor; (3) the Contractor's Drawing Review; and (4) the Mechanical Run Test Review. Additional checklists would encompass such items as Field Storage and Handling which can also influence reliability.

Two different approaches or structures are sometimes used for the forms which document this machinery systems completeness and reliability appraisal. Pages 286–309, which represent the form consulted at the P & ID and Spec Review phase of a project, illustrate the "Tutorial" structure or approach. As the term implies, each checklist item is thoroughly analyzed and explained. Note also the extent, or sheer length of this form. In contrast, the forms used for the other project phases are structured in a pure checklist mode.

12

Critical component assessment*

Whenever a single key part, component, or assembly is unproven or could otherwise influence the overall reliability of a machine, or perhaps the projected output of an entire plant, it will be necessary to make a detailed analysis. This analysis is principally aimed at understanding the vendor's design assumptions and methods. Accordingly, much information such as empirical constants must be supplied by the vendor.

A good example of this rigorous assessment would be a Bendix* diaphragm coupling connecting a 19,600-hp steam turbine to a centrifugal compressor operating at 6700 rpm. The reviewing engineer would proceed after carefully noting that the equations, graphs, and data given here are unique to the "standard" catalog coupling made by Bendix and are not applicable to other manufacturers' designs.

Also, we wish to emphasize that such detailed component assessments are only appropriate for new products for which little field data exists. As of 1989, Bendix (Lucas Corp.) couplings have racked up an enviable reliability record and the authors would see no justification to apply the type of scrutiny which we gave the subject in 1970.

Accordingly, we show this critical component assessment example only to convey to the reader an understanding of the degree of detail that is sometimes warranted.

TORQUE EFFECT ANALYSIS

Shear stress

The first effect torque has on the diaphragm is the shear stress in the contoured section of the diaphragm, henceforth called the "profile" (Fig. 12.1). The Bendix diaphragm has a profile shape such that the shear stress is constant from hub to rim, and thus the diaphragm is nowhere thicker than it need

* The cooperation of Allied Signal Inc., Bendix Fluid Power Division, Utica, N.Y., is gratefully acknowledged.

159

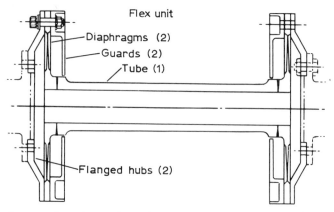

Fig. 12.1   Cross-section views of Bendix contoured diaphragm coupling for turbo-machinery.

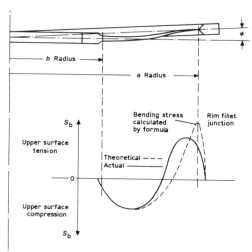

Fig. 12.2   Contoured diaphragm nomenclature and stresses acting in flexible portion.

be to carry the torque. Therefore:

$$S_s = Tq/2\pi a^3(t_a/a) \qquad (12.1)$$

$$t_i = k/r_i^2 \qquad (12.2)$$

where S_s = shear stress (psi),
   Tq = torque (lb-in),
   a = radius "a" (in) (see Fig. 12.2),
   t_a = thickness at radius "a" (in),
   k = constant (in^3),
   r_i = any radius (in),
   t_i = thickness at radius "r_i" (in).

Torsional buckling

A second effect of torque on the diaphragm is the induced compressive stress in the profile as a result of elastic torsional wind-up of the diaphragm. This stress is harmless up to the point at which buckling occurs. This buckling is elastic at the onset but becomes permanent if sufficient overtorque is applied.

$$T_B = \pi E Z a^3 (t_a/a)^3 / 12(1 - p^2) \qquad (12.3)$$

where T_B = buckling torque (lb-in),
E = Young's modulus (psi),
Z = function of b/a ratio (Fig. 12.2). A number such as 2000 is a satisfactory working value (vendor-supplied),
p = Poisson's ratio.

Torsional instability

The third torque effect is not encountered except in multiple diaphragm assemblies. It is a torsional instability of the assembly in which the very flexibility of the diaphragms permits a corkscrew deflection of the torque axis to occur.

$$T_I = 360 K_b / n \qquad (12.4)$$

where T_I = torsional instability (lb-in),
K_b = diaphragm bending spring rate (lb-in/deg),
n = number of diaphragms in each assembly, e.g. for a double-ended coupling with a total of four diaphragms with two at each end, $n = 2$.

MISALIGNMENT EFFECT ANALYSIS

Bending stress

The parallel offset and angular misalignment of the shafts of connected machines cause bending of the flexible diaphragms. The stresses resulting from this bending are fully reversing cyclic fatigue stresses which occur at the rate of 1 cycle per revolution. Therefore, the diaphragm must be designed and manufactured to survive an infinite number of bending cycles.

$$S_b = K_1 E(t_a/a)\phi K_6 \qquad (12.5)$$

where S_b = bending stress (psi),
K_1 = function of b/a (see graph in Fig. 12.2),
E = Young's modulus (psi),
ϕ = bending angle per diaphragm (deg),
K_6 = 0.77, profile correction, vendor-supplied.

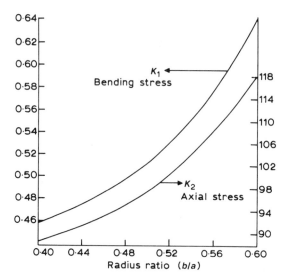

Fig. 12.3   Stress coefficients as a function of diaphragm geometry (i.e. radius ratio *b/a*).

The working limit for the bending stress calculated with equation 12.5 is 24,000.

Axial stress

Changes in length of the coupling can be accomplished by flexurally extending or compressing the "flex unit". This deflection from the free length causes a bending of the diaphragm profile. The stresses resulting from this bending are steady-state stresses, not fatigue stresses, except when the axial deflection is oscillatory and occurs at a frequency more than 10% of the rotational speed. Under the latter conditions the axial stress is fatigue stress and should be added directly to the bending stress.

$$S_a = K_2 G(t_a/a)(x/a)K_7 \qquad (12.6)$$

where S_a = axial stress (psi),
   K_2 = function of b/a (see graph in Fig. 12.3; vendor-supplied),
   G = shear modulus (psi),
   x = deflection per diaphragm (in),
   K_7 = 0.80, profile correction (vendor-supplied).

The working limit for steady-state axial stress for these couplings is 60,000 psi.

Centrifugal stress

The rim of the diaphragm expands under the influence of centrifugal force to a greater extent than the profile portion; thus the profile portion is put

Table 12.1 Coefficients used in conjunction with centrifugal stress

Series	K_5	Series	K_5
305	16.4	405	13.7
306	12.7	406	10.4
308	10.3	408	7.7
310	9.0	410	6.9
312	7.6	412	6.0
314	7.5	414	5.8
316	6.6	416	5.5
318	6.7	418	5.0
322	5.5	422	4.4
326	4.6	426	3.5
331	4.5	431	3.3

into tension as it restrains the rim. The location of the highest radial centrifugal stress is unfortunately near the point of highest bending stress. Therefore, this centrifugal stress must be considered in the combined stress equation 12.8. The shape of the rim and hub, the b/a ratio and the t_a/a ratio affect the centrifugal stress. A listing of coefficients of representative diaphragms is shown in Table 12.1.

$$S_c = K_5 a^2 n^2 \tag{12.7}$$

where S_c = centrifugal stress (psi; maximum at the thin point),
  K_5 = coefficient (vendor-supplied; Table 12.1),
  n = speed (krpm),
  a = radius at A.

Material	*Fatigue strength* *psi*
AISI 4340 VM	70,000
Ti6A14V	72,000

Combined stress

The profile section of the diaphragm is subject to a combined stress which is made up of the various stresses presented on the previous pages. These stresses are broken down into a combined mean stress and a combined alternating stress.

The mean stress is calculated as follows:

If $S_s < \sqrt{p}\,(S_a + S_c)$

$$S_m = (1 + p)/2 + \sqrt{[(1 - p)/2]^2 \times (S_a + S_c)^2 + S_s^2} \qquad (12.8)$$

where $p = 0.32$

$$S_m = 0.66(S_a + S_c) + \sqrt{0.1156(S_a + S_c)^2 + S_s^2}$$

If $S_s > \sqrt{p}\,(S_a + S_c)$

$$S_m = \sqrt{(1 - p + p)^2 \times (S_a + S_c)^2 + 3S_s^2} \qquad (12.9)$$

where $p = 0.32$

$$S_m = \sqrt{0.7824 \times (S_a + S_c)^2 + 3S_s^2}$$

The alternating stress is calculated as follows:

$$S_{\mathrm{alt}} = \sqrt{(S_b + \text{alternating } S_a)^2 + 4(\text{alternating } S_s)^2} \qquad (12.10)$$

CONSTANT-LIFE FATIGUE DIAGRAM

The preceding stresses are plotted on the constant-life fatigue diagram (Fig. 12.4). The limiting parameter is the 10^7 cycle life line. The profile section is designed so the combined stress point, when plotted on this curve, falls on or below a line half way between the origin and the 10^7 life line (Fig. 12.4).

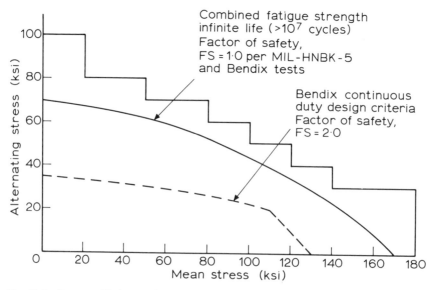

Fig. 12.4 Constant life fatigue diagram for AMS 6414 steel diaphragms in 170,000 uts heat treat condition.

Bending spring rate

The moment which must be applied to a coupling to cause it to deflect through a given angle is a function of the bending spring rate of the diaphragm or diaphragms used in the coupling. If the coupling has a single diaphragm at each end of the unit, as do standard industrial couplings, then the spring rate for the coupling is the same as that of a single diaphragm. However, for a multiple diaphragm coupling the spring rate at each center of articulation, or diaphragm pack, is equal to that of one diaphragm divided by the number of diaphragms in the pack. The spring rate of a single diaphragm is:

$$K_b = [K_3 \pi E a^3 (t_a/a)^3] K_8 \qquad (12.11)$$

where K_b = bending spring rate (lb-in/deg),
K_3 = function of b/a (vendor-supplied, Fig. 12.5),
K_8 = 1.0, profile correction (vendor-supplied).

Reaction loads

The reaction loads on the connected equipment due to this bending spring rate are usually sufficiently low so as to have little or no effect upon these machines. These loads may be readily calculated.

The primary reaction moment for the standard two-diaphragm coupling is equal to the bending spring rate of the diaphragm times the bending angle on that diaphragm.

The secondary reaction moment is at 90° to the primary and is equal to the transmitted torque times the tangent of the diaphragm bending angle. These two moments should be combined by using the square root of the sum of their squares.

The transverse shear load on either machine is equal to the sum of the combined moments on each machine (which can be different) divided by the distance between the diaphragms.

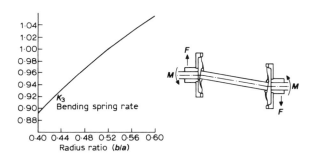

Fig. 12.5 Coefficients used in calculating bending spring rate are dependent on radius ratio b/a.

Transverse loads due to unbalance conditions must be considered separately.

Lateral spring rate

A multiple diaphragm coupling has a distinct spring rate perpendicular to the coupling axis. This can affect the critical speed. Note that a standard industrial two-diaphragm coupling (Fig. 12.6) does not have lateral flexibility unless the center tube flexes. Such a coupling mounted on rigid machines has a critical speed N_c as noted.

Figure 12.7 depicts a coupling with two diaphragms at each end of a center tube. Here it can be seen that a transverse load can cause deflection of the center tube without any flexing of the tube. The radial spring rate (K_r) is given as a function of the per diaphragm spring rate (K_b) and the distance (L) between the profile centerlines of the diaphragms. Should the tube flex the effect of the diaphragm, bending upon total deflection would be reduced. Figures 12.8 and 12.9 show the cases for three- and four-diaphragm packs.

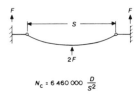

$$N_c = 6\,460\,000\,\frac{D}{S^2}$$

Fig. 12.6   Critical speed formula for industrial two-diaphragm coupling.

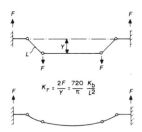

$$K_r = \frac{2F}{Y} = \frac{720}{\pi}\frac{K_b}{L^2}$$

Fig. 12.7   Radial spring rate (K_r) as a function of the per-diaphragm spring rate (K_b) and the distance (L) between the profile centerlines of the two diaphragms at each end of the coupling.

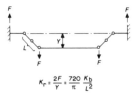

$$K_r = \frac{2F}{Y} = \frac{720}{\pi}\frac{K_b}{L^2}$$

Fig. 12.8   Spring rate relationships for coupling with three diaphragms per coupling end.

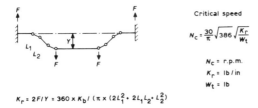

$$K_r = 2F/Y = 360 \times K_b / (\pi \times (2L_1^2 + 2L_1L_2 + L_2^2)$$

Fig. 12.9 Spring rate relationships for coupling with four diaphragms per coupling end.

The critical speed of multiple diaphragm couplings, with non-flexing center tubes, mounted on rigid machines, can be computed from equation 12.12.

$$N_c = 30/\pi \times \sqrt{386} \times \sqrt{K_r/W_t} \qquad (12.12)$$

where N_c = lateral critical speed (rpm),
$\pi = 3.14159$,
K_r = radial spring rate (lb/in),
W_t = suspended weight (lb).

Axial spring force

Changes in length of the coupling by flexure of the diaphragms require a force proportional to the deflection from the neutral length.

$$F = [K_9 E(t_a/a)]X^3 + K_4 G_a(t_a/a)^3 X \qquad (12.13)$$

where F = axial force (lb),
K_4 = function of b/a (see Fig. 12.10; vendor-supplied),
K_9 = function of b/a (see Fig. 12.10; vendor-supplied),
X = deflection per single diaphragm (in).

The total extension (or compression) of a two-diaphragm coupling (as pictured opposite) is twice the deflection (X) of a single diaphragm. Multiple diaphragm couplings have proportionally larger deflections for the same axial force.

Axial preloading

For installations where the axial deflection required for a given coupling is greater than would be permissible from a stress or axial load viewpoint, it may be possible to install the coupling in a preloaded condition. This will enable the axial movement to move the coupling from the preload position through neutral to a load position, thus doubling the axial deflection capability. This technique is especially valuable in connecting machines whose shafts grow towards each other, a known amount, due to thermal expansion.

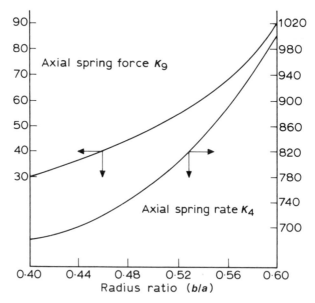

Fig. 12.10   Coefficients used in calculating axial spring force of diaphragm couplings.

Axial vibration

The center tube of the standard industrial coupling is connected only to the two diaphragms; thus it is a mass freely suspended on springs. The total spring rate in this case is twice the single diaphragm rate. The center tube will respond to an axial input vibration and will vibrate at its resonant frequency.

Generally, this vibration does not cause trouble unless a significant amount of vibration at exactly the resonant frequency is applied axially to the coupling. Operation at the rotational speed corresponding to the resonant frequency will cause an axial vibration and, although it is normally not severe, such operation should not be continuous.

A graph of the axial resonant frequency in cycles per minute for the different standard industrial couplings is given in Fig. 12.11. The length referred to on the graph is the length of the tube. This graph is plotted with the diaphragm in its neutral plane (no axial displacement). Operation of the diaphragm with axial displacement can effect a shift in the axial resonant frequency. For operation under these conditions, the vendor's engineering department would have to be consulted.

Balance and critical speed

Determining the amount of unbalance a given set of connected machines will tolerate is a thorny problem. All too often one of the lateral critical

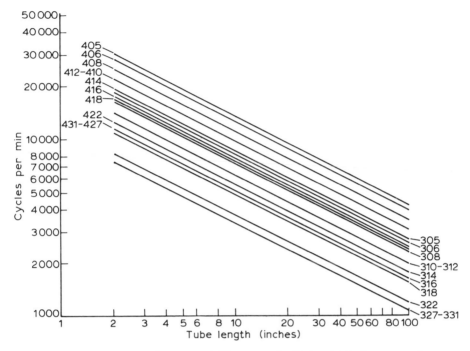

Fig. 12.11 Axial resonant frequency of different standard industrial diaphragm couplings.

speeds of the drive train is near enough to the operating speed to cause any unbalance to be excessive. The best preventive action to avoid any vibration linked with rotational speed is to calculate the various lateral, axial, and torsional resonant frequencies before the design of all the elements of the system is completed. For multi-element systems these calculations are best done on computers.

Coupling balance

The standard industrial diaphragm coupling is designed in such a manner that the "flex unit" may be removed and either of the coupled machines may be operated independently. For this reason, among others, the coupling is balanced in at least four places. The work holding arbor is balanced first of course and then each of the hubs is balanced with the material being removed from the places marked 1, 1a, 2, and 2a (see Fig. 12.12). The "flex unit" is then installed on the arbor with its weight balanced bolts and nuts and it is balanced by material removal at 3 and 4. If a spare "flex unit" is made at the same time as the original unit, it can also be balanced on the arbor-mounted hubs so that it would be ready for use without trim balancing. All units are match-marked to ensure balance after disassembly and reassembly.

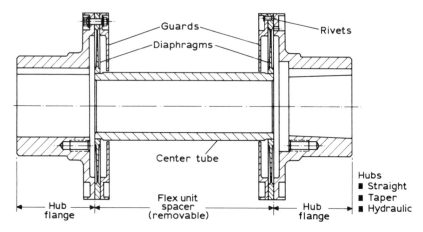

Fig. 12.12   Balance planes for material removal on diaphragm couplings.

Centrifugal force

The centrifugal force acting upon the bearings of a rigid rotating body is a function of unbalance and speed.

$$U = 8We \tag{12.14}$$

$$F_c = 1.778Un^2 \tag{12.15}$$

where U = unbalance (oz-in),
W = weight of body (lb),
e = eccentricity, distance between center of bearings and center of gravity (in),
F_c = centrifugal force (lb),
n = rotational speed (krpm).

The nomograph (Fig. 12.13) will help correlate the above two equations and also aids to point up the effect that weight and eccentricity have upon unbalance. Most balancing machines have a finite balance limit that is reached when the measured eccentricity drops below about 0.00002 in. Below that the noise from the bearings supporting the arbor mask the signal, and the location of the heavy side cannot be determined.

STANDARD INDUSTRIAL COUPLINGS

The standard diaphragms for the industrial couplings have parameters which vary according to their diameter and flexibility. The radius ratio for all of this series of diaphragms is 0.52. This ratio was chosen as a compromise of several design criteria. The thickness ratios (t_a/a) for the 300 and 400 series

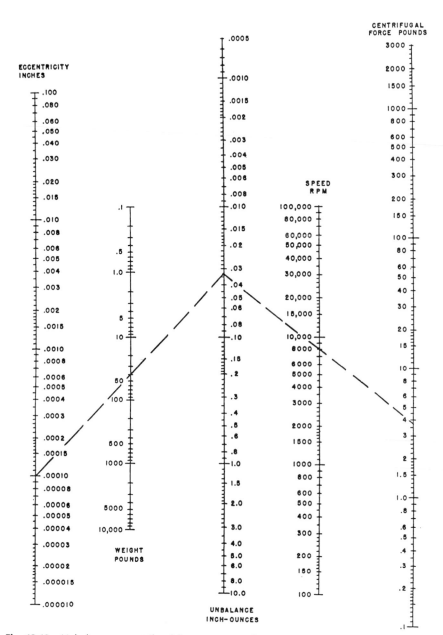

Fig. 12.13 Unbalance – centrifugal force nomograph.

couplings are 0.006 and 0.008, respectively. The major profile radius (*a*) is:

Coupling diameter (in)	Radius a (in)
5.750	2.200
6.750	2.760
8.875	3.605
10.625	4.445
12.750	5.430
14.500	6.200
16.500	7.250
18.000	8.000
22.000	9.930
26.000	11.750

MATERIALS

Bendix standard couplings are made from all wrought steel parts. The diaphragms which are the working elements of the coupling are made from a vacuum melted grade of 4340 alloy steel, AMS 6414. They are produced from either pancake forgings or by cross-rolling into plate, thus giving isotropic properties. The diaphragms are heat-treated to a minimum ultimate tensile strength (UTS) of 170,000 psi.

The center tube is made from cold drawn AMS 6371 tubing or AMS 6415 forgings. They are heat-treated to 130,000 psi UTS minimum.

The guards are made from AISI 4140, which is heat-treated to 130,000 psi UTS minimum. Any flanges for adapting to machinery flanges (such as on steam turbines) are made from AISI 4140, 4340, or AMS 6415. They are heat-treated to 130,000 psi UTS minimum.

The shaft-mounted hubs are made from AMS 6415 forgings which are heat-treated to 130,000 psi UTS minimum.

The diaphragms are welded to the center tube by electron beam welding. This type of welding of diaphragms has been used by Bendix since 1962. The weld is given a thermal treatment after welding to stress relieve the weld zone and temper the heat-affected zone.

RELIABILITY VERIFICATION

We are now ready to focus on the task at hand or, putting it another way, the engineer will now be equipped to perform the actual component reliability assessment by performing a series of calculations.

Noting that a Type 416 coupling has been proposed by the compressor manufacturer, the reviewer proceeds by calculating stress levels, factors of

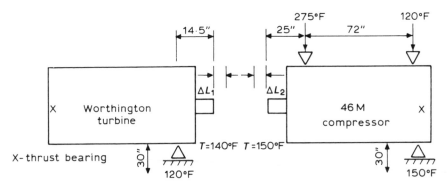

Fig. 12.14 Bendix coupling – thermal growth sample calculation.

safety, coupling deflections, and angular misalignment. Many of these calculations are extremely straightforward. For instance, steady-state stresses are calculated using the well-known formulas given on p. 174, while alternating stresses acting on the compressor/steam turbine shaft system are safely assumed not to exceed 10% of these steady-state stresses. The results of these calculations are then plotted in Fig. 12.4. Other calculations will take into account the anticipated thermal movement of the machinery, using the customary expression $\Delta L = 6.5 \times 10^{-6}$ in/in/°F as a basis.

The engineer making the reliability assessment will draw the following conclusions:

1. Taking into account the various stresses acting on the proposed coupling, the resulting application safety factor will be 6.3. Since experience shows many couplings with lower safety factors in successful long-term operation, the Type 416 coupling will be satisfactory for this service.
2. In the hot state, each of the two machine shafts will effectively grow from the thrust bearing location toward the shaft end. The combined total thermal growth will be partially compensated by installing the coupling with an initial pre-stretch of 0.050 in. The residual deflection will be 0.033 in. In the transient state, the residual deflection will be 0.056 in. Even this worst-case deflection is within the vendor's allowable value.
3. The anticipated total adjusted vertical growth (Fig. 12.14) of the two machines is 0.010 in, and the horizontal misalignment (offset) is expected not to exceed 0.004 in. Using the vector sum and dividing by the distance between coupling flexure planes results in the tangent of the misalignment angle and hence a misalignment angle of 0.106 degrees versus a vendor-allowed angle of 0.25 degrees.

On the basis of the above, the reliability assessment engineer would accept this coupling as meeting reliability standards.

13

Screening of shaft strength

Shafts are usually designed and reviewed for infinite life keeping our four agents force, time, temperature, and reactive environment in mind. Reference 1 provides information on finite life design.

Shafts are primarily designed on the basis of the torsional moment which they must transmit. This torque may be found from the equation:

$$T = 63{,}025 \text{ hp}/N \tag{13.1}$$

where $T =$ torque (lb-in),
  hp = horsepower transmitted,
  $N =$ shaft rpm.

The shear stress developed for a given transmitted torque is given by the formula:

$$S_s = \frac{16T}{\pi d^3} \tag{13.2}$$

where $d =$ shaft diameter (in).

Figure 13.1 can be used to determine shaft diameter in a first approach and is based on a shear stress of 10,000 psi for a given horsepower and speed. For other design stresses, the shaft diameters found in Figure 13.1 may be corrected using the graph in Figure 13.2.

Example: 550 hp is to be transmitted at a shaft speed of 3600 rpm. The shaft diameter as read from Figure 13.1 for a shear stress of 10,000 psi is 1.7 in. If a lower stress of 6000 psi is desired, a shaft diameter of 2 in can be determined from Figure 13.2.

Most shaft failures originate at stress concentrations caused by discontinuities of shaft geometry and loads. Stress concentration factors* for

---

*
$$K_t = \frac{\sigma_{max}}{\sigma_o}, \quad K_{ts} = \frac{T_{max}}{T_0}$$

$K_t =$ stress concentration factor, tension;
$K_{ts} =$ stress concentration factor, torsion;
$\sigma =$ stress (psi);
subscript $o =$ operating condition.

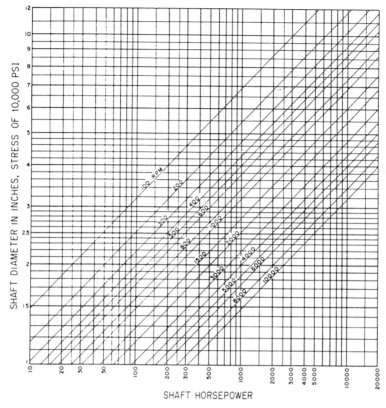

Fig. 13.1 Shaft horsepower rating curve.

shoulder radii and grooves are shown in Figures 13.3 and 13.4. For ANSI keyways, Peterson [2] indicates stress concentration factors of 3.0 and 3.1 for end-milled keyways in torsion and bending, respectively. Orthwein [3] cites stress concentration factors for sled runner keyways 10–30% lower than end-milled keyways. Ways of reducing stress concentration factors are shown in Figure 13.5.

Possible shaft loading in terms of type, amount, and source should be considered as shown in the following checklist when reviewing a given shaft design:

1.0 Forces and moments
   1.1 Bending
   1.2 Torque
2.0 Amounts
   2.1 Mean
   2.2 Peak
   2.3 Alternating

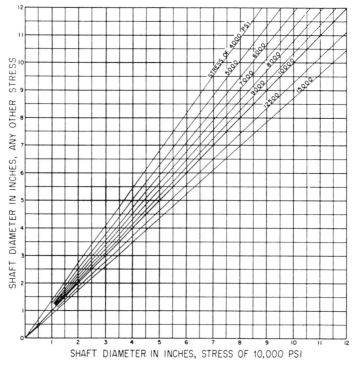

Fig. 13.2 Correction curve for shaft diameters at different stresses.

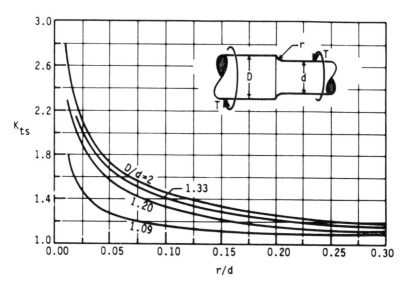

Fig. 13.3 Round shaft with shoulder radius in torsion (from ref. 2).

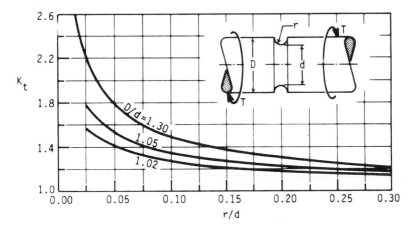

Fig. 13.4 Grooved shaft in torsion (from ref. 2).

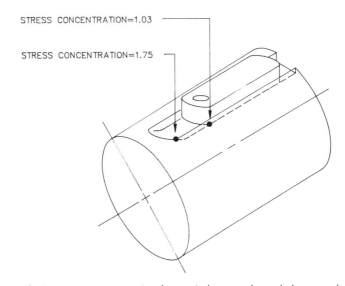

Fig. 13.5 Reducing stress concentration factors in keyways by a sled runner design.

3.0 Sources – FRETT

   3.1 Forces/stresses/impact

      3.1.1 Intended operating loads

      3.1.2 Possible operating loads

      3.1.3 Start-up and shutdown loads

      3.1.4 Weight of shaft

      3.1.5 Unbalance effects

      3.1.6 Internal misalignment effects

3.2 Reactive environment
    3.2.1 Chemical attack
    3.2.2 System resonance – lateral/torsional
    3.2.3 Misalignment between machines
3.3 Thermal effects
    3.3.1 Thermal growth limits
    3.3.2 Cold embrittlement
3.4 Time effects
    3.4.1 Material endurance limits
    3.4.2 Wear from shaft-sealing devices

Shafting material below 3.5 in diameter is generally cold-rolled, cold-drawn, or cold-drawn and ground. Larger shafts are usually hot-rolled and then machined to remove the decarburized surface. Shafts over 5–6 in are usually forged and machined to size. Table 13.1 gives average values of mechanical properties of the most common shaft materials.

FINAL EVALUATION BASED ON REVERSED BENDING AND CONSTANT TORQUE

The most common shaft loading condition is constant torque in combination with reversed bending. For this condition the recently confirmed ANSI/ASME* equation 13.3 may be used:

$$d = \left\{ \frac{32}{\pi} n \left[\left(\frac{M}{S_e} \right)^2 + \frac{3}{4} \left(\frac{T}{S_y} \right)^2 \right]^{1/2} \right\}^{1/3} \qquad (13.3)$$

where M = bending moment (lb-in),
      n = desired factor of safety, e.g. 1, 1.5, etc.,
      S_y = yield strength (psi),
      S_e = fatigue endurance strength of the component.

S_e is arrived at by modifying the laboratory bending endurance strength (Fig. 13.6) with the fatigue strength modification factors as shown in equation 13.4:

$$S_e = k_a k_b k_c k_d k_s k_f S_e' \qquad (13.4)$$

These modification factors take into account:

1. Surface condition: k_a
2. Size: k_b
3. Reliability goal independent of the design factor n: k_c
4. Temperature: k_d
5. Stress concentration(s): k_s
6. Miscellaneous effects: k_f.

---

* B106.1 M-1985.

Table 13.1 Allowable stresses in common steels

Material	Size rounds, in.	Tensile strength	Yield point	Elongation, %	Reduction area	Brinell hardness
SAE 1020 hot-rolled	1	65,000	40,000	30	55	130
SAE 1020 hot-rolled	6	60,000	35,000	30	40	120
SAE 1020 forged	12	55,000	30,000	20	30	110
SAE 1040 hot-rolled	1	94,000	58,000	27	52	187
SAE 1040 hot-rolled	6	84,000	46,000	19	30	160
SAE 1040 forged	12	82,000	44,000	16	28	160
SAE 1040 hot-rolled, water-quenched, tempered at 1200°F	1	100,000	70,000	27	60	200
SAE 1040 hot-rolled, water-quenched, tempered at 1200°F	6	82,000	52,000	25	48	160
SAE 1040 forged, water-quenched, tempered at 1200°F	12	78,000	44,000	23	44	155
SAE 2340 hot-rolled, oil-quenched, tempered at 1200°F	1	112,000	85,000	25	63	230
SAE 2340 hot-rolled, oil-quenched, tempered at 1200°F	6	104,000	75,000	27	58	210
SAE 2340 forged, oil-quenched, tempered at 1200°F[a]	12	100,000	70,000	21	48	200
SAE 2340 forged, normalized, tempered at 1200°F	12	100,000	65,000	20	45	200
SAE 4140 hot-rolled, oil-quenched, tempered at 1200°F	1	145,000	125,000	17	56	293
SAE 4140 hot-rolled, oil-quenched, tempered at 1200°F	6	108,000	80,000	21	54	220
SAE 4140 forged, oil-quenched, tempered at 1200°F[a]	12	103,000	65,000	20	45	210
SAE 4140 forged, normalized, tempered at 1200°F	12	95,000	57,000	21	48	190
SAE 4340 hot-rolled, oil-quenched, tempered at 1200°F	1	150,000	130,000	20	58	302
SAE 4340 hot-rolled, oil-quenched, tempered at 1200°F	6	125,000	100,000	18	54	250
SAE 4340 forged, oil-quenched, tempered at 1200°F[a]	12	110,000	90,000	17	50	230
SAE 4340 forged, normalized, tempered at 1200°F	12	95,000	70,000	20	45	220

[a] It is not generally considered good practice to liquid-quench solid diameters larger than 10 in.

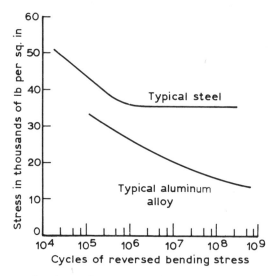

Fig. 13.6   Laboratory bending endurance strength for steel.

Fatigue strength modification factors are described in detail by Mischke [1].

The reader can see that the analysis of stress in stepped, round shafts could involve tedious handbook equations or even the application of finite element programs. Relatively inexpensive PC software is available for the task.*

A typical program requires that the user identify the distance from one end of the shaft to each bearing, load, diameter change, and stress concentration. The software would then assign a node to each of these locations. For example, a program would use three nodes to model a shaft having two diameters, supported at the end by bearings, and carrying a load at the diameter change.

The user would enter program information onto separate spreadsheets. He would have to know the shaft material properties, whether the shaft is hollow or solid, dimensions, loads, restraining conditions, bearing locations, boundary conditions for displacement, and rotational freedom. The latter conditions model support movement and misalignment.

On a good program, values can be assigned for fillet radii, concentrated forces and moments, and stress concentration factors. If stress concentration factors are not supplied for keyways and shoulder radii, they will have to be calculated.

---

* Engineering Software Co., Dallas, Texas.

REFERENCES

1. *Standard Handbook of Machine Design*. Edited by J. E. Shigley and C. R. Mischke. New York, Toronto: McGraw-Hill, 1986.
2. Peterson, R. E., *Stress Concentration Design Factors*. New York: John Wiley, 1953.
3. Orthwein, W. C., A new key and keyway design. *Society of Mechanical Engineers*, **101** (2), 1979, pp. 338–341.

14

Selection of flexible couplings, including disengaging drives and transmission brakes

There are relatively few machines which do not incorporate couplings. Generally, couplings are used to connect sections of shafts or to connect the shaft of a driving machine to the shaft of a driven machine. A coupled connection may be considered permanent, whereas a connection via clutches will provide for engagement or disengagement at will.

There is no coupling which will satisfy the requirements and actual operating conditions of every application encountered on machinery. Coupling selection is thus often a compromise, or an endeavor where trade-offs have to be weighed and considered.

It is our premise that couplings for critical applications should be selected on the basis of reliability, maintainability, and surveillability. Hence, for a given situation, a *reliable* coupling will be the one that is unaffected as much as possible by the four principal agents of component failure – force, reactive environment, time, and temperature. In making a selection or appraising the adequacy and suitability ranking of a given coupling, we must keep in mind that *forces* and thus stress levels encountered are a function of coupling speed, deflection, stretch, compression, offset, power input, shock loading, and perhaps additional parameters. All of these vary for different applications and must be taken into account. A reliable coupling may also be one which will not age. Remember, some elastomers will age, and elastomeric couplings are often popular because of price and other real or perceived advantages. For obvious reasons, elastomeric couplings must be analyzed keeping the agents *time* and *temperature* in mind.

The maintainability of competing coupling types will be reflected in ease of installation, accessibility in the machine, and availability of replacement components. It will be obvious to the engineer or technician making the overall suitability assessment that these factors must also be examined and answered in the affirmative if couplings are to meet our criteria.

Surveillability of couplings varies greatly. A gear-type coupling may be considered surveillable because progressive wear and deterioration will show up in analyzing for wear particles in the lube oil of continuously lubricated couplings, or in vibration of the connected machine elements, or perhaps heat generated at the gear mesh. In contrast, a diaphragm coupling may perhaps not be considered surveillable because its load-carrying component, a thin metal membrane, could fail abruptly and without warning. While this fact in itself should by no means be cause for disqualifying diaphragm couplings from being considered for high-reliability applications, the assessment process must establish strength – in the form of low stress levels – as a compensating feature. Alternatively, the coupling could be designed to revert to a "quasi gear-coupling equivalent" as already described in Chapter 6 (see Fig. 6.1).

In the presence of a *reactive environment*, special attention must be given to the materials of construction of couplings. Table 14.1 illustrates the likely performance of two types of elastomeric couplings in certain chemical plant

Table 14.1 Performance of elastomeric couplings in a chemical environment

	Relative performance of:	
Environment	*Polyurethane*	*Rubber (polyisoprene)*
Abrasion	Excellent	Excellent
Acids (dilute)	Fair	Good
Acids (concentrated)	Poor	Good
Alcohols	Fair	Good
Aliphatic hydrocarbons	Excellent	Poor
Aliphatic gasoline, fuel	Excellent	Poor
Alkalies (dilute)	Fair	Good
Alkalies (concentrated)	Poor	Good
Animal and vegetable oils	Excellent	Fair
Aromatic hydrocarbons	Excellent	Poor
Aromatic benzol, toluene	Poor	Poor
Degreaser fluids	Good	Poor
Heat aging	Good	Good
Hydraulic fluids	Poor	Poor
Low-temperature embrittlement	Excellent	Good
Oil	Excellent	Poor
Oxidation	Excellent	Good
Ozone	Excellent	Poor
Radiation	Good	Good
Silicate and phosphate	Poor	Poor
Steam (hot water)	Poor	Unknown
Sunlight aging	Excellent	Poor
Synthetic lubricants	Poor	Poor
Water swell	Excellent	Good

ambients. Similar concerns must be addressed in selecting diaphragm and disc pack-type flexible membrane couplings in chloride-laden atmospheres.

GEAR VERSUS FLEXIBLE MEMBRANE COUPLINGS

Gear-mesh couplings have for decades been the most frequent type chosen by industry. Their allowable load capabilities are readily calculated by standard strength-of-materials methods, and misalignment tolerance levels and lubrication needs have also been quantified [1]. Using reliability, maintainability, and surveillability criteria, gear couplings do not fare badly. However, their reliability is strongly influenced by the adequacy of lubrication. With continuous lubrication, the oil must be free from water because water will act as the catalyst for sludge formation in lube oils. This sludge is composed of corrosion products and fine dust, generally too fine to be removed by filtration. Besides, ultra-fine filtration would tend to remove oxidation inhibitors from premium-grade lubricants and render the oil unsuitable for long-term, reliable service.

Grease-lubricated gear couplings require premium-grade greases which will not separate from their respective oil and soap constituents while undergoing the high-speed centrifugal force action common to couplings. More importantly, grease-packed couplings must be periodically relubricated. They are one of the few remaining process machinery components which will require preventive maintenance in the truest sense of the term. Although lubrication deficiencies might be considered surveillable, the life-expectancy of incorrectly lubricated gear couplings is too short to advocate anything other than preventive maintenance via periodic shutdowns.

Flexible membrane couplings with such a diaphragm and disc pack couplings are not usually afflicted with these drawbacks. The basic operating principle and design approach to the diaphragm coupling shown in Figure 14.1 is that it must be thick enough to carry the torque load required and must be thin enough to bend through the misalignment angle needed. Since the diameter of the diaphragm also affects how much torque it can carry, diameter, torque, misalignment, and thickness are the main variables to consider in the design of the diaphragm. The interaction of these factors was explained earlier in this text (see pp. 159–173).

Disc pack couplings are somewhat similar to diaphragm couplings inasmuch as both rely on the flexure of metallic elements to accommodate misalignment and axial movement in shafts. The diaphragm coupling employs one or more metallic plates attached to the outside of a drive flange and transfers torque through the diaphragm to a suitable connecting tube or spool piece into the driven flange. Disc pack couplings (Fig. 14.2) usually

DIAPHRAGM

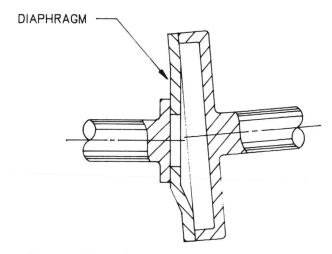

Fig. 14.1   Diaphragm coupling: working principle.

consist of several flexible metallic annuli which are alternately attached with bolts to opposite flanges.

COUPLING RATING BASIS

It is quite typical for couplings manufacturers to base their ratings on horsepower and speed. However, limitations on misalignment capability must be recognized as well. If non-lubricated couplings are grossly misaligned, their flexing components will fail in bending independent of application factors utilized. In contrast, badly misaligned gear couplings will undergo accelerated wear and exhibit progressively increasing vibration levels.

Most manufacturers of membrane couplings design for a stress limit of roughly 70% of the maximum values allowed by a modified Goodman analysis and the reliability assessment effort for critically important membrane couplings should verify that this practice has been adhered to.

SAFETY FEATURES

Since the abrupt failure of a coupling diaphragm cannot be anticipated by conventional monitoring or surveillance instruments, some users have opted to require safety or redundancy features. These would permit continued drive for a limited period via gear tooth engagement (Fig. 6.1). With disc pack couplings, drive engagement would be taken over by the bolt heads or some special disc pack containment lugs. These features would protect turbine drives where the instantaneous removal of driven load could lead to turbine

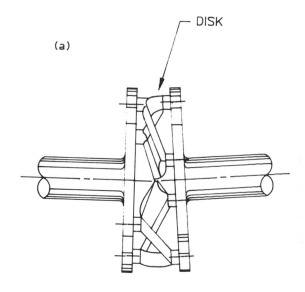

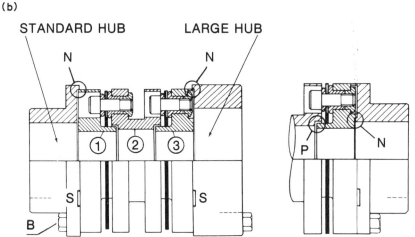

Fig. 14.2 (a) Disc pack coupling: working principle; (b) disc pack coupling with nesting arrangement "N" and floating piloted spacer engagement "P" which prevents spool piece "2" from being flung out in case of failure of the disc pack (from ref. 2).

runaway. During normal operation of these membrane couplings, neither safety feature would be activated.

RELIABILITY, MAINTAINABILITY, SURVEILLABILITY

The reliability of the various coupling types being considered for a project must be assessed on the basis of strength calculations. As indicated earlier,

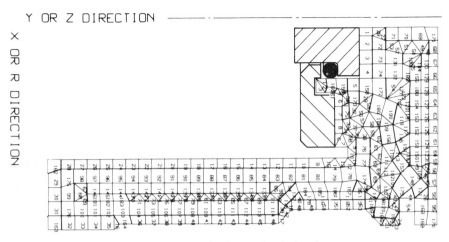

Fig. 14.3   Finite element stress contour plot for mechanical seal.

strength is influenced by force, reactive environment, time, and temperature. Steady stresses are due to the applied torque, centrifugal forces, axial displacement, and built-in membrane pre-stretch. Alternating stresses, at running speed frequency, are due to angular misalignment and are considered to be the more destructive. These are proportional to the square of the angle through which the membranes are deflected. The life of a coupling is thus largely dependent on alignment accuracy.

Stress values in hubs and membranes as well as recommended values of shrink fit for hub to shaft engagement are best modelled by finite element analysis. Similar in appearance to the plot shown in Fig. 14.3, this modern analysis tool makes it possible to describe the complex stress patterns existing in an individual disc pack membrane to which a load has been applied. An analysis of this plot allows the designer to determine an optimum membrane shape. The alternating and steady stresses are usually plotted on the modified Goodman diagram mentioned earlier.

Resistance to a reactive environment can be imparted by the choice of suitable construction materials. Many such materials are available and have been tested. Of all these, AISI 301d high-chrome was found to offer the best combination of strength, corrosion resistance, and stamping characteristics for disc pack couplings at reasonable cost. For the more exotic duties where chloride stress corrosion cracking is thought to be a problem, Inconel 718 has been found to offer the best combination of properties. Incidentally, this material is also chosen where excessively high temperatures have to be tolerated.

High strength steels are generally used for coupling hubs, and also for single contoured diaphragm couplings. These parts are then protected by suitable phosphate primers and durable epoxy paints.

The general maintainability of couplings is very favorably influenced by overall simplicity and low number of parts. Gear couplings rank well in these categories, whereas diaphragm couplings must be considered less maintainable due to the relative care that has to be exercised so as not to nick, burr, or scratch the precisely machined diaphragm contours. The maintainability of all large couplings is enhanced by hydraulic fit-up provisions which should certainly be used whenever possible. Hydraulically engaged coupling hubs represent maximum maintainability.

It should be noted that disc pack couplings are inherently more surveillable than diaphragm couplings. When observed with a stroboscopic light source, defective disc packs will generally exhibit a bow or delamination condition, whereas diaphragm couplings may only exhibit small scratches or microcracks which will be difficult to spot. However, all of this still means that each coupling has its place: A properly engineered diaphragm coupling will outperform everything else when it comes to a 50,000-hp, 4200-rpm compressor; a captured center member disc pack coupling is ideal for a 300-hp, 3600-rpm refinery pump; and the lowly gear coupling may still be the right choice at locations where the equipment must, for some reason or other, be frequently disassembled and serviced. Table 14.2 presents a review of various coupling quality parameters.

DISENGAGING DRIVES AND TRANSMISSION BRAKES

It has been estimated that in the United States alone there are close to 100 different manufacturers of disengaging drives and transmission brakes. Many of these are self-contained, fully assembled units covering over 60 product categories from clutch couplings for duplicating machine roll drives to units on gigantic mining machines and cranes.

The operating and actuating principles of disengaging drives (clutches) and brakes encompass every conceivable mode and method. Tables 14.3 and 14.4 are an attempt to bring order to the profusion, although it makes no claim to accuracy.

A brake can be an electrical, mechanical, hydraulic or pneumatic device, or a combination of any of these. The function of a brake is to absorb the kinetic energy of a moving body, convert the energy to heat, and reject the heat through the body of the brake to the ambient air.

There are very few mechanical power transmission applications which do not need some form of braking – whether for stopping or for holding a load when the motive power is cut off, or for positioning, quick stopping in an emergency, automatic speed control, to provide tension, or simply bringing equipment to rest in a reasonable time. Each application has its own particular requirements which are met by a variety of available designs.

Figure 14.4 will assist in the process of assessing a brake by type, actuator,

Table 14.2 Coupling quality parameters

	Disc	Diaphragm	Gear	Elastomer
Reliability	High	Very high	Moderate	Moderate
Relative maintainability	High	Moderate	High	Moderate
Relative surveillability	Low	Moderate	High	Low
Relative initial cost	Moderate	High	Low	Low
Speed capacity	High	High	High	Moderate
Power-to-weight ratio	Moderate	Moderate	High	Moderate
Lubrication required	No	No	Yes	No
Misalignment capabilities at high speed	Moderate	High	Moderate	Very high
Inherent balance	Good	Very good	Good	Moderate
Overall diameter	Low	High	Low	Moderate
Normal failure mode	Abrupt (fatigue)	Abrupt (fatigue)	Progressive (wear)	Varies
Overhung moment on machine shafts	Moderate	Moderate	Very low	Low
Generated moment when applying torque while misaligned	Moderate	Low	Moderate	Low
Axial movement capacity	Low	Moderate	High	Varies
Resistance to axial movement:				
Suddenly applied	High	Moderate	High	Low
Gradually applied	High	Moderate	Low	Low

Table 14.3 Clutches by engagement method

Mechanical	Via linkages, balls, wedges, cams
Pneumatic	Via actuating pistons, or direct force
Hydraulic	Via actuating pistons, or direct force
Electrical	Electromagnetic force action
Speed-dependent (centrifugal)	Fluid or pellet action
Load-dependent (centrifugal)	Torque demand regulates slip

operating environment, and mounting position. The following list shows the calculation steps needed to assess braking heat load:

1. Torque: prime mover full load torque is given by torque $\propto$ power/speed

$$T = 974 \times kW/N$$

where T is in kg,
         N is in rpm.

Relate brake torque to above by arbitrary service factor.

Table 14.4   Clutches by basic operating principle

Type	*Configuration and operation*
Positive contact	Interlocking cogs, teeth or splines form solid connection
Friction contact	Could be axial, radial, or cone type. Employs two opposing surfaces which are pressed together
Overrunning	Roller, wrap-spring, or sprag type available. Two members run freely relative to each other in one direction and will lock in reverse direction
Fluid	Hydraulic fluid centrifugally flung into contact with two hemispheres which will become locked by trapped oil
Pellet	Steel shot centrifugally flung outward will lock an outside housing, which is part of the input shaft, to an inside housing, which is part of the output shaft
Magnetic	Magnetic coil actuates friction pad via clamping mechanism or by direct flux lines

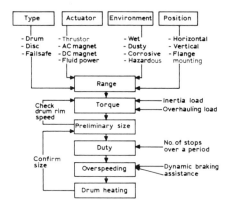

Fig. 14.4   Brake assessment process.

2. Braking time: basically torque ∝ mass × deceleration

$$T = M\alpha = I\omega/t$$

$$t = J/g \times 2\pi/60 \times N/T$$

$$t = \frac{J \times N \times 0.01066}{T}$$

where $g = 9.81$,
   t is in seconds,
   J is in kg^2 (WR^2).

When calculating braking time, each variable must be separately assessed. J is the total polar moment of inertia at the brake shaft.

To transpose linear mass consider equal energies.

$$KE \text{ mass} = KE \text{ polar}$$

$$\tfrac{1}{2}MV^2 = \tfrac{1}{2}I\omega^2$$

$$J = W\left(\frac{V}{2TTN}\right)^2$$

where W is in kg,
   V is in m/min,
   N is the actual speed at the instant of brake application,
   T is the actual decelerating torque which may include the effects
   of gravity.

For hoist applications consider the load driveback torque.

$$T_L = W\left(\frac{V}{2\pi N}\right) \times \eta$$

where T_L is positive to assist the brake when hoisting; and negative,
   detracting from the brake when lowering,
   η is the back efficiency of gears and reeving.
3. Braking distance: basically stop distance $\propto$ mean velocity $\times$ time

$$\theta = \tfrac{1}{2} \times N/60 \times t \qquad \text{where} \quad \theta \text{ is the number of revolutions}$$

or for a linear machine

$$S = \tfrac{1}{2} \times V/60 \times t \qquad \text{where} \quad S \text{ is the distance in meters}$$

Coasting distance during brakeset time must be added, if significant.
4. Brake dissipation: Heating of the brake drum or disc is a complex function
   of kinetic and potential energy and frequency of stopping; compared with
   drum/disc cooling surfaces and permitted lining skin temperature for
   reasonable rate of wear. Heat released into drum/disc during a stop from
   N rpm to rest in t sec with brake torque T is given by:

$$\text{Energy} \propto \text{Mean power} \times \text{Duration}$$

$$KE = \tfrac{1}{2} \times \frac{T \times N}{974} \times t \qquad \text{where} \quad KE \text{ is in kilojoules}$$

When analyzing the adequacy of clutches and brakes from the reliability
point of view, we may resort, if necessary, to applicable calculation methods
from textbooks. Many of the more typical devices respond to standard
calculations while the more sophisticated units are perhaps best reviewed by
enlisting the help of the manufacturer. Complex models have inevitably been
tested and will respond to semi-empirical calculation routines. Particular
attention has to be paid to material selection and wear-related criteria. Also,

heat transfer verification will be in order and FRETT – the reliability impact of Force, Reactive Environment, Time, and Temperature – must certainly be considered.

Maintainability considerations follow the usual pattern and merit a good deal of coverage because clutches and brakes are frequently subject to abuse. Easy access and simple replacement of wear parts are appreciated by service personnel and will reduce downtime in the event of repair.

Surveillability of clutches and brakes can be important if the devices perform safety-related functions and wear indicators, heat sensors, and other surveillance assist components are available. These may incorporate audible, visible, or electronic-remote annunciation features.

REFERENCES

1. *Proceedings of an International Conference on Flexible Couplings*, 29 June–1 July 1977, University of Sussex, Brighton, UK. Farnham, UK: Michael Neale and Associates.
2. Phillips, J. and Umbrich, P., Pump couplings in chloride laden atmospheres. *World Pumps*, 1983, p. 199.
3. Bloch, H. P., Improve safety and reliability of pumps and drivers. *Hydrocarbon Processing*, Feb. 1977.

15

Gear life assessment

Gears are power transmission components designed to transfer rotary motion from one shaft to another by either increasing or decreasing the speed of the driven component. The preferred choice is for gears on shafts with parallel axes. A second choice are gears on shafts with axes intersecting at right angles. Helical teeth are preferable to straight teeth; however, if the latter are significantly less expensive they may be used for low speeds. Table 15.1 shows the general fields of applications of the main classes of gears.

As expected, Force, Reactive Environment, Time, and Temperature (FRETT) can significantly influence the reliability and life expectancy of gears. Further, maintainability should be considered at the initial design, review, and procurement stages. Gears generally lend themselves well to predictive maintenance due to their excellent surveillability. During the design stage, thought should be given to inspection covers and lubricant sample points. A well-designed gear will not usually fail in a sudden failure mode but through gradual wear. Also, we should consider what a gear expert once said: "They wear in and then wear forever."

SPUR GEARS

Spur gears are used to transmit power between parallel shafts. They are cylindrical and their teeth are straight and parallel to the axis of rotation. For slow-speed gearing applications, Lewis' equation to determine *tooth strength* modified by a factor to account for the pitch line speed is generally satisfactory. For high-speed applications, the ability of the teeth to withstand wear is of greater importance. Consequently, the teeth are sufficiently strong and the Lewis equation can no longer be the most important criterion [1].

Using the pitch line velocity

$$V = \frac{\pi d n_p}{12} \quad \text{ft/min} \tag{15.1}$$

193

Table 15.1   Scope and torque capacity of common types of gears

Type	Relation between shaft axes	Gear ratio (m_g)	Max. tooth speed, V (ft/min)	Type of tooth	Max. wheel torque (lb-in)
1	Parallel	up to 10	1000	Helical or straight	80×10^6
			5000	Helical	20×10^6
				Profile ground straight	2×10^6
			40,000	Helical	5×10^6
2	Intersecting	up to 7	500	Spiral bevel or straight bevel	0.8×10^6
			12,000	Spiral bevel	0.4×10^6
3	Nonintersecting Crossed at 90°	up to 50	10,000	Worm and wormwheel	2.5×10^6
				Crossed helical	1.5×10^6
4	Nonintersecting Crossed at 80° to 100° but not 90°	up to 50	10,000	Worm and wormwheel	1×10^6
				Crossed helical	1.5×10^6

Note: These figures are for general guidance only. Any case that approaches or exceeds the quoted limit needs special consideration of details of available gear-cutting plant.

the tangential tooth load W_t (lb) is given by the equation:

$$W_t = \frac{33,000 \text{ hp}}{V} \quad \text{lb} \tag{15.2}$$

The load-carrying capability of the teeth must be based on the stress developed at the mesh or contact lines. This capability is therefore a function of the curvature of the tooth surfaces which in turn vary with the pitch diameters of the gears of the speed changer.

A formula based upon the maximum compressive stress between two parallel cylinders, as developed by Hertz, can be applied to a pair of contacting spur gear teeth:

$$S_{max} = 4582\left(\frac{W_t(1 + m_g)}{FD \sin 2\phi_L m_g}\right)^{1/2} \tag{15.3}$$

The value of S_{max} may not be the actual contact pressure in psi but should be considered to be an experience factor. We encourage our readers, who choose to review an unfamiliar application, to request these factors from their particular vendor or vendors as the case may be.

HELICAL GEARS

The development during the early part of the century of high-speed drivers such as the steam turbine, the electric motor, and the internal combustion engine made a parallel development of high-speed gears necessary. In some industries this development led to the establishment of special-purpose gear standards. Concurrently, the American Gear Manufacturers' Association (AGMA) created unique bodies of standards that are serving gear users well. In the following we will make use of this experience.

LIFE ASSESSMENT OF CRITICAL GEARS

In order to be able to make such an assessment, the information shown in Figure 15.1 has to be solicited from the potential vendor or vendors. It is advisable to do so at any rate, even though one might not use the information immediately. It is much more difficult to obtain this information once the equipment has been acquired or even installed.

Vendor experience with the design and fabrication of special-purpose gearing is no less important than vendor experience with any other critically important machinery category. Questions to be asked relate to pitch-line velocities, gear-blank (web) construction, horsepower levels, bearing design, gear-speed ratios, etc. When vendor experience has been established to the reviewer's satisfaction, he or she is ready to proceed with a comparison of competing bids. This comparison is aimed at determining which of the various offers may represent a stronger, potentially less failure-prone gear.

Design appraisals can be complex and time-consuming if efforts are made to use the full complement of AGMA rating formulas. Moreover, cycles to failures calculated with some of these rating formulas can be drastically influenced by minor changes in the assumed or anticipated surface roughness, tooth spacing, etc. A sensible approach to gear design appraisals would not, therefore, use calculated probable cycles to failure in an *absolute* way. The reviewer would utilize the data only to make a comparison of competing offers and to assign a ranking order.

In the late 1960s, Robert H. Pearson, then chief engineer of the Sier-Bath Gear Company, equated the mathematical expression for estimated gear-tooth compressive stress to that for allowable fatigue stress. Compressive stress is a measure of surface durability and pitting. Pearson's work, summarized in an article published by *Machine Design* magazine in 1968 determined:

$$N_c = 3.8 \times 10^{-10} \left[\frac{dFI(H + 150)^2}{C_d W_t} \right]^{8.77} \tag{15.4}$$

Project . . . .
Gear vendor . . . .
Compressor vendor . . . .
Driver vendor . . . .

Gear data

Type/model
Serial number
Gear ratio
Service horsepower
**Input/output r.p.m.
AGMA service factor
**Operating pitch diameter (in)
 & no. of teeth
 Pinion
 Gear
**Effective face width (in)
**Helix angle (degrees)
**Normal pressure angle (degrees)
**Transverse pressure angle (degrees)
Transverse pitch
**Diametral normal base pitch
Whole depth (in)
Addendum (in)
Dedendum (in)
Circular tooth thickness
**Brinell hardness
 Pinion
 Gear
Oil temperature in/out
Apex leading or trailing

Center distance (in)
Distance gear center to sump bottom
Backlash, min/max (in)
Surface finish, RMS & AGMA number
Weight and WK^2 (lb & lb-ft^2)
 Pinion rotor
 Gear rotor

*Thrust-bearing type, active/inactive side
Design capacity (lb)
Effective area (in^2)
Journal bearing: length/dia./clearance
 Pinion
 Gear
Upmesh or downmesh layout
Oil flow rate (gpm)
 To each H.S. bearing
 To each L.S. bearing
 To thrust bearing
 To mesh
Lateral critical speed of H.S. pinion
**Geometry factor J
 Pinion
 Gear
Calculated scoring index

*Must have identical load carrying capacity.
**Essential input for life & horsepower calculations, all other data required for more general appraisal.

Fig. 15.1   Data required for the assessment of high-speed helical and herringbone gears.

In this expression:

N_c = life in cycles to failure,
d = pinion operating pitch diameter (in),
F = face width (in),
H = hardness (bhn),
I = geometry factor,
W_t = tangential driving force (lb).

The geometry factor I is obtained by dividing durability factor C_3 from Figure 15.2 by a materials factor $(S_{sc}/C_p)^2$ from Figure 15.3. W_t is readily calculated by dividing the pinion output torque by the pinion pitch radius.

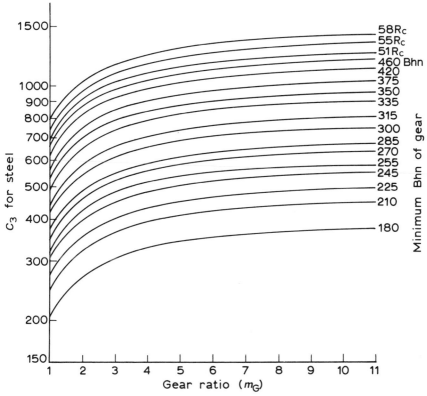

Suggested gear and pinion hardness combinations

	Minimum Brinell hardness												
Gear	180	210	225	245	255	270	285	300	315	335	350	51R$_c$	55R$_c$	58R$_c$
Pinion	210	245	265	285	300	315	335	350	365	385	400	51R$_c$	55R$_c$	58R$_c$

Fig. 15.2 C_3 – values for high-speed gear units (from ref. 1).

C_d, however, is a good deal more difficult to obtain. Five factors make up C_d and are defined as follows:

1. C_o: The overload factor allows for momentary torques and overloads imposed by the driver or driven load. Though best established by field experience, it can be estimated from Table 15.2. Since this factor can derate gear life by about 2000 times, care should be taken in selecting C_o. Uniformity and frequency of torque fluctuation, particularly as related to the tooth mesh frequency, are important. If the rate is low and not a specific harmonic, a low value of C_o can be used safely.

2. C_v: The dynamic factor has to do with the action between mating teeth. Its magnitude depends on spacing accuracy, pitch-line velocity, inertia, and stiffness of connected masses. Lubricant viscosity comes into play at

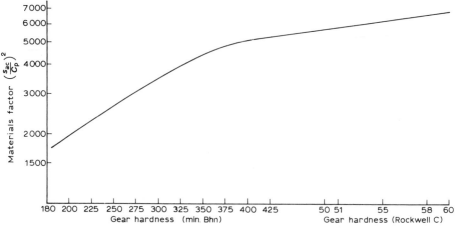

Values are to be taken from the above curve for the minimum hardness specified for the gear.
Values for suggested gear and pinion hardness combinations are tabulated below for convenience.
Pinion and gear may be the same hardness when the gear ratio is "1 : 5" or less.

	Minimum Brinell hardness													
Gear	180	210	225	245	255	270	285	300	315	335	350	51 R_c	55 R_c	58 R_c
Pinion	210	245	265	285	300	315	335	350	365	385	400	51 R_c	55 R_c	58 R_c
$\left(\frac{s_{ac}}{C_p}\right)^2$	1750	2100	2300	2560	2700	2950	3200	3460	3750	4200	4460	5800	6200	6600

Fig. 15.3   Materials factors for high-speed helical and herringbone gear units (from ref. 1).

Table 15.2   Overload factor, C_o

	Characteristic of driven load		
Characteristic of driver	*Uniform*	*Moderate shock*	*Heavy shock*
Uniform load	1.00	1.25	1.75 and up
Light shock	1.25	1.50	2.00 and up
Medium shock	1.50	1.75	2.25 and up

pitch-line velocities of about 10,000 fpm and higher. Until being
questioned in the mid-1970s, commonly used values of C_v ranged from
1.0 to 0.5 and lower. The minimum value of C_v would derate gear life by
about 440 times. Tooth spacing accuracy exerts the greatest influence on
C_v. For AGMA Quality 12 and very low stiffness and inertia, C_v could
approach 1.0.

3. C_s: The size factor allows for gear and gear-tooth size, tooth-contact
pattern, and reliability of material heat treatment. Generally, this factor
is taken as 1, although it may reach about 1.25 when all influencing

conditions are unfavorable. A value of 1.25 would derate gear life about 7 times.

4. The surface-condition factor takes into account tooth surface finish and C_f = residual stress. Generally, it is taken as 1. However, it may go to about 1.25 for rough finishes that would cause localized contacts, or when residual stresses are expected to be high or unpredictable. Both conditions can warrant a value as high as 1.50; this would derate gear life about 35 times.

5. C_m: The load distribution factor allows for anything that might prevent 100% gear-tooth contact—lead and profile errors, stiffness of gears and mountings, deviations from true alignment of the gear axis, and uneven thermal expansion during operation. If all such conditions were ideal, C_m would be 1.00, but practical conditions may drive this value to 3.00 or more. For a value of 3.00, gear life would be derated 530 times.

Experience shows that wide-face gears require special considerations to offset net misalignment and to obtain good load distribution at full torque. Generally, a face width equal to the pinion diameter is the best compromise. It should be remembered that misalignment cannot be readily absorbed by the gear teeth except through plastic deformation, which is difficult to predict in terms of life. Extreme profile modifications in a helical mesh can reduce instantaneous lines of contact to nearly point contacts; this is as bad as a severe misalignment. Usually a few ten-thousandths addendum relief is enough to assure a smooth load transfer from tooth to tooth.

The combined derating factor C_d could reasonably vary from 1.00 to 20.25, with the latter reducing life about 2.87×10^{11} times. AGMA Standard 411.02 is very helpful in determining limits for C_d. Table 15.3, taken from AGMA Standard 411.02, gives derating factors for aircraft-quality gears: these may be used as guides in establishing minimum limits for C_d for any quality level. From a practical point of view, quality speed-increasing gears supplied to the petrochemical industry for instance most often exhibit life-cycle

Table 15.3 Derating factor, C_d

Application	AGMA gear quality number			
	9	10	11	12
Main propulsion drive gears				
Continuous	—	1.8	1.5	1.2
Takeoff	—	1.5	1.2	1.0
Power-takeoff gears	2.4	2.1	1.8	1.5
Auxiliary power supply units	—	2.1	1.8	1.5

Source: AGMA. By permission.

characteristics relating to a C_d of approximately 2.0. For double-helical gears using turbine oil as a mesh lubricant, N_c should be calculated on the basis of this derating factor.

Let us illustrate by example. You receive two proposals for a double-helical gear-speed increaser with an input speed of 1785 rpm, an output speed of 11,415 rpm, and a rated power output of 3900 hp. Given the following data, which gear should you buy?

Pinion data	Vendor A	Vendor B
Pinion diameter, d	6.491 in	6.250 in
Face width, F (total)	10.000 in	10.140 in
Minimum Brinell hardness	350	372

For this design assessment, we use $C_d = 2.0$. Next, for Vendor A, calculate:

$$W_{tA} = \frac{T}{d/2} = \frac{(63,025)(3900)(2)}{(11,415)(6.491)} = 6635 \text{ lb}$$

Similarly, for B:

$$W_{tB} = \frac{T}{d/2} = \frac{(63,025)(3900)(2)}{(11,415)(6.250)} = 6890 \text{ lb}$$

Then

$$N_c = 3.8 \times 10^{-10} \left[\frac{(6.491)(10)(0.2133)(500)^2}{(2.0)(6635)} \right]^{8.77}$$

$$= 5.9 \times 10^{11} \text{ cycles} = 98 \text{ years (Vendor A) at } 6.0 \times 10^9 \text{ cycles/year}$$

and

$$N_c = 3.8 \times 10^{-10} \left[\frac{(6.250)(10.14)(0.202)(522)^2}{(2.0)(6890)} \right]^{8.77}$$

$$= 4.5 \times 10^{11} \text{ cycles} = 76 \text{ years (Vendor B)}$$

This example calculation establishes that Vendor A's offer ranks somewhat ahead of Vendor B's offer, but would probably not merit much of an additional expenditure.

CHECKING FOR SCORING SUSCEPTIBILITY

The susceptibility of a given design to scoring damage in service can be checked by calculating a so-called scoring index:

$$SI = \left(\frac{W_t}{F} \right)^{0.75} \frac{(N_p)^{0.5}}{(d)^{0.25}}$$

In this expression, the previously defined parameters are joined by N_p, the pinion speed (rpm).

The following may be assumed:

$$SI < 14{,}000: \text{ slight probability of scoring}$$
$$14{,}000 < SI < 18{,}000: \text{ moderate probability}$$
$$SI > 18{,}000: \text{ high probability of scoring}$$

Reviewing our example shows

$$SI_A = \left(\frac{6635}{10}\right)^{0.75} \frac{(11{,}415)^{0.5}}{(6.491)^{0.25}} = 8750$$

and

$$SI_B = \left(\frac{6890}{10.14}\right)^{0.75} \frac{(11{,}415)^{0.3}}{(6.25)^{0.25}} = 8993$$

Both offers are completely acceptable from the point of view of scoring.

The reader is reminded that the above screening calculations should be looked at as, again, a first approach in a rather involved decision-making process.

REMAINING LIFE ASSESSMENT OF HIGH-SPEED GEARS

Sometimes we will have to assess reliability of a gear set in mid-life. A good example is the following case.

A major petrochemical company had operated several of their motor-gear-driven process gas compressors (Fig. 15.4) for a number of years at loads above their design rating. An analysis showed that the high-speed

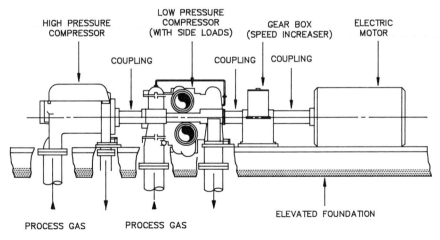

Fig. 15.4 Typical major gas compression train.

gears were particularly vulnerable because they exceeded industry standard (API) durability ratings, occasionally by 30%. Gearbox inspections revealed no visible signs of surface fatigue after continuous operation for 10 years. The owner decided nevertheless to assess remaining life of his gear sets.

The possibility of surface fatigue failure in the form of pitting or spalling can be predicted quite well by analytical techniques [2]. The equation generally used for these predictions appears in AGMA 218.01. The value of surface compressive stress number S_c is calculated as follows:

$$S_c = C_p[CW_t C_a/C_v)(C_a/dF)(C_m C_f/I)]^{0.5} \tag{15.5}$$

This is compared to an allowable value of surface compressive stress number:

$$S_c' = S_{ac}(C_L C_H/C_T C_R) \tag{15.6}$$

The equations are discussed in detail in AGMA 218.01 [3]. The term S_{ac} is the fatigue stress at which the material can be operated for 10^7 cycles. To make this analysis one should calculate several S_c values, one for each combination of load intensity and operating cycles at that intensity.

Typical rating data for the gear sets were:

Original design	Avg. actual	Full load motor
hp	hp	hp
		(OL^a alarm setting)
5800	6080	6670

^{a}OL means overload.

Surface endurance allowable stresses for the gear sets were calculated using C_R, the reliability factor, as variable. Curves I, II, and III in Fig. 15.5 represent gear S_c' values for $C_R = 1.0, 1.25$, and 1.5, respectively. The reliability factor C_R accounts for the effect of the normal statistical distribution of failures found in materials testing. Table 15.4 describes the meaning of C_R. Two S_c values were plotted: point (1), the stress number associated with the past ten years' average actual load; and point (2), the stress number calculated with the overload alarm setting which was assumed to occur five times per day for five minutes during the next ten years.

$$\text{Cycles} = 6917 \text{ min}^{-1} \times 5 \text{ min} \times 5 \times 365 \times 10 = 6.3 \times 10^8$$

Point (2) affords a first assessment of the effect of a temporary overload. The stress number is below the most conservative allowable stress line and therefore it would be safe to permit this temporary overload alarm setting as far as gear life was concerned. Remaining gear life as a function of past and future normal average or temporary maximum load applications was evaluated using Miner's rule as proposed in Appendix D of AGMA 218.01. The results are plotted in Fig. 15.6. The calculated life of gears according to

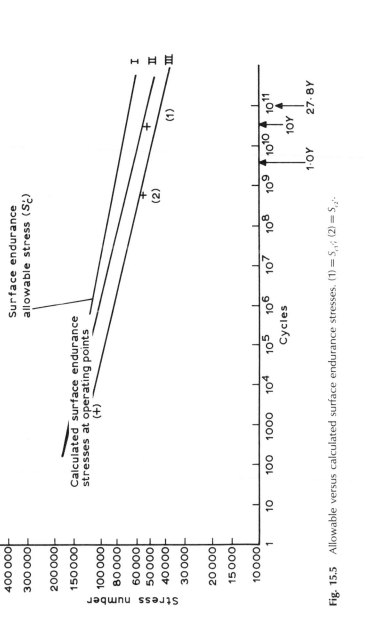

Fig. 15.5 Allowable versus calculated surface endurance stresses. (1) = S_{c1}; (2) = S_{c2}.

Table 15.4   Reliability factor, C_R

Requirements of application	C_R
Fewer than one failure in 10,000	1.50
Fewer than one failure in 1000	1.25
Fewer than one failure in 100	1.00
Fewer than one failure in 10	0.85

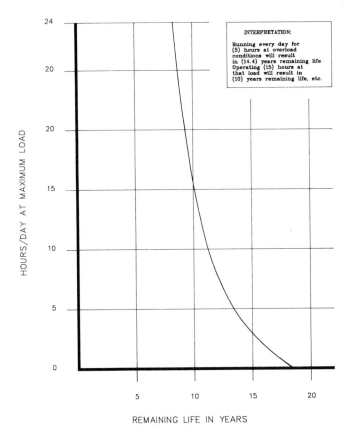

Fig. 15.6   Hours per day at maximum load versus remaining life – high-speed gear unit. *Interpretation:* Running every day for 5 h at overload conditions will result in 14.4 years remaining life. Operating 15 h at this load will result in 10 years remaining life, etc.

Miner's rule is:

$$\text{Gear life (years)} = \frac{X}{\sum_{i=1}^{p} \frac{n_i}{N_{fi}}} \tag{15.7}$$

where X = total years in the duty cycle,
  p = total number of different stress levels,
  n_i = number of cycles at ith stress level,
  N_{fi} = number of cycles to failure at ith stress level.

Miner's rule assumes that the percentage of life used is equivalent to the number of applied cycles at a given stress level divided by the number of cycles to failure at that stress level. Failure occurs when:

$$\sum_{i=1}^{p} \left(\frac{n_i}{N_{fi}} \right) \geqslant 1.0 \qquad (15.8)$$

The following calculations were made:

$X = f(n_2 \text{ and } n_1)$
$p = 2$ (2 conditions)
$n_1 = 10^{11}$ (intersection of normal actual stress number with S'_{cII}) cycles
$n_2 = $ variable from 6.3×10^8 to 3.1×10^{10} (24 h/day equivalent) cycles
$N_{f1} = 10^{11}$
$N_{f2} = 3.24 \times 10^{10}$ (intersection of OL alarm stress number with S'_{cII}) cycles

S'_{cII} was chosen as allowable surface endurance stress line because it appeared reasonable to postulate that since no failures had occurred during operation for 10 years at stress point (1), S'_{cII} would be a sufficiently conservative limit.

REFERENCES

1. AGMA Standard Practice 421.06 for High Speed Helical and Herringbone Gear Units. Arlington, VA: American Gear Manufacturers Association, 1969.
2. Dean, P. M., *Introduction to Gear Failure Analysis*. Paper presented at the National Conference of Power Transmission, Chicago, 1979.
3. AGMA Standard Practice 218.01 for Rating the Pitting Resistance and Bending Strength of Spur and Helical Involute Gear Teeth. Arlington, VA: American Gear Manufacturers Association, 1982.

16

Screening of belt drives

Belt drives are classified as positive and non-positive. Non-positive belt drives are friction drives in which forces are transmitted from belt to pulley and vice versa by friction. V-belts and flat belts belong to this category. The positive belt drive transmits the peripheral force by positive locking of transverse elements such as teeth on the belt and pulleys. Timing belts belong to this last category of belts.

V-BELTS

V-belts are chosen in North America from a selection of standard narrow belt cross-sections (3V, 5V, and 8V) and wider "classical" B, C, and D sizes. The choice of belt section is based on design horsepower of the drive, the speed of the fastest shaft, and the fastest shaft sheave diameter. The fastest shaft is in nearly every case the motor shaft. As a first check one would look at belt speed, service factors, and the need for dynamic balancing sheaves and pulleys.

3V, 5V and 8V belt speeds are optimally in the 7000–8500 fpm range. Conventional belts (A, B, C, and D) have an optimum speed range of 5000–6000 fpm. Service factors can be checked using Table 16.1.

The need for balancing the drive becomes important in order to reduce disturbing forces that might give rise to bearing damage and other machinery distress. All sheaves are statically balanced by the manufacturer which is usually enough to prevent unbalance forces to become a problem. However, at the maximum speeds used today, single plane, or static, balance may not be sufficient. Two-plane dynamic balancing should be specified for sheaves exceeding the limits set by Figure 16.1.

The design and analysis of V-belt drives often involves tedious catalog selection work. Frequently, there are so many combinations of belts and pulleys that it is difficult to optimize a V-belt drive design. Personal computer-based selection tools have been developed to address this problem. These programs evaluate all possible combinations of standard sheaves and allow an optimal selection meeting design needs [1].

206

Table 16.1 V-belt service factors

	Electric motors										Engines
	Alternating current						Single phase		Direct current		Gas and diesel
	Squirrel cage			Wound rotor (slip-ring)	Synchronous		Repulsion and split phase	Capacitor	Shunt wound	Compound wound	4 or more cylinders above 700 rpm
	Normal torque line start	Normal torque compensator start	High torque		Normal torque	High torque					
Agitators – paddle-propeller	1.2	1.0	1.4	1.2	..	..	..	..	1.4	..	..
Brick and clay machinery	1.5	1.3	1.8	1.5	..	..	..	1.0	..	..	1.2
Bakery machinery	1.2	..	..	..	1.5	..	1.2	1.2	..	..	..
Compressors	1.4	1.4	..	1.5	..	..	1.2	..	1.2	..	1.2
Conveyors	..	1.6	1.8	..	1.4	..	..	..	1.6	1.6	1.6
Crushing machinery	1.6	1.4	1.6	1.4	1.4	1.6	..	..	..	..	1.6
Fans and blowers	1.4	1.6	2.0	2.0	2.0	2.0	..	..	1.4	..	1.6
Flour, feed, and cereal-mill machinery[a]	1.4	1.4	1.6	1.4	1.4	..	..	..	..	..	1.8
Generators and exciters	1.2	..	..	..	..	..	..	..	1.2	1.2	2.0
Laundry machinery	1.2	..	..	..	1.4	..	..	..	..	1.4	..
Line shafts	1.4	1.4	..	1.4	..	2.0	1.4	1.4	1.4	1.2	1.6
Machine tools	1.2	..	..	1.4	..	..	1.2	1.2	1.2	1.4	..
Mills	..	1.6	1.6	1.4	1.6	..	..	..	..	1.4	..
Oil-field machinery	1.2	1.2	1.4	..	1.6	1.8	..	..	1.4	1.4	1.4
Paper machinery	1.2	1.4	1.8	1.5	..	..	..	..	1.5	1.5	..
Printing machinery	1.2	1.2	..	1.2	..	..	..	..	1.2	1.2	..
Pumps	1.4	1.4	1.4	1.6	1.6	1.8	1.2	1.2	1.2	..	2.0
Rubber-plant machinery	1.4	1.4	1.4	1.4	..	1.8	..	..	..	..	..
Screens	1.2	1.2	1.4	..	..	..	..	..	..	..	..
Textile machinery	1.6	..	1.8	..	..	..	..	..	..	..	..

[a] The use of a service factor of 2.0 is recommended for equipment subject to choking.

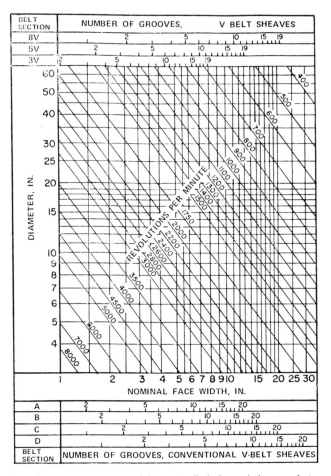

Fig. 16.1 Chart shows maximum speed for statically balanced sheave of given diameter and face width. To exceed this speed, sheave should be dynamically balanced.

In Europe, the narrow V-belts with ISO profiles SPZ, SPA, SPB, and SPC, in accordance with DIN 7753, have been gaining increasing popularity. Belt manufacturers can provide detailed information pertaining to these profiles.

Because of the extreme variety of V-belt applications it is difficult to come up with general belt life data. V-belt life depends largely on the application conditions. The factors affecting belt life are influenced by FRETT:

- Age (T in FRETT),
- Assembly (Force F in FRETT),
- Operating Conditions (Temperature T and Force F in FRETT),

- Environmental conditions, i.e. oil, dust, and climatic conditions (RE in FRETT).

Power transmissions capacities specified in the relevant standards are based on the empirical belt life of 24,000 h, which assumes optimum operating conditions and an excellent maintenance environment.

Multiple V-belts are the preferred choice in critical applications over all other types of belt drives since the probability of sudden failure of a complete belt set is very low. Consequently, belts have excellent surveillability due to their inherent redundancy features. A multiple V-belt failure will normally annunciate itself before a complete drive failure occurs.

TIMING BELT DRIVES

Timing belts, or synchronous belts, operate on the tooth-grip principle and provide precise transfer of power from driver to driven equipment. They consist of four components: a molded backing, a helically wound tension member, teeth, and a nylon tooth facing. The backing and teeth, made of the same neoprene material, are molded integrally. Enclosed within the backing are tension members that carry the load and prevent belt elongation. The precisely formed teeth have their root line located on the pitch line, enabling the tooth spacing to be unaffected by belt flexing. The wear-resistant nylon fabric facing protects the wearing surfaces of the teeth and keeps friction loss to a minimum.

Most timing belt manufacturers offer belt sizes according to engineering standards of publication IP-24 (I-78) of the Rubber Manufacturers Association (RMA). This means that belts are specified by pitch length, tooth pitch, and top width (Fig. 16.2).

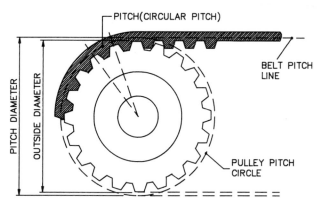

Fig. 16.2 Timing drive pitch measurements.

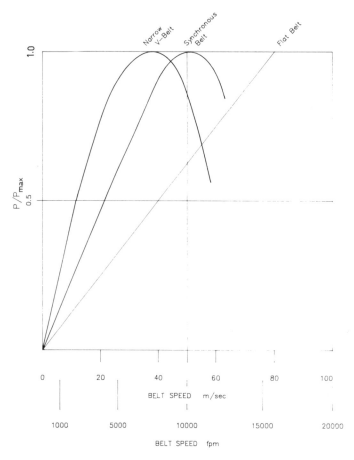

Fig. 16.3 Belt speeds as a function of power ratio.

Belt pitch length is the total length (circumference) of the belt at the cord line in 0.1-in increments. The pitch line is measured where the tensile members run through the belt and remains constant regardless of backing thickness.

Tooth pitch refers to the distance in inches between two adjacent tooth centers. There are five standard belt tooth pitches and each has a code designation which in turn implies the type of service (i.e. XL = extra light with $\frac{1}{5}$-in pitch; L = light with $\frac{3}{8}$-in pitch; H = heavy with $\frac{1}{2}$-in pitch; XH = extra heavy with $\frac{7}{8}$-in pitch; and, finally, XXH = double extra heavy with $1\frac{1}{4}$-in pitch).

The top width of a synchronous belt is measured in hundredths of an inch and is indicated by the third part of the belt number. Accordingly, a belt with a part number 750 H100 has a pitch length of 75.0 in, a tooth pitch of 0.5 in, and a top width of 1.00 in.

Timing belts are not recommended where excessive or extreme shock

Table 16.2 Belt quality ratings

	Flat belts	Standard V	Narrow V	Timing belts
Reliability index	2	1	2	3
Maintainability	10	5	5	10
Surveillability	5	10	10	1
Power per volume (kW/cm³)	0.8	0.7	1.8	1.9
Efficiency (%)	98	95	96	98
Temperature range (°C)	−40 to +80	−55 to +70	−55 to +70	−50 to +120

loads are expected. This is due to their positive drive characteristics and high-modulus cord material in the tension members. Other belt drives are able to absorb heavy shock loads by allowing slippage and belt elongation.

Timing belts require no maintenance once they are properly aligned. Their failure mode, however, is often sudden and unpredictable which makes them an inferior choice in critical drive applications.

There are reliability differences between V-belts and timing belt design criteria. Consequently, substantial differences between operating lives of the drives may exist.

Timing belts should not be applied where debris is present unless the drive can be adequately protected. Debris caught between the timing belt and its pulley will destroy the tensile cords of the belt. Oil and petroleum derivatives may change the belt polymers and adhesion components.

Timing belt drives are selected from engineering manuals available from timing belt manufacturers. One major manufacturer suggests a five-step selection routine that uses driver and driven sheave diameters and approximate center distances as input.

Figure 16.3 allows timing-belt drive screening and Table 16.2 shows quality ratings of belt drives for comparison.

REFERENCE

1. *V-belt Editor*. Houston: DESMAC Corp., 1987.

17

Assessment of bearing life, load ratings, sliding bearings, and shaft sealing devices*

ASSESSMENT OF BEARING LIFE AND LOAD RATINGS

How long a ball bearing will last under load depends on two groups of variables. First, there are physical characteristics of the bearing which include how it is designed, the material from which it is made, and how it is manufactured. Second, there are the conditions under which it is applied, such as load, operating speed and temperature, the way it is mounted, and the way it is lubricated.

Even if a ball bearing is operated under ideal conditions – where it has been properly mounted, lubricated, protected from foreign particles, and not subjected to extreme temperature or speed – it will ultimately fail due to either material fatigue or wear. Fatigue failure results from the repeated stresses that are developed in the contact areas between the balls and raceways. Failure shows up as spalling of the load-carrying surfaces. Excessive wear occurs when operating conditions are other than ideal. These conditions are generally those which cause high friction and/or heat within the bearing.

PREDICTING BEARING LIFE

It is not possible to predict the exact fatigue life of an individual bearing. Instead, the designer of a system incorporating ball bearings must rely on the results of extensive research and testing done on the life of groups of identical bearings operated under identical conditions. Tests show that lifetime ratings of bearings operating under these identical conditions vary due to intricate differences between individual bearings. These lifetimes, however, follow definite statistical distributions. Load ratings, boundary

---

* *Source:* General Bearing Corporation, Blauvelt, New York. Adapted by permission.

dimensions, and tolerances for ball bearings and cylindrical roller bearings are computed from AFBMA and ISO standards.

Such statistical distributions can be represented by equations which relate predicted bearing life to factors such as the load it must bear, its operating speed, and the physical characteristics of the bearing. It is up to the designer then to determine which bearing is best for a particular application by use of these equations, and it is certainly appropriate for the reviewing engineer or technician to ascertain that the proposed design incorporates all relevant factors.

L_{10}, or rating life, is the life most commonly used in load calculations. We mentioned in Chapter 2 that this is the life in units of either hours or millions of revolutions that 90% of a group of apparently identical ball bearings will complete or exceed. Another accepted form is L_{50} or median life. This is the life which 50% of a group of bearings will complete or exceed. L_{50} is usually not more than five times L_{10}.

Another important definition is that of the basic dynamic load rating, C. For a radial ball bearing, the basic dynamic load rating is the constant radial load which a group of identical bearings with a stationary outer ring can theoretically endure for 500 h at $33\frac{1}{3}$ rpm.

The relationship between bearing life and applied load can be expressed as *life in revolutions*:

$$L_{10} = 3\left(\frac{C}{P}\right)^3 \times 10^6 \qquad (17.1)$$

or *life in hours*:

$$L_{10} = 3\left(\frac{C}{P}\right)^3 \frac{16,667}{N} \qquad (17.2)$$

where L_{10} = the rating life,
C = the basic dynamic capacity as shown in the catalog,
P = the equivalent radial load on the bearing (lb),
N = speed in rpm,
3 = life improvement factor for vacuum degassed 52100 steel.

EQUIVALENT RADIAL LOAD

Bearings often must carry a combination of radial and thrust loads. The equations stated above are based solely on radial loaded bearings. Therefore, when radial and axial loads are present, an equivalent radial load (P) must be calculated. The equivalent radial load is the greater of:

$$P = XF_r + YF_a$$
$$P = F_r$$

where P = equivalent radial load (lb),
  F_r = applied radial load (lb),
  F_a = applied axial load (lb),
  X = radial load factor = 0.56,
  Y = axial load factor dependent on the magnitude of F_a/C_o,
  C_o = catalog static load rating (lb).

F_a/C_o	Y
0.014	2.30
0.028	1.99
0.056	1.71
0.084	1.56
0.11	1.45
0.17	1.31
0.28	1.15
0.42	1.04
0.56	1.00

STATIC LOAD RATING

C_o, the static load rating, is the non-rotating radial load which produces a maximum contact stress of 580,000 lb/in^2 at any point within the bearing. When static load exceeds the catalog rating, a significant decrease in bearing smoothness and life can be expected when rotation is resumed.

As with dynamic load ratings, static loads are usually a combination of radial and thrust loads. Equivalent static load must therefore be calculated.

The static equivalent load for radial ball bearings is the greater of:

$$P_o = 0.6F_r + 0.5F_a$$

or

$$P_o = F_r$$

where P_o = equivalent static radial load (lb),
  F_r = applied radial load (lb),
  F_a = applied axial load (lb).

EXAMPLES OF LIFE AND LOAD CALCULATIONS

Example 1

Determine the L_{10} life hours of a 6203 ball bearing operating at 800 rpm with a radial load of 250 lb. The basic dynamic capacity from the catalog is

$C = 1653$:

$$L_{10} = \text{unknown}$$
$$C = 1653 \text{ lb}$$
$$F_r = P = 250 \text{ lb}$$
$$N = 800 \text{ rpm}$$

$$L_{10} = 3\left(\frac{C}{P}\right)^3\left(\frac{16{,}667}{N}\right)$$

$$L_{10} = 3\left(\frac{1653}{250}\right)^3\left(\frac{16{,}667}{800}\right)$$

$$L_{10} = 18{,}067 \text{ h}$$

Example 2

Determine the minimum static and dynamic load ratings required to carry a 300-lb radial load and 75-lb axial load for 3500 h at 650 rpm:

$$C = \text{unknown}$$
$$C_o = \text{unknown}$$
$$P = \text{unknown}$$
$$P_o = \text{unknown}$$
$$Y = \text{unknown}$$
$$X = 0.56$$
$$F_r = 300 \text{ lb}$$
$$F_a = 75 \text{ lb}$$
$$N = 650 \text{ rpm}$$
$$L = 3500 \text{ h}$$

$$P_o = 0.6F_r + 0.5F_a = 217.5 \text{ lb}$$

or

$$P_o = F_r = 300 \text{ lb}$$

Therefore

$$P_o = C_o \text{ minimum} = 300 \text{ lb}$$

$$F_a/C_o = 75/300 = 0.25$$

Then by interpolation $Y = 1.19$.
    Equivalent radial load

$$P = XF_r + YF_a = 0.56(300) + 1.19(75) = 257.3 \text{ lb}$$

$$P = F_r = 300 \text{ lb}$$

Therefore $P = 300$ lb

$$L_{10} = 3\left(\frac{C}{P}\right)^3\left(\frac{16{,}667}{N}\right)$$

or

$$C = \left(\frac{L_{10}N}{3(16,667)}\right)^{1/3}(P)$$

$$C = \left(\frac{(3500)(650)}{50,000}\right)^{1/3}(300) = 1071 \text{ lb}$$

Answer: C_o minimum = 300 lb
C minimum = 1071 lb

Simplified load calculations
A simplified version of the above equations can be used as follows:

$$L_f = \frac{C \times S_f}{P}$$

where $\quad L_f$ = life factor,
C = basic dynamic load rating from catalog,
S_f = speed factor,
P = equivalent radial load (lb).

Speed and life factors are given in Table 17.1.

Table 17.1   Speed and life factors for anti-friction bearings

				Speed factor (S_f)				
rpm	10	11	12	13	14	15	16	17
S_f	1.49	1.44	1.40	1.36	1.33	1.30	1.27	1.25
rpm	18	19	20	21	22	23	24	25
S_f	1.22	1.20	1.18	1.16	1.14	1.13	1.11	1.10
rpm	26	27	28	29	30	31	32	33.3
S_f	1.08	1.07	1.06	1.04	1.03	1.02	1.01	1.00
rpm	34	36	38	40	42	44	46	48
S_f	0.99	0.97	0.95	0.94	0.92	0.91	0.89	0.88
rpm	50	55	60	65	70	75	80	85
S_f	0.87	0.84	0.82	0.80	0.78	0.76	0.74	0.73
rpm	90	95	100	110	120	130	140	150
S_f	0.71	0.70	0.69	0.67	0.65	0.63	0.62	0.60
rpm	160	170	180	190	200	220	240	260
S_f	0.59	0.58	0.57	0.56	0.55	0.53	0.51	0.50
rpm	280	300	320	340	360	380	400	420
S_f	0.49	0.48	0.47	0.46	0.45	0.44	0.43	0.43
rpm	440	460	480	500	550	600	650	700
S_f	0.42	0.41	0.41	0.40	0.39	0.38	0.37	0.36
rpm	750	800	850	900	950	1000	1050	1100
S_f	0.35	0.34	0.34	0.33	0.32	0.32	0.31	0.31

Table 17.1 (cont.)

				Speed factor (S_f)				
rpm	1150	1200	1250	1300	1400	1500	1600	1700
S_f	0.30	0.30	0.29	0.29	0.28	0.28	0.27	0.27
rpm	1800	1900	2000	2100	2200	2300	2400	2500
S_f	0.26	0.26	0.25	0.25	0.24	0.24	0.24	0.23
rpm	2600	2700	2800	2900	3000	3200	3400	3600
S_f	0.23	0.23	0.22	0.22	0.22	0.21	0.21	0.21
rpm	3800	4000	4200	4400				
S_f	0.20	0.20	0.19	0.19				

				Life factor (L_f)				
L_{10} (h)	300	600	900	1200	1500	1800	2100	2400
L_f	0.58	0.73	0.84	0.92	1.00	1.06	1.12	1.17
L_{10} (h)	2700	3000	3300	3600	3900	4200	4500	4800
L_f	1.21	1.26	1.30	1.34	1.37	1.41	1.44	1.47
L_{10} (h)	5100	5400	5700	6000	6600	7200	7800	8400
L_f	1.50	1.53	1.56	1.59	1.64	1.69	1.73	1.77
L_{10} (h)	9000	9600	10,200	10,800	11,400	12,000	13,500	15,000
L_f	1.81	1.85	1.89	1.93	1.96	2.00	2.08	2.15
L_{10} (h)	16,500	18,000	19,500	21,000	22,500	24,000	25,500	27,000
L_f	2.22	2.29	2.35	2.41	2.47	2.52	2.57	2.62
L_{10} (h)	28,500	30,000	31,500	33,000	34,500	36,000	37,500	39,000
L_f	2.66	2.71	2.76	2.80	2.85	2.89	2.93	2.96
L_{10} (h)	40,500	42,000	43,500	45,000	46,500	48,000	49,500	51,000
L_f	3.00	3.40	3.07	3.11	3.14	3.18	3.21	3.24
L_{10} (h)	52,500	54,000	55,500	57,000	58,500	60,000	63,000	66,000
L_f	3.27	3.30	3.33	3.36	3.39	3.42	3.48	3.53
L_{10} (h)	69,000	72,000	75,000	78,000	81,000	84,000	87,000	90,000
L_f	3.58	3.63	3.68	3.73	3.78	3.82	3.87	3.91
L_{10} (h)	93,000	96,000						
L_f	3.96	4.00						

Simplified version
Use of Table 17.1 is shown below by resolving Examples 1 and 2.

Example 3 (see data from Example 1)

Speed factor $S_f = 0.34$

$$L_f = \left(\frac{C \times S_f}{P}\right) = \left(\frac{1653(0.34)}{250}\right) = 2.25$$

Then L_{10} from table = approximately 18,000 h.

Example 4 (see data from Example 2)

P and C_o must be solved in the same fashion shown in Example 2.
  From tables:
  $S_f = 0.37$
  $L_f = 1.32$

$$C = \frac{L_f(P)}{S_f} = \frac{1.32(300)}{0.37} = 1070 \text{ lb}$$

RADIAL INTERNAL CLEARANCE

Radial internal clearance is a measure of the radial looseness, or play, between the inner and outer rings. Precision bearings are available in four classes of looseness. The amount of looseness necessary is dependent on many factors, including shaft alignment, shaft and housing fits, bearing speed, etc. As rpm, shaft misalignment, and press fits increase in magnitude, so should radial play. Radial clearances are listed in Table 17.2.

Table 17.2   Radial internal clearance: Single row, radial contact, ball bearings

Basic bore diameter, d (mm)		C-2 acceptance limits		Standard acceptance limits		C-3 acceptance limits		C-4 acceptance limits	
Over	Incl.	Low	High	Low	High	Low	High	Low	High
2.5	10	—	3.0	1	5	3	9		
10	18	—	3.5	1	7	4	10	7	13
18	24	—	4.0	2	8	5	11	8	14
24	30	—	4.5	2	8	5	11	9	16
30	40	—	4.5	2	8	6	13	11	18
40	50	—	4.5	2	9	7	14	12	20
50	65	—	6	3	11	9	17	15	24
65	80	—	6	4	12	10	20	18	28
80	100	—	7	5	14	12	23	21	33
100	120	—	8	6	16	14	26	24	38
120	140	—	9	7	19	16	32	28	45
140	160	—	9	7	21	18	36	32	51
160	180	—	10	8	24	21	40	36	58
180	200	—	12	10	28	25	46	42	64

Note: Tolerance limits in 0.0001 in.

ADJUSTMENTS FOR UNUSUAL OPERATING CONDITIONS

The above short-cut calculation routines cover the majority of conditions which are typically encountered by rolling element bearings. These calculations are ideally suited to the type of rapid screening study which is made in the course of a design appraisal for the machinery categories listed in Table 17.3. However, adjustment factors would have to be applied for unusual operating conditions. For instance, at elevated temperatures the hardness of the bearing materials is reduced and the dynamic load-carrying capacity is also reduced as a consequence. The reduction in dynamic load-carrying capacity at different temperatures is taken into account by multiplying the basic dynamic load rating C by a temperature factor obtained from the following table:

Bearing temperature (°F)	302	392	482	572
(°C)	150	200	250	300
Temperature factor	1.00	0.90	0.75	0.60

Table 17.3   Guide to values of recommended basic life rating for different classes of machines

Class of machine	L_{10} operating hours
Domestic machines, agricultural machines, instruments, technical apparatus for medical use	300 to 3000
Machines used for short periods or intermittently: electric hand tools, lifting tackle in workshops, construction machines	3000 to 8000
Machines to work with high operational reliability during short periods or intermittently: lifts, cranes for packaged goods or slings of drums, bales, etc.	8000 to 12,000
Machines for use 8 h per day but not always fully utilized: gear drives for general purposes, electric motors for industrial use, rotary crushers	10,000 to 25,000
Machines for use 8 h per day and fully utilized: machine tools, woodworking machines, machines for the engineering industry, cranes for bulk materials, ventilator fans, conveyor belts, printing equipment, separators and centrifuges	20,000 to 30,000
Machines for continuous use 24 h per day: rolling mill gear units, medium-sized electrical machinery, compressors, mine hoists, pumps, textile machinery	40,000 to 50,000
Water works machinery, rotary furnaces, cable stranding machines, propulsion machinery for ocean-going vessels	60,000 to 100,000
Pulp and paper making industry, large electric machinery, power station plant, mine pumps and mine ventilator fans, tunnel shaft bearings for ocean-going vessels	> 100,000

The satisfactory operation of bearings at elevated temperatures also depends on whether the bearing has adequate dimensional stability for the operating temperature, whether the lubricant selected will retain its lubricating properties, or whether the materials of the seals and cage, etc., are suitable. For any unusual conditions, it would be appropriate to request further information from one of the principal bearing manufacturers.

SLIDING BEARINGS

Experience assists vendor and user in the design and selection of high reliability sliding bearings. While the average-speed, moderately-loaded sliding or sleeve bearing is generally unsophisticated and forgiving, considerably more care is required for high-speed, heavily-loaded bearings. And again, the specifying or reviewing engineer must make his overall selection based on his perception of FRETT, the combined action of Force, Reactive Environment, Time, and Temperature.

Reliably designed turbomachinery bearings for instance require oil film thicknesses which increase as the shaft peripheral velocity increases. This does not necessarily mean, however, that the as-designed or permissible clearance ratio is increased in the same manner. A reasonable rule of thumb assumes a bearing clearance of [(0.001 in/in diameter) + 0.002 in], that is a 3-in journal would have a nominal diametral clearance of 0.005 in and a 12-in journal would have a nominal diametral clearance of 0.014 in.

The load and speed capacity of sliding bearings is quite heavily influenced

Table 17.4   Speed and load ranges of typical materials

Material	Typical speed range (fpm)	Typical load (psi on projected area)
Bronze	250	200
Leaded bronze	300	200
Cast iron	250	100
Zinc	250	70
Steel	200	50
Babbitt	1000	50
Babbitt-steel	1000	75
Babbitt-bronze	1000	75
Aluminum	1200	3500
Aluminum–steel	1600	3600
Overlay aluminum–steel	2000	4100
Overlay copper-lead–steel	2200	3400
Copper-lead–steel	1900	2800
Copper-lead	1500	2300

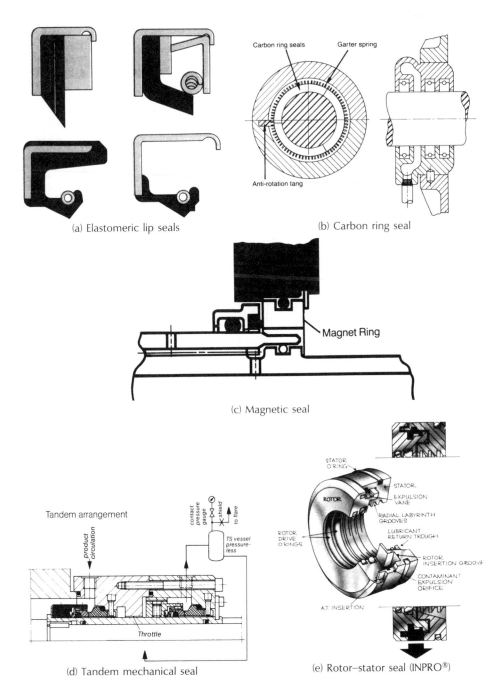

(a) Elastomeric lip seals

(b) Carbon ring seal

Carbon ring seals   Garter spring

Anti-rotation tang

(c) Magnetic seal

Magnet Ring

(d) Tandem mechanical seal

Tandem arrangement

product circulation

contact pressure gauge

shield

to flare

TS vessel pressure-less

Throttle

(e) Rotor–stator seal (INPRO®)

STATOR O'RING

ROTOR

STATOR

EXPULSION VANE

RADIAL LABYRINTH GROOVES

LUBRICANT RETURN TROUGH

ROTOR DRIVE O'RINGS

ROTOR INSERTION GROOVE

CONTAMINANT EXPULSION ORIFICE

A.T. INSERTION

Fig. 17.1 Some shaft sealing devices typically applied in rotating machinery

Table 17.5 Principal attributes of shaft sealing devices

	Bearing housing seals									Process containment seals															
	Hermetic sealing	Least expensive	Shaft wear likely	High temperature	Easily available	Emergency bearing	Accommodates ΔP	Avg. life < 3000 h	Avg. life > 8000 h	Extreme temperature	Coking fluid	Flashing fluid	Abrasive fluid	Poor lubricant	High loss potential	Low leakage	Medium leakage	High leakage	Low HP draw	Slow failure	Sudden failure	Good surveillability	High maintainability	High reliability	High speed
General-purpose seals																									
Elastomeric lip seal		●	●		●			●																	
Labyrinth seal, rotating				●					●																
Magnetic face seal	●						●		●																
Rotor – stator seal						●			●																
Labyrinth seal, stationary		●	●	●	●				●																
Pump seals																									
Metal bellows design										●						●					●				
Isolated spring design											●	●	●			●					●				

222

Hydraulic balance design

Double seal

Tandem seal

Stationary seal head design

Cartridge design

Compressor seals

Labyrinth seal

Carbon rings

Dry running gas seal

Bushing seal

Pumping seal

Mechanical face seal

(a)

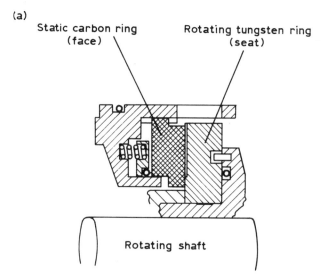

Fig. 17.2a   Gas seal cross-section. The seal typically consists of an O-ring sealed carbon ring (face), located within the casing, spring loaded against a rotating tungsten carbide ring (seat) fastened to the shaft (courtesy John Crane – Houdaille Co.).

by the material composition. Although metal alloy or composite bearings are of primary importance in most machinery, a large number of other materials are available for special applications and cannot be overlooked by the serious designer. For an overview of typical materials and their speed and load ranges refer to Table 17.4, which should be used in screening studies to determine if the vendor's selection is in the right range.

In general, it may be said that bronzes are high-strength bearing materials with varying amounts of copper, tin, and perhaps zinc, which impart high load carrying capacity. It must be noted that bronze bearings may not resist seizure at high speeds. Certain aluminum-base alloys allow for high loads and speeds and resist corrosion in hot oxidized oil, whereas copper-lead with high lead content will not resist hot oxidized oil.

Babbitt is, of course, one of the most prevalent bearing materials. It is often used because dirt particles in the lube oil will not easily destroy the bearing surface. However, thick babbitt will easily distort and may not be suitable for machines undergoing shock loading or heavy vibrations.

Plastic bearings are rapidly replacing the more mundane metallic sleeve bearings and new compositions enter the market every year. The user should not be afraid to accept them as long as prior experience can be demonstrated and as long as proper thought has been given to FRETT.

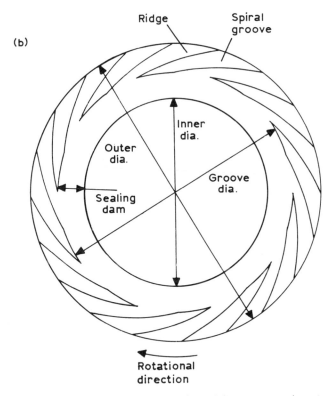

Fig. 17.2b Sealing surface of rotating seat. Sealing of the compressed gas is achieved at the radial interface of the rotating and the stationary rings in a unique and ingenious way. The sealing surfaces are lapped to a high degree of flatness, but the rotating tungsten carbide ring has spiral grooves machined into its surface. Rotation gas is pumped inwards towards the internal diameter of the groove called the sealing dam. The sealing dam provides resistance to gas flow, and pressure increases. The generated pressure lifts the carbon ring surface out of contact with the tungsten carbide ring, thus providing a controlled sealing gap (courtesy John Crane – Houdaille Co.).

SHAFT SEALING DEVICES

One important, yet often overlooked source of machinery distress is lube oil contamination due to the intrusion of water or air-borne dust and debris. These contaminants generally enter at locations where the machine drive shaft protrudes through the bearing housing. At other times, outward leakage of bearing lube oil causes loss of lubricant and contamination of the surrounding area.

In either case, the concept of machinery reliability enhancement again brings to mind why an assessment of these sealing devices is so important. Some of them are clearly subject to potential degradation due to the by now

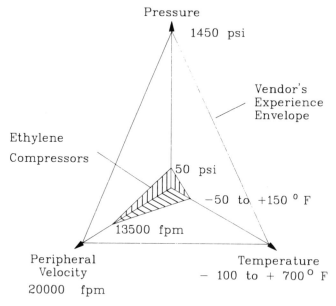

Fig. 17.3   Gas seal parameter triangle.

familiar agents force, reactive environment, time, and temperature; others may be less susceptible. Some are highly surveillable and some less so, and the same is true for maintainability.

Figure 17.1 a–e gives a brief overview of the most commonly encountered shaft sealing devices or closures. Their principal attributes are shown in Table 17.5.

An interesting screening exercise has been done for dry running gas seals. These seals are simpler than many other sealing devices in so far as they are not subject to wear during normal operation. Their maximum rotational speed is therefore only limited by the stress exerted on their rotating parts through centrifugal force. A typical cross-sectional view of a gas seal is shown in Figure 17.2. This seal was proposed for retrofit in a process gas compressor owned by a petrochemical company. The seal manufacturer's literature indicates three parameters as influencing seal reliability. These parameters are:

- peripheral shaft or seal velocity (ft per min),
- pressure (psig),
- temperature (°F).

The maximum vendor experience is shown as 18,000 fpm, 1450 psig and 750°F. This can be illustrated in a parameter triangle (Fig. 17.3). The user's application was 13,500 fpm, 50 psig sealing pressure, and 150°F gas temperature. Entering these values into the parameter triangle provides a first view of how the planned application fits into the seal manufacturer's experience envelope.

18

Bolt loading assessment

In the preceding chapters we directed our attention time and again at the powerful contribution made by the principle of redundancy to machinery reliability. This principle is applied as a matter of course when it comes to fixing elements. Fixing elements are, for example, unthreaded and threaded fasteners. Because fixing elements are usually applied in a redundant or load sharing fashion they are normally not considered critical components in the context of machinery reliability assessment. Yet, from time to time, a review is required. As an example, we would like to show how one particular aspect of threaded fasteners may be evaluated.

EVALUATING BOLTS IN TENSION

There are distinct theoretical and practical aspects to the topic. Quite often there is sufficient disagreement between both aspects to justify a thorough investigation of the design assumptions as well as of the actual field application in a particular situation [1].

To get the right clamping force, one tightens a bolt or screw by applying torque to the head. If resistance is met, the bolt will continue to rotate until a balance is achieved between the torque applied and the sum of bolt tension and friction. There is usually a wide distribution between these factors [2].

Other concerns include:

- selection of a safe design load level,
- loosening tendencies of bolted joints,
- prevention of premature loosening.

Theoretical investigations into bolted connections that are dynamically stressed in tension can be made by evaluating the maximum bolt load.

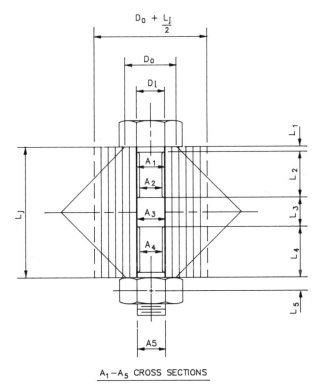

Fig. 18.1 Dynamically stressed bolted joint in axial tension.

Maximum bolt load is defined as:

$$F_{max} = F_P + \Delta F_B \qquad (18.1)$$

where F_P = initial preload (lb, N),
   ΔF_B = change in bolt load or differential load (lb, N).

This would apply to a typical bolted joint as illustrated in Figure 18.1. A graphical presentation of the joint forces is shown in Figure 18.2.

The differential load may be determined by trigonometry from Figure 18.2 and is defined as:

$$\Delta F_B = L_X \frac{\Delta l_B}{\Delta l_B + \Delta l_J} \qquad (18.2)$$

where L_X = external tension load (lb, N),
   Δl_B = elongation of bolt (in, mm),
   Δl_J = shortening of joint (in, mm).

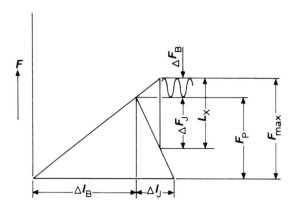

Fig. 18.2   Joint tension diagram.

Another relationship is

$$\Delta F_B = L_X \frac{K_B}{K_B + K_J} \qquad (18.3)$$

where K_B = spring constant of bolt,
K_J = spring constant of joint.

The fastener spring constant can be calculated from:

$$K_B = \frac{E_B}{\dfrac{l_1}{A_1} + \dfrac{l_2}{A_2} + \cdots} \qquad (18.4)$$

The joint spring constant may be evaluated from:

$$K_J = \frac{\pi E_J \left(D_0 + \dfrac{l_B}{2} \right)^2 - D_i^2}{4 l_B} \qquad (18.5)$$

where $E_{B,J}$ = modulus of elasticity of bolt resp. joint material (psi, GPa),
l_1, A_1, l_2, A_2 = length and cross-sectional area of individual bolt and thread
parts (Fig. 18.1; in, mm; in², mm²),
l_B, D_0, D_1 = dimensions in Figure 18.1.

Once F_{max} is known we can proceed to determine σ_{tmax}, the maximum
acceptable tensile stress:

$$\sigma_{tmax} = \frac{F_{max}}{A_s} \quad \text{(psi, MPa)}$$

where A_s = cross-sectional area of the threaded portion of the bolt (in², mm²).

NEW IDEAS FOR FASTENERS*

From the preceding pages we can easily see how threaded fasteners can become a maintainability problem due to the fact that calculated bolt pretension has somehow to be achieved and maintained. Many approaches to bolt pretensioning have been tried and are in fact being practiced. All methods clearly have their advantages and disadvantages. In the following we present a new solution to an old problem.

The strength of a screw type fastener increases with the square of its diameter. However, to make use of that strength it is necessary to torque the fastener until it is prestressed to a certain level. A reasonable level of prestress for a high strength bolt is 40,000 psi or 275.6 MPa in the minor thread area. This results in a safety factor of approximately 3. Unfortunately, while the strength of the fastener increases by the second power of the diameter, the torque required to prestress it increases by the third power of its diameter. Figure 18.3 shows the strength and torque requirements of high strength bolts prestressed to 40,000 psi (275.6 MPa).

The torque that can be exerted by ordinary torquing devices, such as hand wrenches, torque wrenches, and air impact wrenches is rather limited. A torque of 1000 lb-ft or 1360 N-m is about the maximum that can be obtained. However, that torque already requires wrench extensions of 10 ft

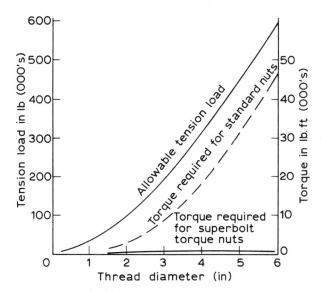

Fig. 18.3   Tensioning torque for bolts.

---

* By permission of Superbolt, Inc., Carnegie, Pennsylvania.

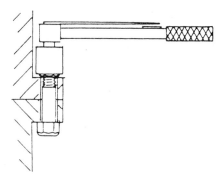

Fig. 18.4   Applying a torque wrench to a small nut and bolt.

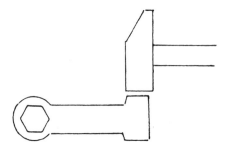

Fig. 18.5   Sledgehammer method of tightening nuts.

or 2.5 m and a pull of 100 lb or 444.8 N at the end of the extension. As can be seen from Figure 18.3, a 1-in or 25-mm diameter bolt is about the largest that can still be torqued to capacity directly. Anything larger needs more than 1000 lb-ft of torque. A 2-in (50-mm) thread needs three times what a torque wrench can handle. A 4-in (100-mm) thread needs 25 times, and a 6-in (152-mm) thread needs 90 times the torque available from a hand wrench.

Over the years a number of methods or devices have been used to prestress large diameter screws. Although they help to some extent, they all have severe limitations. Figures 18.4–18.7 show the various methods employed to tighten large screws. Figure 18.4 shows the direct torquing of a screw by a hand wrench of some sort, a box wrench, socket wrench, torque wrench, etc. Perhaps 99% of all screws are tightened this way. The upper size limit for screws that can be tightened to capacity directly is about 1 in thread size.

Figure 18.5 shows a crude but often used method to tighten screws above 1 in diameter, the sledge hammer method. It works reasonably well in some applications, but there is little control over the amount of torque. There is also still a size limitation.

Figure 18.6 shows a long practiced method to tighten fasteners. The bolt is heated, quickly inserted, and the nut tightened as much as possible. When

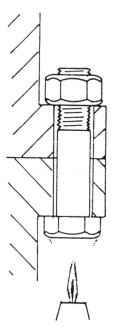

Fig. 18.6   Tightening bolt by heating.

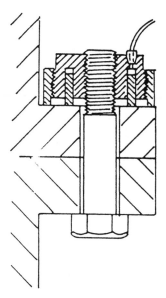

Fig. 18.7   Hydraulic bolt tensioning device.

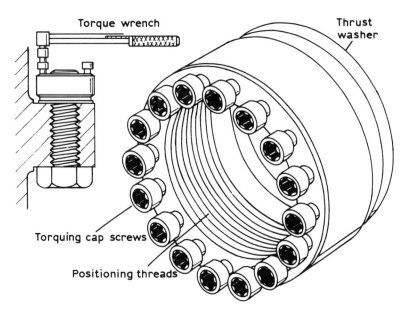

Fig. 18.8   Multiscrew nut lowers bolt torquing forces.

the bolt cools, it shrinks and pulls the workpieces together. Although there is no size limitation, there is still no direct control over the amount of tightness required – it all depends on calculated data. The other problem for heat tightened screws is how to loosen them.

Figure 18.7 shows a popular method to tighten large screws. The details vary with many patented devices, but basically one uses a hydraulic ring piston to stress the bolt. After the stress is hydraulically induced, a mechanical take-up device ensures that the stress is maintained after the hydraulic pressure is released. There are many problems with "hydraulic nuts". They are expensive, need a lot of room, and have relatively low tightening power. The only reason for their popularity is that they are easy to use and have no size limitations.

Figure 18.8 shows a new type of torque nut. This torque nut is inexpensive, easy to tighten, easy to loosen, light, uses no more room than an ordinary nut, gives good torque control, has 10 times the tightening power of a hydraulic nut, and will not come loose if properly torqued. There is no size limitation for these torque nuts. To date nuts have been designed up to 24 in (610 mm) thread size with 25,000,000 lb (111.25×10^6 N) work load. Even for torque nuts of this size the torque on individual jack bolts is only 1000 lb-ft (1360 N-m).

For process machinery the obvious application for the described torque nuts is at high-pressure flanges, particularly at housing splits. At these places they can make it a lot easier for the manufacturer to design and build the

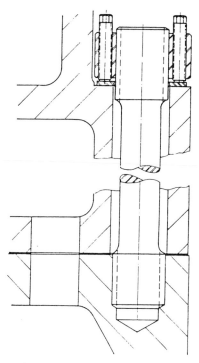

Fig. 18.9 Installation example: torque nut.

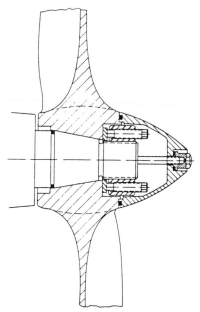

Fig. 18.10 Installation example: torque nut.

Table 18.1 Tension/torque data for torque nuts

Thread size (in)	Allowable tension load (lb)	Torque required standard nut (lb-ft)	Torque required "Superbolt" torquenut (lb-ft)	Outside diameter "Superbolt" torquenut (in)
$1\frac{1}{2}$	62,000	1,900	40	$2\frac{3}{4}$
$1\frac{3}{4}$	62,000	2,100	40	3
2	115,000	4,700	100	$3\frac{1}{2}$
$2\frac{1}{4}$	115,000	5,100	100	$3\frac{3}{4}$
$2\frac{1}{2}$	175,000	9,100	185	$4\frac{1}{2}$
$2\frac{3}{4}$	175,000	9,600	185	$4\frac{3}{4}$
3	260,000	16,400	185	$5\frac{1}{4}$
$3\frac{1}{4}$	260,000	17,100	185	5
$3\frac{1}{2}$	260,000	17,800	185	$5\frac{3}{4}$
$3\frac{3}{4}$	260,000	18,500	185	6
4	350,000	28,700	185	$6\frac{1}{2}$
$4\frac{1}{2}$	350,000	30,800	185	7
5	500,000	50,100	320	$7\frac{1}{2}$
$5\frac{1}{2}$	500,000	53,200	320	8
6	500,000	57,000	320	$8\frac{1}{2}$

Fig. 18.11 Installation of torque nuts on a boiler feedwater pump.

machinery. Maintenance people in particular will benefit from the ease of taking the machinery apart. Torque nuts are ideal for replacing the original nuts during overhaul periods, since they take no more room.

For designers of process machinery the new torque nuts offer new possibilities. With the tremendous tightening forces available it is possible to use heavy, long anchor bolts to assemble huge machines from several sections. This can have advantages in both manufacturing and shipping. It is also possible now to force rotors of all kinds on to steeply tapered shafts. The connection has the holding power of a shrink fit, but is self-releasing after removal of the torque nut. Figures 18.9 and 18.10 show examples of the use of the new torque nuts.

Table 18.1 shows a basic comparison between the torque requirements for standard nuts and the new torque nuts. Also, outside diameters are listed. More powerful torque nuts can be designed; however, diameters will increase somewhat.

Finally, Figure 18.11 shows a torque nut installation on a boiler feedwater pump in a major utility company.

REFERENCES

1. Bickford, J. H., *An Introduction to the Design and Behavior of Bolted Joints.* New York: Marcel Dekker, 1981.
2. Haviland, G. S., Designing with threaded fasteners. *Mechanical Engineering*, Oct. 1983, pp. 17–31.

19

Lubrication systems

GENERAL CONCEPTS

Reliability, maintainability, and surveillability are our stated objectives for a machine, an assembly, and a component part. And, we have repeatedly accepted the fact that force, reactive environment, time, and temperature can have an adverse influence on the achievement of our objectives. With virtually every machine incorporating moving parts and with every moving part requiring some form of appropriate lubrication, we must address this issue with considerable urgency.

It has been shown that lubrication systems design for many process machines is often only an afterthought and that improvements are necessary to make sure that lubrication does not become the weak link in the component chain. Lube systems deficiencies in large turbomachinery for refining and chemical plant installations have prompted the American Petroleum Institute (API) to develop standards for the design and fabrication of these systems. The principal requirements of these standards are aimed at upgrading the selection of materials, instruments, and auxiliary equipment such as pumps, accumulators, and the like. Many of these so-called API requirements are equally applicable or beneficial in non-petrochemical machinery.

As can be expected, force, reactive environment, time, and temperature will surely influence the overall reliability of a lube oil system. Undue *force* is exerted on piping or tubing if service personnel stand on thin-walled, or unsupported fluid lines. Similarly, undue force is created by fluid flow pulsations in these lines. Any reliability assessment task will thus have to address pipe size, pipe support, and pulsation suppression. And, while this may rightly be considered the vendor's responsibility, he may extricate himself from this implied obligation by claiming that he had never meant for the user's personnel to stand on the piping, or that pulsations are always present and nothing lives forever; hence, says the vendor, be prepared to accept the need for repairs.

Looking at the contributing agent *time*, we should note that lube oils may in time become contaminated or that elastomeric components such as rubber diaphragms in control valves or bladders in accumulators may degrade with time. Of course, we would want to draw certain conclusions from this observation and look for ways to purify the lube oil, preferably through the use of vacuum dehydration methods [1]. While this may be beyond the scope of the review or appraisal, the engineer may nevertheless ask for provision of suitable drain valves which, at some future date, would make it possible to retrofit such devices with relative ease. The other time-affected elements (i.e. rubber diaphragms and accumulator bladders) will again merit our attention when the question of maintainability comes up for discussion.

Lube oil *temperature* deviations will have a significant influence on equipment reliability from two different points of view:

1. High or low ambients may require special provisions for maintaining acceptable oil temperatures in hot or cold climates. In a hot climate the lube oil viscosity may drop below acceptable limits unless the facilities are suitably designed. The opposite concern should prompt an investigation in cold climates where an increase in oil viscosity would mandate the installation of heaters or similar provisions.
2. High or low operating temperatures will require that all lubricated parts be designed to accept the specially compounded lubricants which may be required for the protection of machinery whose component parts are surrounded by hot or cold working fluids. The review engineer may wish to investigate oil film strength and thickness relationships in these cases.

It would also be fruitful to understand the acceptable temperature versus viscosity range of a chosen lube oil and its condemnation limits in terms of deviations in flash point, viscosity, acid number, water, and oxidation inhibitor content [1]. Of course, these considerations move towards maintainability and surveillability concerns which should always be kept in sight.

Our next and last agent is a *reactive environment*. Lube oil systems may be subjected to this agent from either external (ambient) or internal exposure. The principal reactive agent is water, which can cause greatly accelerated corrosion of system reservoirs, piping, auxiliary components, instruments and, of course, major machine internals. One authoritative source [2] has documented that a water content of 0.002 to 6% can reduce the fatigue life of anti-friction bearings from 48 to 83%. For the sake of reliability, the lube oil system may have to be hermetically sealed, or purged with an inert gas such as nitrogen. Alternatively, other means may have to be found or devised in efforts to prevent the ingress of free water and atmospheric moisture into the oil reservoir or oil system.

External protection will often be required in the form of primer paints, protective paints, or stainless steel. Experience shows that in the majority of cases the use of stainless steel for lube oil reservoirs and piping is cost-justified, while instruments and control valves could be made of carbon steel. The petrochemical industry has made it a practice not to allow non-ferrous alloy components in lube oil systems associated with large rotating machinery in an effort to limit failure risks in cases of fire. However, this may not be an important concern for all machinery categories.

A reactive environment could also be created by lube oils which are not properly inhibited or which lose their inhibitors when contacted by certain process gases or liquids reaching the bearing or seal areas. Here is where the person engaged in reliability assessment will have to decide whether it will be appropriate to treat the root cause or the symptom. In other words, should a seal system be devised so as to preclude the movement of contaminants into the lube oil, or should the lube oil be continuously purified, or would it be possible to obtain a lubricant which will not easily degrade when exposed to the contaminant? A general approach to the reliability assessment of critical lube oil systems has already been presented in Chapter 1.

DETAILED APPRAISALS

We had made the statement earlier that frequent deficiencies with large lubricating oil systems had prompted the petrochemical industry to devise a standard specification with clauses and requirements aimed at reliability improvement. This standard, API 614, should be consulted whenever large, critically important systems are needed. One such system is depicted in Figure 19.1.

However, regardless of whether this or any other specification is invoked, a detailed appraisal should be made. This appraisal should result in verifying and, if necessary enhancing, the various components which constitute a given lube oil system. Remember, again, that the user's design appraisal is aimed not only at reliability but also maintainability and surveillability.

The review should start at the main component, the storage tank or reservoir illustrated in Figure 19.2. Its reliability is influenced by the strength of the construction materials and by the geometric shape and material properties selected. The assembly method is important also; evidently, welding will impart a certain resistance to leakage and a given rigidity which may be different if other assembly methods had been chosen. Reliability is also affected by the size and elevation of the vent or breather opening of the lube oil tank. Both must be sized to prevent the accidental over-pressurization of the lube oil tank.

The tank must be maintainable, perhaps through the judicious use of access ways or clean-out ports, or by easy access to its periphery and the many appurtenances entering and leaving it. The tank must also be surveillable. Oil levels and sometimes pressures and temperatures must be read from gauges or meters, and provision must be made for sample taking.

Tank heaters may be incorporated in the tank or below the slightly sloped

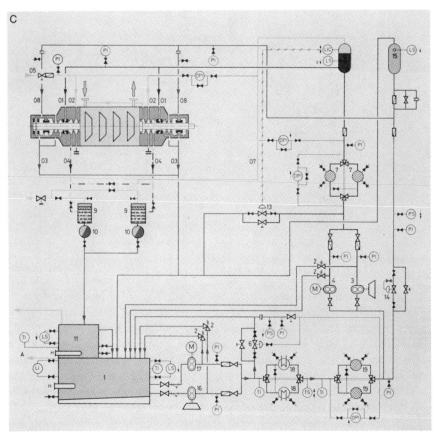

Fig. 19.1 Lube oil system with seal oil booster system for floating oil rings. *Source:* Mannesmann-Demag, Duisburg, West Germany. 1, Oil tank; 2, relief valve; 3, main oil pump (high pressure); 4, auxiliary oil pump (high pressure); 5, oil cooler (high pressure); 6, regulating valve; 7, high-pressure filter; 8, seal oil overhead tank; 9, demister separator; 10, condensate trap; 11, degassing tank; 12, sour oil tank; 13, level control valve; 14, pressure-reducing valve; 15, lube oil overhead tank; 16, main oil pump (low pressure); 17, auxiliary oil pump (low pressure); 18, oil cooler (low pressure); 19, low-pressure filter; 01, seal oil supply; 02, buffer gas supply; 03, outer drain; 04, inner drain (seal oil/buffer gas); 05, buffer gas supply; 07, reference pressure line; 08, lube oil supply; PI, pressure indicator; PS, pressure switch; DPI, differential pressure indicator; TI, temperature indicator; TS, temperature switch; LI, level indicator; LS, level switch; LIC, level controller; H, heater; A, reservoir vent.

bottom of the tank. To be maintainable, it should be possible to withdraw or otherwise remove these heaters without requiring equipment shutdown or tank drainage. However, the heaters should have sufficient surface area to keep the skin temperature at a moderate level, thus preventing the lube oil from overheating and carbonizing.

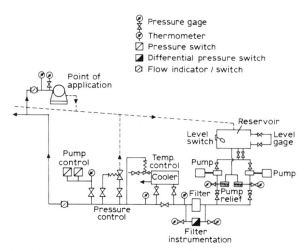

Fig. 19.2 Basic lubrication system with warning and protective devices.

Lube oil return lines should deposit the return oil below the liquid level and should be liberally sized so as to fill at most one-quarter of the cross-sectional area of the pipe. Lube oil supply lines leaving the tank should have strainers between the tank outlet and pump inlet if the pump is a close-clearance positive displacement type. Remember that the strainers must be maintainable. A small sampling or blowdown valve will make the strainers surveillable (see Figure 19.3 for guidelines on lube oil tank design).

Pumps should be arranged as main and spare sets. The spare is fed from a different power supply or is perhaps driven by an electric motor, whereas the main pump would be driven by a steam turbine, or vice versa. On installations with two electric motor-driven pumps, a selector switch may be used occasionally to reverse the designation main versus spare between the two pumps.

The engineer making the appraisal must, in fact, completely think through all the possible operating modes or modes proposed by the lube system vendor. For instance, if one of the two pumps is to be slow-rolled so as to facilitate instantaneous start-up, will this cause deterioration of the mechanical seals in the pump? Will these seals overheat due to slow-roll? What prior experience exists with this equipment? Should he verify this experience by contacting other users? The answers to all these questions may vary from case to case.

Oil coolers frequently exist in many circulating lube oil systems. There are inexpensive ones and expensive ones; maintainable ones and others very difficult to maintain properly. The materials of construction for these coolers should vary with the water quality that can be ascertained for a given location and air coolers may be appropriate at times. Reliability, maintainability, surveillability – these factors must constantly be kept in mind and, since

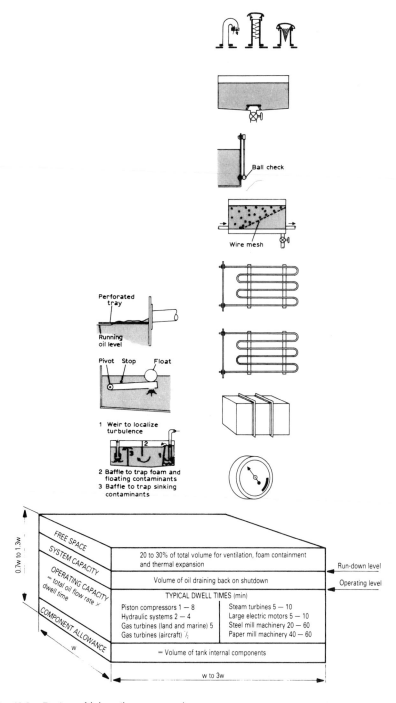

Fig. 19.3 Design of lube oil storage tanks.

Table 19.1 The function of major lubrication system components and required control devices

System component	Function of component	Information or control required to maintain function	Device required	Comment on application
Reservoir	Maintenance of required volume of oil	Level indication Level control	Level gauge Level switch	Visual indication of contents 1. High level warning of overfilling or ingress of water 2. Low level warning of system leak or normal consumption
Pumps	Delivery of the required quantity of oil at prescribed pressure	Pressure indication	Pressure gauge	1. Situated at pump discharge, is necessary when adjusting pump valve 2. Situated at a point downstream of system equipment, is necessary when adjusting pressure control valve
		Pressure control	1. Pressure switch(es) 2. Pressure control valve or other method	To switch in standby pump on pressure. Give warning of falling pressure To spill-off surplus oil and regulate pressure variations due to temperature fluctuations and changes in system demand
		Flow indication and control	Flow switch	Indicates satisfactory flow is established
		Pump protection	Spring relief valve	Protection from overpressure due to system malfunction

244

Filter(s)	Maintenance of cleanliness of oil	Construction of filter	1. Pressure gauges	Visual indication of pressure drop across filter
			2. Differential pressure switch	To signal when pressure drop has increased to a predetermined value and filter requires cleaning
Cooler(s)	Maintenance of prescribed temperature of oil supply	Temperature indication	Thermometer	Visual indication of temperature of oil into and out of cooler. Is necessary when adjusting temperature control valve
		Temperature control	Temperature control valve	Regulates temperature of oil leaving cooler
Point of application to lubricated parts	Dispense correct quantity of oil at prescribed pressure and temperature	Oil pressure, oil temperature, oil flow	Pressure gauge Thermometer Flow indicator	Local visual indication or fitted with contacts to give malfunction signal

Table 19.2 Protective devices for lubrication systems: Selection and installation

Device	Some types	Comments on selection	Installation
Level gauge	1. Dip-stick	Cheap, simple to make	Through reservoir top
	2. Glass tube	Direct reading, simple in design, requires protecting with metal tube or cage	On reservoir side complete with shut-off valves
	3. Dial. Float actuated	Direct reading. For accuracy of calibration reservoir dimensions, shape and any internal obstructions, also specific gravity of oil must be considered	On reservoir side or top
	4. Dial. Hydrostatic, pneumatic operated	Remote reading. Comments as for (3.). Absolutely essential capillary joints are positively sealed against leak	Through reservoir top or side with dial remote panel mounted
Level switch	1. Float actuated	Usually magnetic operation, glandless, therefore leakage from reservoir into switch housing not possible. Some types are level adjustable	Through reservoir top or side, or, on reservoir side in float chamber complete with isolating valves
	2. Sensing probe	Techniques generally used: capacitance, conductivity or resistance. Complete lubricant characteristics must be provided to supplier for selection purposes. Not adjustable after fitting	Through reservoir top or side

246

Component	Type	Description	Mounting / Location
Pressure gauge	1. Bourdon tube	Available for front flange, back flange or stem mounting. Any normal pressure range	Local stem mounted or remote panel mounted, complete with shut-off valve
	2. Diaphragm actuated	More robust construction, more positive indication than Bourdon tube type, mainly stem mounting. Withstand sudden pressure surges, overload pressures, overheating. Any normal pressure range	As for (1.)
Pressure switch	1. Bellows actuated 2. Piston actuated 3. Bourdon tube	Select switch to satisfy pressure and differential range requirements, also oil temperature. (1.) and (3.) suitable for oil and air. (2.) limited to oil applications; will withstand higher pressures	Local stem mounted or remote panel mounted
Pressure control valve	1. Standard high-lift spring-loaded relief	Simple design, cheap. Is viscosity sensitive but normally satisfactory for control of small simple systems	Piped-in branch from main delivery after filter. Discharge back to reservoir
	2. Direct operated diaphragm	Used on larger systems, this valve will maintain control within acceptable limits provided viscosity remains reasonably stable	Fit as for (1.). Pressure signal transmitted to diaphragm from tapping downstream of filter and cooler
	3. Pneumatically controlled diaphragm	Employs same type of valve as in (2.) but the diaphragm is actuated by low pressure air via a control instrument, providing more positive control. Used in place of a direct operated valve, where large diaphragm loads might occur	Fit as (1.). Air control remote panel mounted. Pressure signal transmitted as (2.)

247

Table 19.2 (cont.)

Device	Some types	Comments on selection	Installation
Header tank		Control by use of static-head and provide emergency oil supply on failure of pumps	Install at height equivalent to system pressure requirement
Flow switch	1. Vane actuated	Simple design, robust construction. Can be obtained to give visual as well as electrical indication. Vane angle should be between 30° to 60° for maximum sensitivity	Pipe in-line
	2. Orifice with differential pressure switch	Used when very precise control is required. Orifice size and positioning of pressure tappings is critical	Pipe in-line
Spring relief valve	1. Standard safety valve 2. High lift	Simple design, either chamfered or flat seat Skirted flat seat, designed to give small difference between opening pressure and full flow pressure. Use where pressure accumulation must be minimal	Integral with pump arranged for internal relief, or, fitted remote for external relief
Differential pressure switch	1. Opposed bellows actuated	Comments as for pressure switches	Local or remote panel mounted complete with shut-off valves
	2. Opposed piston actuated	Comments as for pressure switches	Pressure taps close to filter inlet and outlet

248

Thermometer	1. Bi-metallic	Dial calibration, almost equal spacing	Local vertical or co-axially mounted in separable pocket
	2. Vapor-pressure	Dial calibration, logarithmic-scale. Changes of ambient have no effect on reading. Allowance must be made if difference in height between bulb and dial exceeds 2 m. Maximum capillary length about 10 m	Bulb in separable pocket. Dial either local or remote panel mounted
	3. Mercury-in-steel	Dial calibration, equally spaced. Ambient changes have minimal effect on reading. No error of importance results from difference of height between bulb and dial. Maximum capillary length about 50 m	Install as for (2.)
Temperature control valve	1. Direct operating with integral bi-metallic element	Reverse acting. Control is not instantaneous, but is generally acceptable for many systems. Restricts flexibility of pipe routing. Care necessary in adjusting when control is operating	Valve in cooling water supply, operating element in the cooler oil outlet
	2. Vapor-pressure actuated, plunger operated	Reverse acting, reliable, protected against over-temperature. Can adjust when control is operating	Valve in cooling water supply. Remote sensing bulb in separable pocket in cooler oil outlet
	3. Pneumatically controlled diaphragm	Employs valve similar to pressure control, reverse acting	Valve as (2.). Air controller remote panel mounted. Remote sensing

249

they are relative terms, they must be assessed on a case-by-case comparison basis. What about resistance to abuse? What about our old area of concern, FRETT? We should use this approach on filters, and then on everything from check valves to control valves and protective devices. Refer to Tables 19.1 and 19.2 for guidance and direction.

REFERENCES

1. Bloch, H. P. and Geitner, F. K., *Machinery Failure Analysis and Troubleshooting*. Houston: Gulf Publishing, 1983, pp. 189–210.
2. Armstrong, E. L., *et al.*, Evaluation of water-accelerated bearing fatigue in oil-lubricated ball bearings. *Lubrication Engineering*, **34** (1), 1977, pp. 15–21.

20

Rotor dynamics audits*

Machinery reliability is immensely important to the petrochemical industry because the output of an entire plant may stop for days when critical components fail in a major turbomachine. In 1982, we reported that when ethylene plants in the size range of 800,000 metric tons per year experience emergency shutdowns of a few hours' duration, flare losses alone can amount to $400,000 or more [1]. This has prompted many large petrochemical companies to arrange for the performance of dynamic design audits of major machinery for the purpose of identifying and eliminating serious problems before the machine is manufactured. Machinery design audits have also been quite successful in preventing costly project delays.

There are several levels of analysis options, based on the job requirements. The minimum-effort project might be a brief machinery audit, whereas our maximum-effort project would be a full rotor dynamics analysis. The two major divisions of rotor dynamics analysis are lateral and torsional. Lateral analysis is used to optimize radial journal bearings, while torsional analysis addresses the optimization of coupling and shaft elements. This chapter will focus upon lateral rotor dynamics because it specifically addresses the impact of radial journal bearings on turbomachinery vibrations. Also, we consider lateral rotor dynamic analysis quite representative of the extensive reliability assessment efforts applied by the petrochemical industry in the industrial world.

Six key constituent parts make up a full lateral rotor dynamics analysis. Job proposals should be carefully examined to determine that the vendor bids include these essentials unless there is a special reason that warrants a lesser effort.

LATERAL ROTOR DYNAMICS ANALYSIS

Model preparation

The rotor is represented as a series of lumped masses which are connected by elastic beam sections (Fig. 20.1) [2]. The rotating element model is

---

* *Source:* Dana J. Salamone, Salamone Turbo Engineering Inc., Houston, Texas. Adapted by permission of the author.

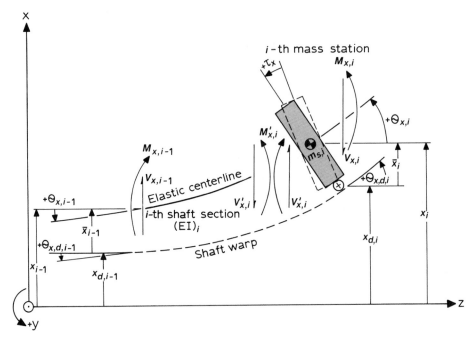

Fig. 20.1   Visualization of rotor behavior in rotor dynamic analysis.

supported by bearings and, in some cases, the seals. The fluid film of the bearings and seals is represented by speed-dependent stiffness and damping characteristics. The stiffness and damping matrices include the principal and cross-coupled components. In cases where foundation effects are significant, the structural support mass, stiffness, and damping can be added under the bearings and seals (Fig. 20.2) [2].

Hydrodynamic bearing stiffness and damping coefficients

The bearing and seal stiffness and damping coefficients are calculated by the finite element method. This method provides a means to model the oil film around the bearing annulus. The pressure profile is integrated around the bearing in order to solve for the resultant forces. This procedure is iteratively applied in order to solve for the equilibrium position. Once the equilibrium position is obtained, the stiffness and damping characteristics can be computed. Figure 20.3 is a typical plot of stiffness and damping versus rotor speed for a four-shoe tilting pad bearing [3].

Critical speed map

The critical speed map (Fig. 20.4) [4, 5] is a plot of the critical speeds of the rotor versus a range of bearing stiffness values. This plot illustrates the

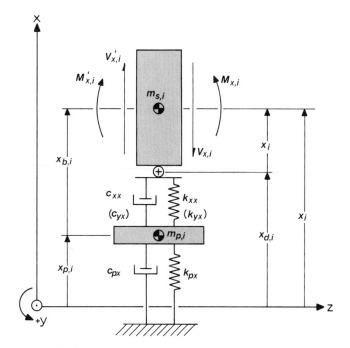

Variable relationships

$x_i = x_{p,i} + x_{b,i}$ (Absolute x-displacement of mass station)

$\bar{x}_i = x_i - x_{d,i}$ (Elastic x-displacement of mass station)

Where

$x_{p,i}$ = absolute x-displacement of pedestal,

$x_{b,i}$ = relative x-displacement of bearing,

$x_{d,i}$ = absolute x-distortion deflection (warp) of shaft.

Fig. 20.2   Typical bearing representation in rotor dynamic analysis.

interaction between the bearing stiffness and the stiffness of the rotor shaft. In addition, if the bearing characteristics are cross-plotted on this map, one can quickly identify possible problem areas. Specifically, the intersections of the bearing stiffnesses with the critical speed curves indicate the undamped critical speeds. These numbers provide an indication of the potential for critical speed infringement.

Mode shapes

The mode shapes at the critical speeds indicate the effectiveness of the bearings and the locations where balancing is most effective. The first two mode shapes for a typical rotor are shown in Figures 20.5 and 20.6 [6].

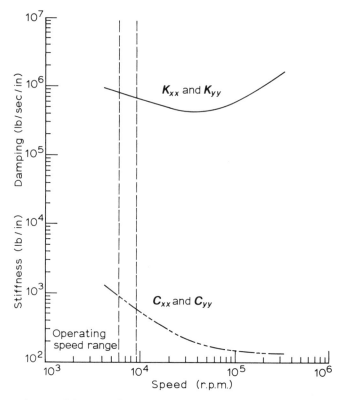

Fig. 20.3   Tilting pad bearing characteristics as represented in rotor dynamic studies. Original governor end bearing, original rotor loads, 4-shoe tilting pad bearing. Arc length $= 55.8°$; pivot angle $= 45°$; offset factor $= 0.5$; preload $= 0.222$; length $= 1.5$ in; diameter $= 4$ in; clearance (dia.) $= 0.009$ in; L/D ratio $= 0.375$; journal load $= 791.8$ lb; $\mu = 1.57\text{E} - 06R_e$.

Synchronous unbalance response

The synchronous unbalance response analysis [7, 8] is a simulation of the rotor response during a start-up or coast-down. The response analysis will indicate the location of the damped peak responses and their sensitivities to unbalance excitation. In addition to unbalance, it is sometimes necessary to examine the effects of a bowed shaft and skewed wheel [2]. Wheel skew can result from imperfect shrink fits or couple unbalance. Figures 20.7 and 20.8 illustrate typical amplitude and phase angle plots versus rotor speed for a two-bearing turbomachine [6]. These two plots are often referred to as a "Bodé plot" [9]. Figure 20.9 illustrates a typical Nyquist plot [9] of the same data presented in the Bodé plots [6].

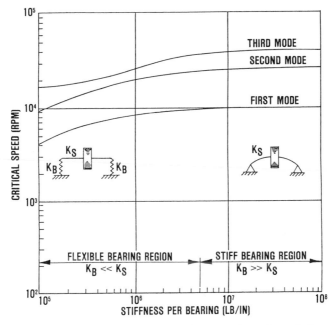

Fig. 20.4   Undamped critical speed map as used in rotor dynamic analysis.

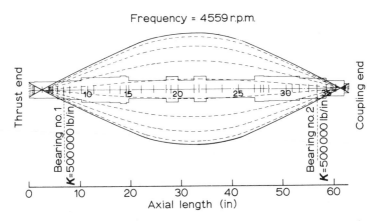

Fig. 20.5   Simulation of compressor rotor mode shape 1 with oil seals removed.

Rotor stability analysis

The rotor stability analysis is often referred to as an "eigenvalue analysis" or "damped critical speed analysis" [8, 10, 11]. The "eigenvalues" are complex natural frequencies of the rotor/bearing system. The real part of the eigenvalue indicates the growth factor and the imaginary part indicates

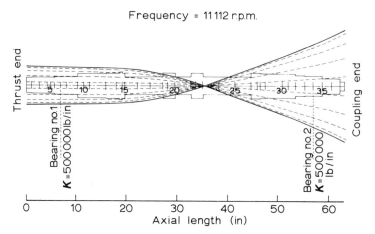

Frequency = 11 112 r.p.m.

Fig. 20.6   Simulation of compressor rotor mode shape 2 with oil seals removed.

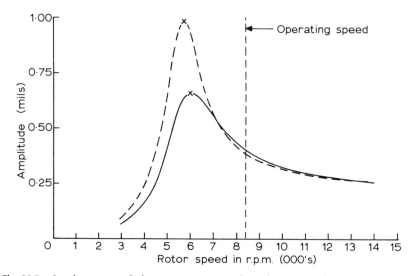

Fig. 20.7   Synchronous unbalance response analysis for a recycle gas compressor: synchronous amplitude versus speed at a given station. – – –, x-direction; ———, y-direction.

the damped natural frequency. These parameters are used to calculate the log decrement, which is the natural log of the ratio of two successive vibration amplitudes [3]. If the log decrement is negative, the system is unstable. Figure 20.10 illustrates a typical damped mode shape [3] at the first damped critical speed (eigenvalue), also referred to as an "eigenvector".

There are dynamic influences that reduce the stability of the rotor and bearing system. These influences are referred to as "cross-coupling coefficients", which originate from fixed-geometry journal bearings, seals,

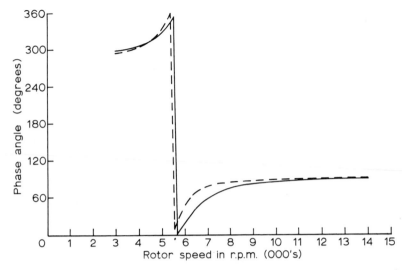

Fig. 20.8 Synchronous unbalance response analysis for a recycle gas compressor: phase angle versus speed at the station examined in Fig. 20.7. – – –, x-direction; ———, y-direction.

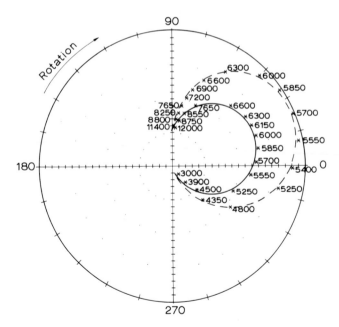

Fig. 20.9 Polar plot of amplitude versus phase for a given station. This is part of a synchronous response analysis used in compressor rotor dynamic studies. Full-scale amplitude = 1 mil; amplitude per division = 0.05 mil. – – –, x-direction; ———, y-direction.

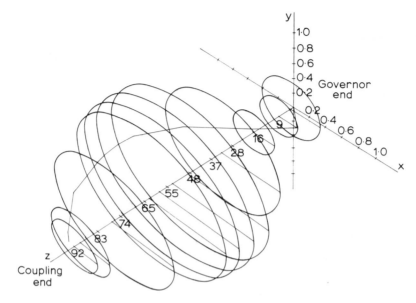

Fig. 20.10　Rotor stability analysis for a centrifugal compressor.

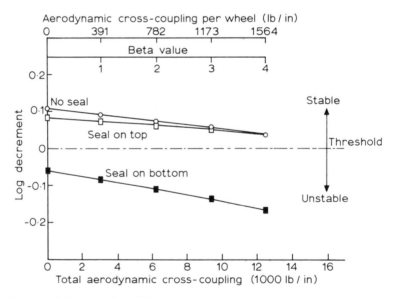

Fig. 20.11　Stability map plotted for a centrifugal compressor.

gas forces in compressor wheels, and steam forces at turbine stages. Figure 20.11 illustrates a typical "stability map" for a centrifugal compressor, which considers the rotor stability for various levels of aerodynamic cross-coupling. Also note that seal effects can be considered in various configurations. The point illustrated in Figure 20.11 is that as cross-coupling is increased, the rotor stability decreases. Therefore, this effect must be considered in a full analysis.

ACTION STEPS RESULTING FROM ROTORDYNAMIC AUDITS

A skilled analyst will have little difficulty studying the effects of changes in bearing geometry, bearing span, coupling mass, etc., on the vibration behavior or stability of the machine. With the powerful computational techniques available today, these alternatives can be assessed and designed before hardware is built. This capability allows plant managers to investigate positive improvements prior to critical shutdowns.

REFERENCES

1. Bloch, H. P., *Improving Machinery Reliability*. Houston: Gulf Publishing, 1982.
2. Salamone, D. J. and Gunter, E. J., Effects of shaft warp and disk skew on the synchronous unbalance response of a multimass flexible rotor in fluid film bearings. *Topics in Fluid Film Bearings and Rotor Bearing System Design and Optimization*, AMSE Book No. 100118, 1978.
3. Majovsky, B. and Salamone, D. J., Dynamic analysis of an 8,000 hp steam turbine operating near its second critical speed. *Proceedings of the Fifteenth Turbomachinery Symposium*. College Station, Texas: Turbomachinery Laboratories, Texas A&M University, 1986.
4. Salamone, D. J., Introduction to hydrodynamic journal bearings. *Vibration Institute Minicourse Notes – Machinery Vibration Monitoring and Analysis*. Clarendon Hill, Illinois: The Vibration Institute, 1985, pp. 41–56.
5. Salamone, D. J., Journal bearing design types and their applications to turbomachinery. *Proceedings of the Thirteenth Turbomachinery Symposium*. College Station, Texas: Turbomachinery Laboratoties, Texas A&M University, 1984, pp. 179–188.
6. Cerwinske, T. J., Nelson, W. E. and Salamone, D. J., Effects of high pressure oil seals on the rotor dynamic response of a centrifugal compressor. *Proceedings of the Fifteenth Turbomachinery Symposium*. College Station, Texas: Turbomachinery Laboratories, Texas A&M University, 1986.
7. Lund, J. W. and Orcutt, F. K., Calculations and experiments on the unbalance response of a flexible rotor. *Journal of Engineering for Industry*, Transactions of ASME, Nov. 1967.
8. Barrett, L. E., Gunter, E. J. and Allaire, P. E., Optimum bearing and support damping for unbalance response and stability of rotating machinery. *Journal of Engineering for Power*, Transactions of ASME, **100** (1), 1978, pp. 89–94.
9. Jackson, C. J., *The Practical Vibration Primer*. Houston: Gulf Publishing, 1979.
10. Lund, J. W., Stability and damped critical speeds of flexible rotor in fluid film bearings. *Journal of Engineering for Industry*, Transactions of ASME, pp. 509–517, May 1974.
11. Gunter, E. J., *Dynamic Stability of Rotor-Bearing Systems*. NASA.

21

Proposal analysis

In the earlier chapters we made the point that it is not the intention of the engineer who makes an assessment of the reliability or adequacy of a machine to actually check or calculate every feature embodied in its design. Not only would it be far too time-consuming to do so, but it would also be unnecessary to go into such detail with the overwhelming majority of machines in order to rank their respective standing among competing offers. We liken our "spot-check" process to a poll taker who needs only a small sampling of the public's opinion in order to determine whether or not the nation agrees that smoking is dangerous to one's health.

In assessing the adequacy or potential vulnerability of a special "single-source" offer, it would be similarly unnecessary to scrutinize every conceivable feature or component. The engineer engaged in machinery reliability assessment should keep in mind that the task could also be described as a screening effort. The assessment effort will be channeled in the right direction by the constant reminder that there are only seven causes of machinery failure:

1. Design deficiencies.
2. Material defects.
3. Processing and manufacturing deficiencies.
4. Assembly error.
5. Off-design or unintended service conditions.
6. Maintenance deficiencies (neglect, incorrect procedures).
7. Improper operation.

In addition, the review process must continually focus on the fact that the basic agents of machinery components and part failure mechanisms are *always* force, reactive environment, time, and temperature.

Having explained the principles of the reliability assessment process, we are now ready to apply the process to a specific machine, that described in US Patent No. 3,913,435 (Fig. 21.1). Moreover, we assume that this machine was offered to us in response to a request for quotations on equipment capable of trimming the edges of metal containers.

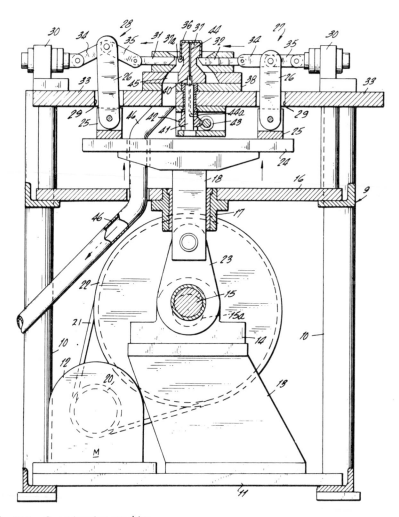

Fig. 21.1 Can trimming machine.

Since this is a pre-procurement screening effort, we look for potential design deficiencies and, using some surprisingly simple short-cut calculations, try to identify where unexpected or uncontrolled levels of force could be the potential failure mechanism. But first, a little more about this machine and its intended use.

BACKGROUND

In the fabrication of electronic enclosures commonly known as cans, one generally employs deep drawing processes. The sheet metal blanks are of sufficient length to leave a certain amount of overhang after drawing is completed. Finishing operations are needed to remove this overhang so as to obtain cans of proper length with clean, burr-free edges.

One of the fastest and possibly most economical methods of trimming is the adoption of sliding cutters which would remove excess stock by positive shearing action. In view of this, a manufacturer offers us a machine which utilizes reciprocating shear blades which modify the shape of the open ends of metal containers. The machine incorporates a mandrel (item 37) for supporting the container or enclosure to be trimmed and this mandrel is executed with shearing edges adjacent to the open end of the container. Eight cutting blades are equally spaced around the mandrel and toggle linkages coupled to each of the cutting blades and the toggle links are operated by reciprocating links secured to a platen. A power transmission unit, including a motor and single-revolution clutch-brake mechanism, is connected to the platen for first moving it in one direction and then in the opposite direction. As the platen moves upward from its idle/rest position, half of the cutting blades move into and out of a shearing position; conversely, as the platen moves downward from its idle/rest position, the other half of the cutting blades move into and out of a shearing position. This double movement, which is accomplished in a single machine cycle, completely trims and shapes the open end of the container.

DESIGN OBJECTIVES VERSUS RELIABILITY ASSESSMENT OBJECTIVES

It is important to recognize that the *reliability assessment* objectives are not as broad and detailed as the *design* objectives for a process machine. In this particular instance, we can assume that the *design* objectives of the machine designer were:

1. To investigate the feasibility of building a trimming machine with toggle-operated sliding cutters.
2. To determine the power requirements to perform trimming operations on steel cans of sizes ranging up to $3 \times 3''$ and $\frac{1}{16}''$ wall thickness.
3. To calculate forces acting and stresses set up.
4. To determine deflections.
5. To determine physical dimensions of machine components.
6. To suggest practical design features and configuration of a toggle acting trimming machine.

The reliability assessment objectives would be somewhat narrower. To the extent deemed appropriate in a screening study, the task seen by the reviewing engineer would be:

1. To calculate forces acting and stresses set up in critical linkages.
2. To determine deflections of selected components.
3. To establish the power needed for trimming operations on steel cans ranging in size to $3 \times 3''$ $(76 \times 76\,mm)$ and 0.062 in (1.5 mm) metal thickness.

The reviewer now has the option of asking the equipment vendor to disclose his design assumptions and calculations. Whenever these data are not available, the reviewer will make his own assumptions and often quite elementary calculations. Dimensions may be scaled from drawings if not actually provided by the equipment manufacturer.

As explained earlier, we know that four cutters will be simultaneously engaged in cutting action. Maximum length of material to be sheared by four side cutters $(L) = 8''$; maximum thickness of material $(T) = \frac{1}{16}''$; average shear strength of material $(S) = 50,000$ psi; intentional overdesign in view of the possibility of using dull cutters and excessive bearing friction $=$ a multiplication factor of 3. Therefore, the total force required is

$$P = 3SLT$$

$$P = 3 \times 50,000 \times 8 \times 1/16 = 75,000\,lb$$

The maximum probable toggle displacement at point of contact between cutters and sheet metal $= 10°$ (Fig. 21.2):

$$F \cos \alpha = 2P \sin \alpha$$

$$F = 2P \tan \alpha$$

$$= 26,000\,lb$$

Experimental evidence points to a material being sheared completely after the cutters have penetrated one-third of its thickness. We may consider it an additional factor of safety not to have taken this into consideration when

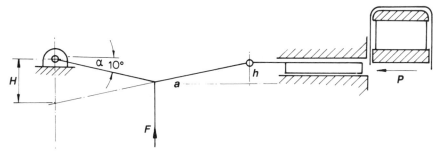

Fig. 21.2 Toggle displacement.

calculating P total. Hence, our reliability assessment will be based on the total input force requirement of $F = 26,000$, as calculated above. This force has to travel through a distance of 0.56 in in the proposed machine, resulting in a torque of 14,600 lb-in on the output side of the drive module which will rotate at 175 rpm. Using a 1750-rpm input motor and a 10:1 gear-reducing unit assumed to be 90% efficient, an input torque of 1622 lb-in is required. Input horsepower:

$$\text{hp} = \frac{1750 \text{ rpm} \times 1622 \text{ lb-in}}{63,025 \text{ lb-in/hp}} = 45 \text{ hp}$$

However, in reviewing the design, we note that the manufacturer of this trimming machine plans to provide only a 1.5-hp electric motor in conjunction with a flywheel which is mounted on the input shaft of the gear speed reducer. To determine whether this is satisfactory, the strength and reliability appraisal effort must first be directed towards calculating the forces which exist in toggle action. The next step would be to assess the adequacy of the flywheel proposed by the machinery manufacturer.

Calculating forces in toggle action (Fig. 21.3)
Width of cut $(L) = 2''$
Thickness of material $(T) = 0.0625''$
Shearing stress $(S) = 50,000$ psi
P = Force applied to face of cutter

$$P = SLT = (50,000)(2)(0.0625) = 6250 \tag{21.1}$$

$$\overset{\curvearrowright}{+} \sum M_A = 0$$

$$Fl \cos \theta - N(2l \cos \theta) = 0$$

$$F = 2N \tag{21.2}$$

$$+\uparrow \sum F_y = 0$$

$$A_v + N - F = 0$$

$$A_v - N = 0$$

$$A_v = N \tag{21.3}$$

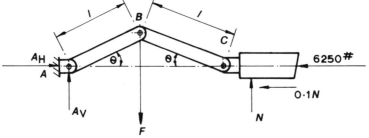

Fig. 21.3 Toggle action and forces.

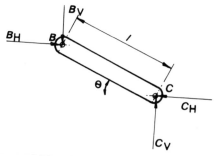

Fig. 21.4 Forces on toggle arm.

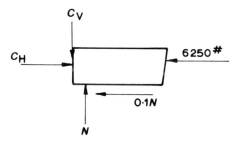

Fig. 21.5 Forces on cutter.

$$+\rightarrow \sum F_x = 0$$

$$A_H - 0.1N - 6250 = 0 \qquad (21.4)$$

$$\sum M_c = 0 \quad \text{(Fig. 21.4)}$$

$$B_H l \sin\theta - B_v l \cos\theta = 0$$

$$B_H \tan\theta = B_v$$

$$C_H \tan\theta = C_v \qquad (21.5)$$

$$\sum F_x = 0 \quad \text{(Fig. 21.5)}$$

$$C_H - 0.1N - 6250 = 0 \qquad (21.6)$$

$$\sum F_y = 0$$

$$N - C_v = 0 \qquad (21.7)$$

Using equations (21.5) and (21.7):

$$C_H \tan\theta = N$$

$$C_H - 0.1N = 6250$$

$$N\left[\frac{1}{\tan\theta}\right] - 0.1 = 6250 \qquad (21.8)$$

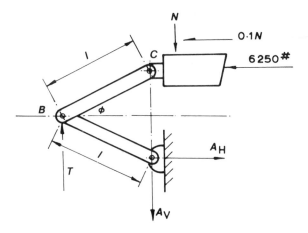

Fig. 21.6   Bearing forces.

$$F = 2N = \frac{12{,}500 \tan \theta \text{ lb}}{1 - 0.1 \tan \theta} \tag{21.9}$$

For the anticipated maximum value of $\theta = 10°$, $\tan \theta = 0.176$:

$$F = \frac{12{,}500(0.176)}{1 - (0.1)(0.176)} = \frac{12{,}500(0.176)}{0.9824} = F = 2240 \text{ lb}$$

This is the maximum force needed to cut the can. Since there are four cutter assemblies of this type operating simultaneously, the total force on the hinge plate is 8960 lb. From Figure 21.6, we note that the cutter is in such a position that the line of action of N is in line with bearing A.

$$+\uparrow \sum F_y = 0$$

$$T - N - A_v = 0 \tag{21.10}$$

$$+\rightarrow \sum F_x = C$$

$$A_H - 0.1N - 6250 = 0 \tag{21.11}$$

$$\curvearrowright \sum M_A = 0$$

$$Tl \cos \phi - 0.1N \times 2l \sin \phi - 6250 \times 2l \sin \phi = 0 \tag{21.12}$$

Figures 21.7 and 21.8 allow us to visualize the action of forces in the horizontal and vertical directions:

$$\vec{+} \sum F_x = 0$$

$$B_H = C_H = 0$$

$$\underset{+}{\uparrow} \sum F_y = 0$$

$$B_v - C_v = 0$$

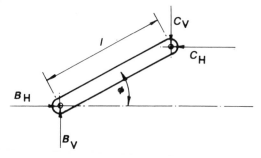

Fig. 21.7 Hinge plate forces.

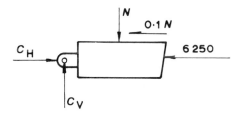

Fig. 21.8 Cutter detail and forces.

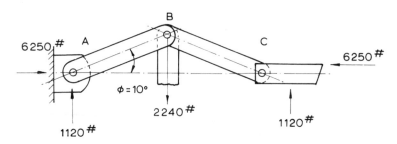

Fig. 21.9 Yoke linkage.

We use Figure 21.9 to indicate summation of moments:

$$\stackrel{\curvearrowleft}{+} \sum M_B = 0$$

$$C_v l \cos \phi - C_H l \sin \phi = 0$$

$$C_v = C_H \tan \phi \qquad (21.13)$$

$$\stackrel{\rightarrow}{+} \sum F_x = 0$$

$$C_H - 0.1N - 6250 = 0 \qquad (21.14)$$

Comparing this equation to equation (21.11) we can seen that $C_H = A_H$.

$$\uparrow + \sum F_y = 0$$

$$C_v - N = 0$$

$$C_v = N = C_H \tan \phi$$

$$C_H = \frac{N}{\tan \phi}$$

$$\frac{N}{\tan \phi} - 0.1N - 6250 = 0 \tag{21.15}$$

$$N\left(\frac{1}{\tan \phi} - 0.1\right) = 6250$$

$$T = N + A_v \tag{21.16}$$

$$Tl \cos \phi - 2l \sin \phi(6250 + 0.1N) = 0 \tag{21.17}$$

$$T = 2 \tan \phi(6250 + 0.1N)$$

From equation (21.14) we see that $6250 + 0.1N = N/\tan \phi$.

$$T = 2N$$

$$T = \frac{12,500 \tan \phi}{1 - 0.1 \tan \phi} \tag{21.18}$$

Thus $T = F$ and the force on the hinge plate is the same for both halves of the cutting cycle.

STRESS ANALYSIS OF YOKE LINKAGE

The preceding equations and resulting force values lead to a stress analysis of the principal linkages in the trimming machine. Figure 21.9 helps us to visualize the process.

Combined force acting on pin A = force along axes of links

$$= \sqrt{(6250)^2 + (1126)^2} = 6350 \text{ lb}$$

This force of 6350 lb will be acting where indicated (A, C, AB, BC):

Total force on pin B $= \sqrt{(6250)^2 + (2240)^2} = 6650 \text{ lb}$

In Figure 21.10, the machine manufacturer shows a cross-sectional area at plate Y-Y $= 0.53 \text{ in}^2$ and will provide a steel with ultimate tensile and

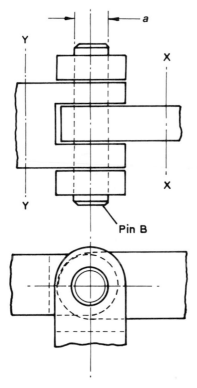

Fig. 21.10 Linkage detail.

compression strengths of 60,000 psi. Therefore, factor of safety (F.S.):

$$F.S. = \frac{\text{cross-sectional area} \times \sigma_{ut}}{\text{force applied}} = \frac{(0.53)(60,000)}{6350} = 5$$

This would be considered an adequate factor of safety, although shock loading and recurring fluctuating stresses are likely to be encountered by the linkage.

Testing for shear failure across section "BB" (Fig. 21.11): The manufacturer proposes to use a steel pin with $d = \frac{3}{8}''$ (0.375"). We note that the total area resisting shear is twice the area of the pin at section "BB". Hence, the factor of safety:

$$F.S. = \frac{(2)(\text{cross-sectional area of pin})(\text{allowable shear stress})}{\text{force applied to pin}}$$

$$= \frac{(2)(\pi)(0.375)^2(45,000 \text{ psi})}{(4)(6650 \text{ lb})} = 1.5$$

This appears to be a rather low factor of safety. We recalculate on the basis

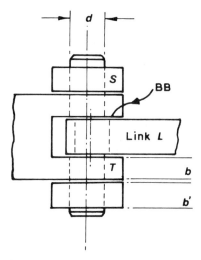

Fig. 21.11   Linkage detail.

of a high alloy tool steel with an allowable shear stress of 150,000 psi:

$$\text{F.S.} = \frac{(2)(\pi)(0.375)^2(150{,}000 \text{ psi})}{(4)(6650 \text{ lb})} = 5$$

and make a note to alert the machine manufacturer to the fact that a high alloy tool steel pin will be required for a reliable design.

Next, we test for the factor of safety associated with the bearing stress between pin and link. Using Figures 21.12–21.14:

$$\text{F.S.} = \frac{(\text{end-area of link})(\text{allowable tensile stress})}{(\text{force applied at extremity of link})}$$

$$= \frac{(m-d)(c)(200{,}000 \text{ psi})}{6350 \text{ lb}} = \frac{(0.16 \text{ in}^2)(200{,}000 \text{ psi})}{6350 \text{ lb}} = 5$$

Note that we have made the decision to use high-strength tool steel for this important component. This fact will have to be discussed with the machinery manufacturer in order to verify concurrence and compliance.

The projected area between pin and yoke is scaled from the manufacturer's proposal drawings: $A = (b)(a) = 0.08 \text{ in}^2$

$$\text{F.S.} = \frac{(2)(200{,}000)(0.08)}{6350} = 5$$

obtained by using high-strength tool steel.

The possibility of the shear failure shown on the sketch must be considered. Shear exists on a length e and a depth c on both sides of the pin. Since by now it appears that the machine manufacturer is often using a factor of safety

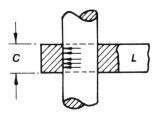

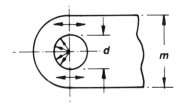

Fig. 21.12 Hinge detail.

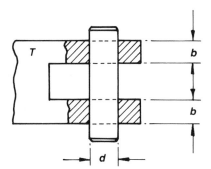

Fig. 21.13 Hinge detail.

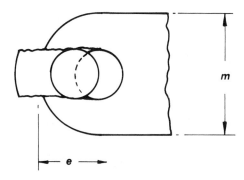

Fig. 21.14 Hinge plate detail.

of 5, we may now opt to rearrange the standard equation and solve for shear stress:

$$\text{Shear stress} = \frac{(\text{F.S.})(\text{force})}{\text{area } 2ce}$$

$$= \frac{(5)(6350 \text{ lb})}{(2)(0.10)} = 158,750 \text{ psi}$$

This would again point towards the need to make this link out of high-strength tool steel.

As described earlier, the manufacturer proposes to utilize a 1.5-hp motor and flywheel combination to power this edge-trimming machine. Our screening study is now directed to this drive combination.

We note that the proposed flywheel is solid steel of 10 in diameter and $1\frac{3}{16}$ in thickness. We also note that on the first cutting stroke, each of the four cutters would exert a force of 6250 lb on the container and would move through 0.0625 in of steel. With the four cutters moving in unison, the total energy requirement is:

$$(4)(6250)(0.0625) = 1562 \text{ lb-in} \qquad \text{or} \qquad 130 \text{ lb-ft}$$

This calculated energy requirement would include a considerable factor of safety because, as stated earlier, shearing action ceases whenever the cutter has penetrated one-third the distance through the steel to be sheared and no work is done during the remaining two-thirds of the stroke.

The electric motor by itself provides some of the required energy while the flywheel must supply the remainder.

$$\text{Torque} = \frac{63,025 \text{ hp}}{\text{rpm}} = \frac{(63,025)(1.5)}{1750} = 54 \text{ lb-in}$$

$$= 4.5 \text{ lb-ft}$$

This is an almost negligible amount and we will have to depend on the apparently small flywheel to supply virtually all of the required energy.

$$E = 130 \text{ lb-ft} = -\frac{I}{2}(\omega_f^2 - \omega_i^2) \tag{21.19}$$

where ω_f and ω_i are the final and initial angular velocities, respectively.

$$I = \frac{1}{2}mr^2, \qquad m = \frac{w}{g} = \frac{(0.286)(\pi)(D^2)(t)}{48}$$

where D and t are the diameter and thickness of the solid steel flywheel.

$$I = \frac{1}{2}\left[\frac{(0.286)(\pi)(100)(1.187)}{(4)(32.2)}\right]\left(\frac{5}{12}\right)^2$$

$$I = 0.0719 \text{ slug-ft}^2$$

Using this result in equation (21.14), we obtain:

$$\frac{(2)(130)}{0.0719} = -(\omega_f^2 - \omega_i^2) = -\omega_f^2 + 33{,}584$$

and

$$\omega_f = [33{,}584 - 3616]^{1/2} = 173 \text{ rad/sec} = 1653 \text{ rpm}$$

This end result tells us that the machine manufacturer chose a motor–flywheel combination which will produce the required power output and, in the process of one rotation, will decay from an instantaneous speed of 1750 rpm to a speed of 1653 rpm. He will have to select an electric motor which will re-accelerate during the next revolution to restore the 1750-rpm speed.

We could, of course, also investigate the strength and acceptability of the clutch-brake device and transmission belts. However, experience tells us that here we can generally depend on the rating charts published by the manufacturers of these mass-produced components. Using the proper service factor in selecting these drive elements will give us additional assurance of the adequacy of the vendor's selection.

Finally, in a competitive bidding situation we would probably make similar screening studies for the range of components, linkages, pins, bearings, etc., offered by the various bidders. This effort could be tabulated and conclusions drawn from the bid comparison as regards the respective strength, vulnerability, maintainability, and reliability rankings of several competing executions. As a concluding reminder, the possible influence of FRETT – Force, Reactive Environment, Time, and Temperature – must be kept in mind during every phase of the appraisal or bid comparison process.

22

Selection process for machinery

A formalized process can make the selection from a large number of alternative and competing proposals easier. In principle, such a selection process should be gone through after each work step during which alternatives appear. One would only pursue items with the following characteristics:

- *Compatible* with our objectives and general goals.
- *Fulfilling the demands* made by our scope of requirements (Fig. 22.1).
- *Allowing to recognize feasibility* with regard to basic features like range, size, etc.
- *Allowing to expect an appropriate effort.*

If there are many remaining alternatives, a preferred choice may be justified by fulfillment of such basic requirements as meeting safety standards, manageable in-house with demonstrated capabilities, materials and/or work processes [1].

THE ASSESSMENT PROCESS

For a more accurate judgment of solutions which have to be pursued after having gone through a basic selection process, we enter into a formalized assessment process. This assessment process has to determine the value of a particular design solution with respect to predefined objectives. Technical as well as economic viewpoints have to be considered at this point. Several methods exist [2, 3]. The general steps of this assessment process are as follows:

Recognition of assessment criteria

A general goal as a rule comprises several objectives. Assessment criteria are directly related to them. In the following we will call them "factors". Factors must be formulated in a positive way, that is "low noise level" instead of

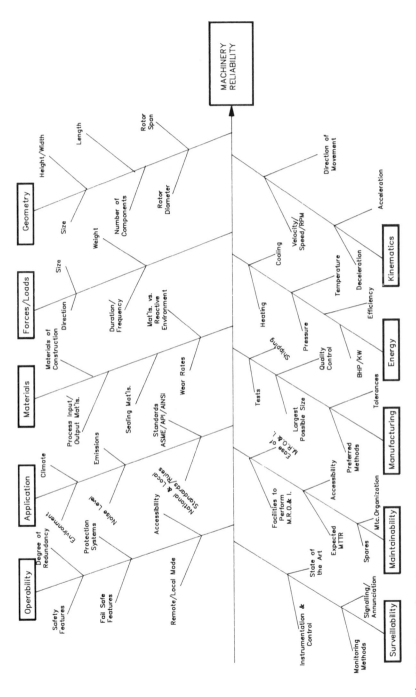

Fig. 22.1 Scope of requirements for process machinery reliability.

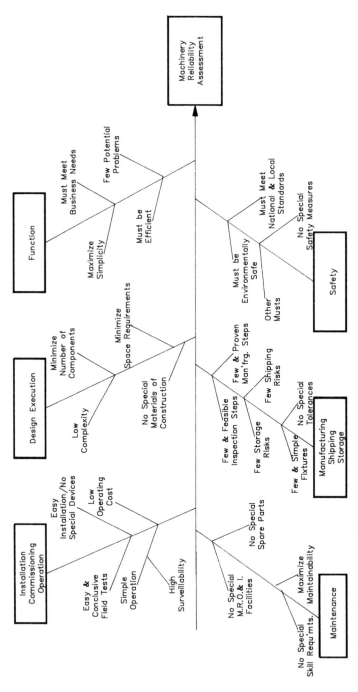

Fig. 22.2 Factors for process machinery assessment.

276

Table 22.1 Machinery assessment and selection list

No. (i)	Factors	Priority number (p_i)	Parameter	Units	Alternative A_1			Alternative A_2		
					Quality (q_{i1})	Grade (g_{i1})	Weighted value ($p_i g_{i1}$)	Quality (q_{i2})	Grade (g_{i2})	Weighted value ($p_i g_{i2}$)
1	Reliability	10	Expected MTBF	Years	2.0	71.0	710.0	7.0	85.0	850.0
2	Cost	9	Project cost	k$	700.0	100.0	900.0	886.1	79.0	711.0
3	Maintainability	8	Expected MTTR/ease of maintenance	Hours	Complex	50.0	400.0	Medium	75.0	600.0
4	Operability	7	Expected degree of operator attention	No. of operations	High	50.0	350.0	Low	100.0	700.0
...		...	...	...	...	...	...	...	...	...
i	......	p_i	......	......	q_{i1}	g_{i1}	pg_{i1}	q_{i2}	g_{i2}	pg_{i2}
...		......			......	......	......	......	......	......
n		p_n			q_{n1}	g_{n1}	pg_{n1}	q_{n2}	g_{n2}	pg_{n2}

"loud". Factors may be divided into minimum requirements (musts) and other requirements (wants) (Fig. 22.2). The assessment criteria must be independent from each other so duplication may be avoided.

The factors are then listed in a way similar to Table 22.1, where each factor in turn receives a weight priority number. The actual assessment is accomplished by assigning grade numbers to each factor by columns that represent different alternatives under consideration. Grade numbers reflect our understanding of relative values and consequently tend to be subjective. It would be well, therefore, to allow a group of different individuals to participate in the grading. Grading should always be performed by proceeding from factor to factor for all alternatives, that is row by row.

All factors, or technical requirements, in our selection process are prioritized on a scale of 1–10; the most important factor receives a 10. Similarly, our grade scale ranges from a poor fit = 0 to a very good fit = 100. One particular weighted factor analysis procedure then performs the following steps [4].

For one possible alternative solution:

1. For all factors, the grade of the alternatives is multiplied by the factor's priority.
2. The grade–priority product of all the factors are needed.

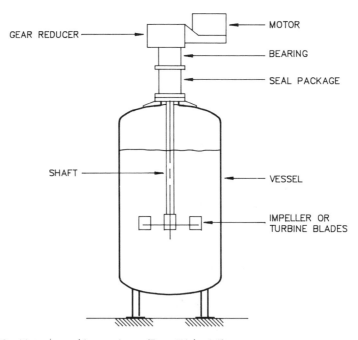

Fig. 22.3 Motor/gear driven agitator (Type "Lightnin").

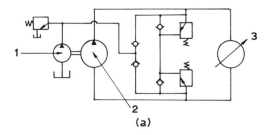

(a)

(b)

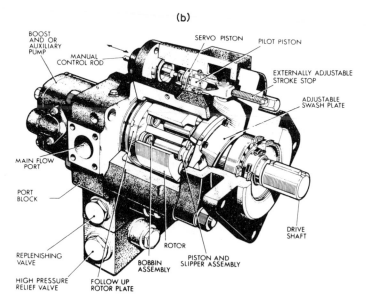

BOOST
AND OR
AUXILIARY
PUMP

MANUAL
CONTROL ROD

SERVO PISTON   PILOT PISTON

EXTERNALLY ADJUSTABLE
STROKE STOP

ADJUSTABLE
SWASH PLATE

MAIN FLOW
PORT

PORT
BLOCK

DRIVE
SHAFT

REPLENISHING
VALVE

ROTOR

PISTON AND
SLIPPER ASSEMBLY

BOBBIN
ASSEMBLY

HIGH PRESSURE
RELIEF VALVE

FOLLOW UP
ROTOR PLATE

Fig. 22.4 (a) Fluid power drive schematic. 1 and 2, Constant capacity pump; 3, variable capacity motor. (b) Variable-capacity axial-piston pump/motor (after ref. 5). (Reprinted from Warring, R. H., *Hydraulic Handbook*, 8th Edition, 1983, p.320, Fig. 1, by courtesy of Trade & Technical Press Ltd.)

3. The priorities of all the factors are added together.
4. The sum of the priorities is divided into the sum of the grade–priority products.

The assessment value V_j is:

$$V_j = \frac{\sum\limits_{i=1}^{n} p_i g_{ij}}{\sum\limits_{i=1}^{n} p_i}$$ (22.1)

Table 22.2   Description of alternative solutions for variable speed agitator drive

Factor	Gearbox options	Hydrostatic option	Motor/pump sets
Reliability	MTBF = 2 years	MTBF = 7 years "prototype"	MTBF = 8.2 years proven technology
Cost of variable speed conversion	435.4 k$	289 k$	217 k$
Cost of mtce. and production losses	49.1 k$ p.a.	10.0 k$ p.a.	10 k$ p.a.
	Lifts/supports	Troubleshooting difficult	Easy
Ease of mtce.	2 alignments	1 alignment	1 alignment
Operator attention	10 oil levels	1 oil level	10 oil levels
Accuracy/speed control	± 3 rpm	± 1 rpm	± 1 rpm
User experience	Like here – not great	Very good – all satisfied	Very good
Reactor room noise	Approx. 90 dBA, exceeds OHSA[a] limits, 80 k$ for noise abatement	Less than 68 dBA (approx. 85 dBA + in pump building)	In separate room, but could be noisier
Vendor support	Historically long lead times on parts	Most items are off the shelf	Most items are off the shelf
Energy costs	33.0 k$ p.a.	78.1 k$ p.a.	< 78 k$ p.a. due to higher efficiency
Initial costs	250 k$	900 k$	900 k$
Engineering effort	Need to solve "cantilever" situation	Need to optimize component selection	"Write purchase order"
Ease of installation	All work to be done in shutdown	Most work can be done before S/D	In between
Back-up power	Electric generator only	Can use electric generator or other direct driver	Electrical

[a] OHSA = Occupational Health & Safety Act (Canada).

Comparison of alternatives

Once the assessment value for each alternative has become available one can start comparing alternatives. It would be well to determine technical assessment values and economic values separately. The latter should take

Table 22.3 Weighted factor assessment and selection of variable speed agitator drive

	Analysis: Reactor agitator drives
Result	Alternative
65.8	Gear box
77.5	Ring main hydrostatic
79.8	Individual M/P set hydros
Priority	factor
10.0	Reliability
9.0	Cost of variable speed
8.0	Maintainability
7.0	Operability
6.0	Reactor room noise
5.0	Energy costs
4.0	First cost
3.0	Engineering effort
2.0	Ease of installation
1.0	Back-up power supply

Factor: Reliability

Grade	Alternative
71.0	Gear box
85.0	Ring main hydrostatic
100.0	Individual M/P set hydros

Factor: Cost of variable speed

Grade	Alternative
100.0	Gear box
79.0	Ring main hydrostatic
86.0	Individual M/P set hydros

Factor: Maintainability

Grade	Alternative
50.0	Gear box
75.0	Ring main hydrostatic
100.0	Individual M/P set hydros

Factor: Operability

Grade	Alternative
50.0	Gear box
100.0	Ring main hydrostatic
50.0	Individual M/P set hydros

Factor: Reactor room noise

Grade	Alternative
10.0	Gear box
100.0	Ring main hydrostatic
100.0	Individual M/P set hydros

Factor: Energy costs

Grade	Alternative
100.0	Gear box
42.0	Ring main hydrostatic
45.0	Individual M/P set hydros

Factor: First cost

Grade	Alternative
100.0	Gear box
28.0	Ring main hydrostatic
28.0	Individual M/P set hydros

Factor: Engineering effort

Grade	Alternative
50.0	Gear box
60.0	Ring main hydrostatic
100.0	Individual M/P set hydros

Factor: Ease of installation

Grade	Alternative
50.0	Gear box
100.0	Ring main hydrostatic
90.0	Individual M/P set hydros

Factor: Back-up power supply

Grade	Alternative
50.0	Gear box
100.0	Ring main hydrostatic
50.0	Individual M/P set hydros

life-cycle costs into account, provided that prices and costs are known. Before one makes a decision as to final selection one should review individual values with regard to uncertainties and unjustified assumptions. Large differences in individual weighted values indicate problem areas and should be thoroughly investigated.

AN EXAMPLE

A process plant using a number of fixed speed motor gear driven agitators (Fig. 22.3), was experiencing reliability problems with them. The technical staff identified an opportunity to increase process reliability as well as product quality by converting the fixed speed agitator gear drives to variable speed hydraulic drives. Two different solutions were offered. One, individual hydraulic motors driving the agitators via a hydraulic ring line that was to be fed and pressurized by a central remote pumping facility. Another design was to furnish individual pumping units for each agitator drive, in close proximity, similar to Figure 22.4. The alternatives offered are described in Table 22.2. Table 22.3 shows the weighted factor analysis results.

REFERENCES

1. Pahl, G. and Beitz, W., *Konstruktionslehre*. Berlin: Springer, 1977.
2. VDI guideline 2225.
3. Kepner, C. H. and Tregoe, B. B., *The Rational Manager*. New York: McGraw-Hill, 1965, pp. 173–206.
4. Cox, J. G., The oracle. *Nibble Magazine*, **4** (5), 1983, pp. 21–31.
5. Warring, R. H., *Hydraulic Handbook*. Morden, Surrey, U.K.: Trade & Technical Press Ltd., 1983.

Appendix A

The coin toss case

The coin toss case follows a "binomial" probability distribution. This distribution can be described by the general function,

$$f(x) = C_x^n p^x q^{n-x}$$

where
- n = number of trials (i.e. six tosses of a coin in our case),
- x = number of successes per trial (i.e. $0 \cdots 6$),
- p = the probability of success per trial (i.e. 50% in our example),
- q = the probability of failure (i.e. 50% or $q = 1 - q$),

$$C_x^n = \frac{n!}{x!(n-x!)}$$

Example. What is the probability of getting 2 heads in 6 throws of a coin?
Solution

$$n = 6$$

$$x = 2$$

$$p = 0.5$$

$$q = 0.5$$

$$C_x^n = \frac{6!}{2!(6-2)!} = \frac{6!}{2!\,4!} = \frac{1 \times 2 \times 3 \times 5 \times 6}{1 \times 2 \times 1 \times 2 \times 3 \times 4} = \frac{720}{2 \times 24} = 15$$

$$P_{head} = 15 \times 0.5^2 \times 0.5^4 = 15 \times 0.25 \times 0.0625 = 0.23 = \frac{15}{64} \quad \text{or} \quad 23\%$$

The binomial function can have many useful applications. One of the most common applications is in lot-by-lot acceptance sampling inspections, where the decision is made to accept or reject an entire lot of parts based on the numbers of defectives found in a sample of these parts.

Appendix B

Mechanical operation and construction of variable inlet guide vanes

The purpose of this section is to explain the mechanical operation and construction of variable inlet guide vanes. Automatic control is accomplished by means of a power cylinder and an automatic controller.

DESCRIPTION AND OPERATION

To best understand the guide vane operation, consider Figure 6.3, the adjustable inlet guide vane assembly. Note also the top of Section B-B.

Torque to move the guide vanes is transmitted along upper drive shaft (1) to the pulley (2), which is keyed to the shaft (see item 3, section A-A) and secured by a cap screw and two drive pins. Guide vane (4) is also secured to shaft (1) and will turn with the pulleys and shaft. Shaft (1) revolves on a set of anti-friction ball bearings (5).

Consider now Section A-A to see how the motion of the drive pulley (2) is transmitted to the other pulleys. Each pulley is connected to adjoining pulleys by a set of two short lengths of $\frac{1}{8}''$ diameter aircraft type cables. Thus pulley (2) is connected to both pulleys (6) and (8). End pulleys (6) and (7) are only connected to one other pulley, as illustrated by the developed view. The cables lie along grooves on the pulleys and are secured by elastic stop nuts. This type of pulley and cable construction removes any danger of slip or backlash. Thus movement in drive pulley (2) causes an equal turning in the same direction of all the other pulleys in the upper half of the guide vanes. The operation of the lower half of the pulleys, as driven by lower drive shaft (10), is identical to the above. Due to the external linkage not described here, both top and bottom halves work as a unit.

Although not shown in our illustration, variable inlet guide vanes are often used at the inlet passage of high-efficiency, capacity-controlled process gas compressors similar to the one depicted in Fig. B.1.

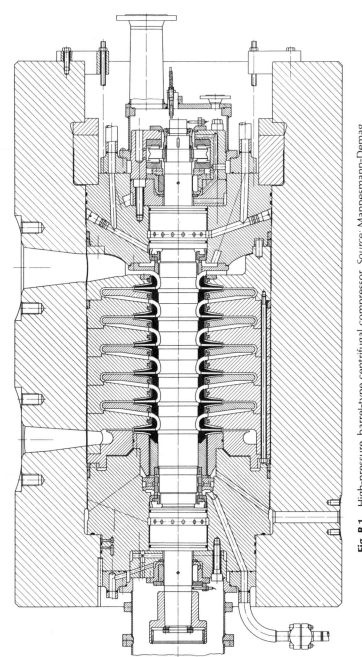

Fig. B.1 High-pressure, barrel-type centrifugal compressor. *Source:* Mannesmann-Demag, Duisburg, West Germany.

Appendix C

Machinery system completeness and reliability appraisal forms

Project Phase: P & ID Specification Review
Machine Category: Centrifugal Compressors

Designation: _____

Location: _____

Service: _____

1. Is there an adequate liquid removal system for compressor suction to each stage?

_____
(Yes or No)

The suction drums must be sized to prevent liquid carry-over that could damage the machine. Worst case operating conditions should be considered. For example: highest suction pressure, highest flow, or two machines operating in parallel using the same suction drum should be considered in drum sizing. Use of crinkled wire mesh in drum outlet (appropriately designed, of course) and tangentially entering inlet nozzle to drum should be considered to provide the most effective liquid separation. A tangentially entering inlet nozzle will effectively increase drum flow capacity at idential inlet conditions by about 40%. This may provide a simple upgrade to increase capacity of existing drums equipped with an inlet nozzle that is perpendicular to drum walls.

Review suction piping layout to be certain that no liquid traps exist between suction drum and compressor. Suction lines must be sloped to drain either to compressor or to suction drum and not be flat. Liquid slugging and compressor damage can occur as a result of liquid trapped in these low points.

2. Is there a gauge glass, LHA alarm, and LHA trip on this drum? _____
(Yes or No)

This instrumentation is extremely critical; catastrophic compressor failure and possible personnel injury can occur if the machine ingests significant amounts of liquid. Failure typically initiates in a thrust bearing and can cause internal rubbing

286

of rotor wheels and diaphragms. Sufficient separation between alarm and trip should be provided so that operators have time to respond to an increase in level before a trip occurs.

3. Is the suction line such a length and size that it can be thoroughly cleaned by water washing? _____

(Yes or No)

If not, what method do you propose?

Construction debris (hats, bolts, welding rods) can be inadvertently left in suction piping. Additionally, long-term storage of piping outdoors can result in dirt, sand, and rust in suction piping interior. Such material will likely be ingested by the machine during operation causing imbalance or damage to components. Close clearance parts are most vulnerable (seals, etc.). As a result suction piping should be thoroughly cleaned and inspected prior to start-up. Visual inspection and removal of larger debris is a mandatory first step for any system cleaning procedure. Piping should be designed to facilitate cleaning.

One relatively inexpensive method of cleaning small debris such as rust, scale, and dirt is to hydroblast with high-pressure water. There are practical limits to lengths of line that can be hydroblasted and limits in piping configurations (hydroblasting typically cannot go through a side outlet "T", for example). Adequate drain piping to remove water after hydroblasting must also be provided. Note that some piping materials may be sensitive to chlorides and other contaminants in water, thus requiring selection of an acceptable source of water.

Another suitable method of cleaning is chemical cleaning. This method is particularly effective for corrosion products and in removing oil-based preservatives. This method, however, is not as effective as hydroblasting in removing larger objects because it often utilizes low-velocity circulation of cleaning solutions.

Access for internal inspection and cleaning by the method selected will likely require additional connections that are for cleaning only (blind flanges at ends of long runs of piping, for example).

4. Does a heat loss calculation indicate that the suction line needs to be traced? If so, is it traced? (allowable condensate is 2% by weight)

_____

(Yes or No)

In general, gases that have no potential for formation of liquids over the range of atmospheric temperatures do not require heat tracing. The ingestion of liquids in excess of 2% by weight of total flow, however, can cause deterioration of thrust bearings and labyrinth seals.

5. If the flow control is by suction butterfly valve, have minimum stops been provided?

_____

(Yes or No)

A centrifugal compressor will surge if the suction butterfly valve is closed. This can damage the compressor. A mechanical stop should be provided. This stop prevents the valve from closing beyond a point that would reduce flow so as to cause surge.

6. Has a discharge check valve been provided?

_____

(Yes or No)

The discharge check valve prevents process gas from flowing backwards through a machine if the machine trips or is shut down. This reverse flow can cause a machine to overspeed, damaging compressor and driver. Bearing damage is also likely. Typically, bearings are designed for one direction of rotation. If a shaft-driven oil pump is used, reverse rotation may occur without lube oil being supplied, leading to more severe damage to lubricated components.

7. Does the antisurge protection come from upstream of a discharge check valve?

_____

(Yes or No)

Compressors may not develop sufficient differential pressure during start-up or process upsets to open discharge check valves. Antisurge protection should be upstream of check valve so that flow can be maintained through a machine, via an antisurge loop, when it cannot discharge into normal system. Lower than design molecular weight gas, for example, could prevent a machine from developing enough differential pressure to open the discharge check valve.

8. Is the antisurge system sized so that it is large enough to prevent compressor surging?

_____

(Yes or No)

Compressor surge systems must be sized so as to prevent compressor surging even if normal process flow is blocked. Vendor-predicted surge flows may not be high enough to use in system sizing. It is suggested that vendor surge flows be increased by at least 50% for surge system sizing purposes. Variable speed machines should have surge systems sized for the highest speed surge flow plus a minimum of 50%.

9. If the antisurge system returns to the suction of the compressor, is it cooled?

_____

(Yes or No)

Does it return upstream of the suction knockout facilities?

_____

(Yes or No)

Compressor operation with gas being continuously recirculated may result in high gas temperatures. Driver horsepower input will essentially be converted into heat energy added to the circulating gas. Higher than design operating temperatures for the compressor can occur in a relatively short period of time when all gas is recirculated. These high temperatures can damage labyrinth seals, and cause rubbing (due to rotor thermal growth relative to the casing).

Liquid that is formed at compressor discharge conditions will be returned directly to the compressor suction and could cause damage. Recirculated gas must first be routed through suction knockout facilities to prevent this problem.

10. Does the system need a discharge safety valve?   _____

(Yes or No)

Consideration should be given to including discharge safety valves on compressors if they can develop discharge pressures in excess of design pressures of downstream equipment. High compressor suction pressure, operation of compressor near surge or at maximum speed, and higher than normal molecular weights are all examples of conditions that will cause higher discharge pressures that may necessitate the use of a relief valve.

11. If the compressor trips out, how high can the suction pressure go even if the discharge check valve does not close tight? Has this pressure been specified?

_____

(Yes or No)

Assume the suction valves are automatic close, isolation-type.

A relief valve or other system design changes may be required to protect upstream equipment if the predicted "settling out" pressure exceeds the pressure rating of piping, vessels, or valves. Also, the compressor suction nozzle rating may be exceeded if suction pressures rise too high. System designers must be aware of compressor casing limitations when selecting pressure relief and pressure control systems.

12. Is there a flow measuring device for the main flow and all other flows?

_____

(Yes or No)

Performance checking after initial start-up and performance monitoring will not be possible without these flow loops. Troubleshooting of machines will be difficult without knowledge of operating point. Surge prevention systems must have data from these flow measurement devices in order to function and to provide necessary protection.

13. Is there a PI and a TI on each stage suction and discharge?

_____

(Yes or No)

This instrumentation is essential when troubleshooting or monitoring machine performance.

14. Has the full range of expected conditions been specified? A centrifugal compressor typically has a head rise of only about 6%. Therefore, with a drop in molecular weight of 6% the capacity drops to about 65% of design if system pressures are as specified.

_____

(Yes or No)

Maximum discharge pressure, minimum suction pressure, maximum suction temperature, and minimum molecular weight are conditions that must be considered when determining the head rise which a compressor must develop at design flow rate. These alternative operating conditions increase head required to develop required discharge pressure. Refrigeration compressors and other compressors that raise gas pressure to condensing pressure are affected by heat sink temperatures (i.e. cooling water, air, or other cooling media). An increase in heat sink temperature will increase compressor head requirements at design flow.

15. If the system has a low dynamic head loss, a high head rise to surge is necessary for stability. This often limits off-design capacity. Has the head rise to surge been specified?

(Yes or No)

Compressor control system will not function in a stable manner unless sufficient head rise to surge is provided. Additionally, slight changes in suction and discharge pressure could cause the compressor to surge if insufficient head rise to surge is provided.

16. How close to surge do you wish to operate?

(Yes or No)

If less than 15% above surge flow, will the control system prevent surge? When operating at only 15% above the surge line, it becomes even more difficult to prevent surge if the machine operates at variable speed.

It is generally difficult to operate closer than 15% to the surge line without special control systems. There often exist significant economic incentives, due to reduced power costs, to permit operation at minimum flow. This is generally true if the compressor is expected to operate at lower flows for any significant portion of time. Computer control schemes that correct predicted surge flow for suction temperature, speed, and molecular weight can be used to operate near surge. This system must also include a surge detection device that is reliable so that surge events do not go unnoticed. This surge detection device should alarm and automatically actuate an antisurge or recycle valve to take the machine out of surge if it should occur.

17. Have block valves been provided to isolate the machine for maintenance?

(Yes or No)

Provision for blinding machines should also be made inside isolation block valves. Blinds must be rated at the same pressure as piping system in which they will be installed. Watch settling out pressure.

18. Has the shaft sealing system been specified?

(Yes or No)

(a) *Labyrinth seals:* Use this type whenever possible. These are simple, proven devices that are inexpensive and require no elaborate seal oil system. These types of seals do leak, however, so that product losses or environmental regulations may preclude their use. Labyrinth seals are commonly seen in services where the compressed gas is not of great value or where the sealing pressure is relatively low. Examples are air, nitrogen, steam, and ammonia.

(b) *Mechanical face seals:* These seals are similar to pump mechanical seals. The seal runs in an oil-pressurized cavity. These seals are susceptible to dirt and similar contaminants, so buffer gas should be used in dirty gas service. This type of seal is also more vulnerable to mechanical damage than other types because it sometimes incorporates relatively brittle components.

This seal does have some advantages over other types. One advantage is the ability to contain process gas in the event of a power failure (the compressor must be down, however). Seal oil pressure is not required to contain gas; seal faces close and leakage is minimal. Another advantage is that mechanical face seals do not need seal oil differential pressure controls as precise as liquid film seals. Mechanical face seals are also less likely to lead to rotor dynamics problems that can occur with liquid film seals. Liquid film seals will sometimes act as bearings.

(c) *Dual or redundant seals or liquid film seals:* Typically used for low molecular weight or toxic hydrocarbons or when no atmospheric leakage can be tolerated. These are the most commonly used seal types for petrochemical and hydrocarbon applications. Moreover, these seals are rugged but require an elaborate seal oil support system and precise control of seal oil differential pressure. Dual seals do not leak gas to the atmosphere. Single seals do leak to the atmosphere but flow is dramatically less than leakage of labyrinth seals.

(d) *Gas seals:* This is a relatively new development that is a modification of mechanical face seal technology as applied in pumps. An elaborate seal oil support system is not required. Due to the recent development and application of these seals, vendor experience should be investigated to verify applicability.

19. If the gas being handled contains H_2S above 10 ppm and water, or can ever do so, specify 90,000 psi yield maximum and state H_2S content on data sheet. Has this been done?

_____
(Yes or No)

Stress corrosion cracking could occur in high stress areas if not in compliance with these limits.

20. If the gas contains over 100 ppm of H_2S, then provision to dispose of contaminated oil from sour oil pots must be made. Has provision of a vacuum dehydrator been considered to recover oil?

_____
(Yes or No)

Internal seal oil leakage will be exposed to process gas, rendering the oil unusable due to high H_2S content. If this oil is reused, it will cause corrosion of some metal components. This oil should be collected in sour oil pots and must either be disposed of or recovered via a vacuum dehydrator. Recovered oil may be reused and could thus provide sufficient economic and environmental incentive to justify a vacuum dehydrator.

21. If the process system cannot tolerate trace quantities of lube oil and seal oil, has provision been made for the trap vents to go to a destination other than the machine suction?

_____
(Yes or No)

Traps are sometimes called sour seal oil pots. They collect oil migrating towards compressor internals from mechanical face seals or liquid film seals. Typically, leakage to traps is about 1–10 gallons per day. Vents from traps will allow small amounts of oil to enter the compressor suction under normal operation and could allow large amounts of oil to enter if level control instrumentation should malfunction. Vents should be routed to another system that has pressure maintained at or below compressor suction pressure if contamination by trace amounts of oil is a concern. Maintaining a separate disposal system at compressor suction pressure is necessary for drainage of oil to trap.

22. If the compressor can operate under a vacuum has provision been made to maintain the seal system above atmospheric pressure by buffering or some other means?

_____
(Yes or No)

If the compressor is allowed to operate under vacuum there is a possibility of air being drawn into the compressor through seals and suction piping leaks unless seal oil is maintained above atmospheric pressure. This condition can be detrimental to certain processes and potentially dangerous if flammable gases are being compressed. Preventing the inward leakage of air may only require an appropriate adjustment of seal oil pressure.

23. If the gas handled contains more than 50 ppm of H_2S or other constituents which will degrade the seal oil, is a clean gas available which could be used to buffer the seals?

If so, has buffer gas injection from a clean source been specified?   _____
(Yes or No)

Degradation of the oil additive package and viscosity can be caused by certain contaminants in process gases. This is more of a concern if the system is designed to reuse oil that is collected in seal traps. Contamination can also occur by diffusion even when sour oil leakage is collected in traps. Internal oil leakage is exposed to process gas and may absorb gas components that can cause degradation. Buffer gas can be injected into the cavity between the seal and the process gas to prevent contact of seal oil with process gas. Buffer gas must be compatible with the process because it will enter into the process stream. As a result, the selected buffer gas may be the major constituent in the process gas. Of course, the buffer gas comes from a source that has the potential contaminants removed.

24. Have flange finish and facings been specified?   _____
(Yes or No)

Flange facing should generally be consistent with the plant or attached process piping. Raised face flanges, for example, should be used for compressor flanges if the piping will utilize raised face flanges. Some prefer special finishes on flange sealing surfaces to enhance gasket sealing. Concentric grooves similar in size and appearance to phonograph grooving are used in some instances to enhance sealing. The compressor flange finish should also be consistent with this piping.

25. If a gear is involved in the compressor drive system, has a torsional analysis

been specified?

_____
(Yes or No)

Torsional resonances in compressor trains (including all drivers, compressors, and couplings) may be excited by forcing functions generated by a gear. A torsional analysis will indicate if potential problems exist and will allow necessary adjustments in torsional stiffness and damping to be made. Generally, necessary adjustments can be made by changes in couplings. Considerable time may be lost during equipment start-up if torsional problems develop (i.e. time required to analyze system and obtain and install parts required to correct problem). Torsional analysis of the system during the design phase will minimize the possibility of start-up delays.

A torsional analysis can be made relatively easily using computer methods if all torsional information is available for the various train components. Determination of optimal corrective action and acceptable differences between significant forcing frequencies and natural torsional frequencies are judgments best made by individuals with experience in this area. A number of qualified consultants are available to assist in verifying vendor calculations if needed.

26. Can the lube/seal oil system for this machine be combined with others? Have dry running gas seals been considered? Can the system be isolated for maintenance? Would the entire installation have to be out of service for seal maintenance?

_____
(Yes or No)

Considerable savings can result if lube/seal oil systems are combined. It is generally not a good idea to combine systems if process gas contains contaminants that could degrade the oil (if bad oil goes undetected it could damage two machines). Use of combined systems is not a good idea if H_2S in process gas exceeds 100 ppm.

Gas seals, if available, and if vendor experience can be verified, might further reduce the complexity and cost of the overall equipment package. The slight amount of atmospheric leakage of process gas, however, could be environmentally unacceptable.

27. Do you want the vendor to supply the lube/seal oil console? If so, has this been specified?

_____
(Yes or No)

It may be necessary to purchase this system from the compressor manufacturer to protect guarantees. (Generally, it is a good idea to purchase this system from the compressor vendor.)

28. Must a separate seal oil console be provided? If so what is the economic justification for this?

_____
(Yes or No)

In some instances it is possible to use lube oil pumps as a source of seal oil; this occurs when compressor suction pressures are relatively low (typically < 200 psig).

In these instances the discharge pressure of oil pumps can be increased to the pressure necessary to supply seal oil, thus eliminating the need for separate seal oil pumps. Alternatively, separate seal oil pumps can be provided that take suction from the main lube oil pumps. Economic justification for a separate seal oil console might include the additional energy required for main oil pumps to boost pressure of all oil to seal oil pressure. The main (lube) oil pumps would have to provide only a pressure of typically 40 psig (3 bar) if a separate seal oil console or separate seal oil pumps are provided.

29. Will shop tests specified be witnessed?

    (a) Shop inspection   _____*

    (b) Hydrostatic   _____*

    (c) Mechanical run   _____*

    (d) Performance (air)   _____

    (e) Performance (gas)   _____

    (f) Auxiliary equipment   _____*

    (g) Driver tests   _____

    * Typically witnessed in vendor's shop.

Usually, centrifugal compressors comprise custom-designed, high-value, long-delivery machinery. The equipment user's exposure to potential problems is thus higher and of more serious consequences than for less expensive standard products. It is prudent to monitor certain key production achievements in the vendor's shop during manufacturing and testing to avoid surprises in the field. This is also a good opportunity to become familiar with the details of construction and idiosyncracies of the machine. All inspection requirements and "hold points" for the customer's inspector should be defined soon after placement of the purchase order. Remember it is much easier and less painful to resolve problems in a vendor's shop than in an operating plant.

30. Are diffusers vaneless or are they of the vane type maximum efficiency design?

                                           _____

                                     (Yes or No)

Vaneless diffusers are less complex and less costly than vaned diffusers. The latter are, however, more efficient and may be justified in non-fouling services.

31. Has the control system been fully described? Will it function correctly under all operating conditions?

                                     _____

                                     (Yes or No)

The mechanical engineer and chemical (process) engineer counterpart must communicate so that the mechanical engineer fully understands all anticipated operating modes of the compressor. This communication should uncover operating conditions that require special controls, alarm, and shutdown devices. The compressor vendor will generally not contribute a great deal to the definition of control systems. The owner's mechanical or chemical engineers who recognize potential problem areas must take the lead in ensuring that control systems are adequately defined. Situations that are likely to lead to problems if not properly addressed in control systems include, but are not limited to, the following:

- *Start-up and shutdown:* off design conditions such as low flow or low molecular weight often occur during these periods. Surge control system should be capable of controlling flow to prevent the compressor from surging. This will require correcting variables as needed to determine how close to surge the machine is operating. A surge detection system should also be considered.
- *Low flow:* due to reduced unit throughput. Suction temperature control may be required if the antisurge valve recirculates a high percentage of compressor total flow.
- *Regeneration service:* may require separate instrumentation for compressor antisurge and flow control. Operation in regeneration service is sometimes quite different from normal process operation. Different gases may have to be handled at different operating pressures and temperatures.
- *Variable speed:* the surge point increases with speed. Also, the compressor performance changes with speed. The control system must compensate for these changes as required to meet process needs. It may be desirable to vary the compressor speed automatically to meet these changing requirements.

32. Unless a pressure-regulated seal has been specified, have all probable suction pressures been specified?

(Yes or No)

Abnormally low suction pressure will affect seal cooling and can affect the seal oil control system. Suction pressure at or above seal oil pressure will allow the process gas to escape if the seal oil pressure is not pressure compensated.

33. If in a single casing the gas is to be removed from the machine, passed through coolers, and/or other processes, and returned, is it essential that the internal bypassing be within certain limits?

If so, has this limit been specified? Or, should you have specified a back to back impeller at the interstage?

(Yes or No)

Cross-contamination of gases with different compositions or at different temperatures will occur due to leakage across interstage labyrinth seal. In some processes this allows clean gas to leak towards the dirty (contaminated) gas stream instead of vice versa. There is a small penalty due to somewhat higher leakage.

34. Are any provisions needed for future requirements?

If so, have these been specified?

_____

(Yes or No)

Future debottlenecks may be anticipated during this phase of project development. At relatively low incremental cost, key equipment can be sized for or easily modified to a debottlenecked condition. Examples are: larger suction knockout drum, larger piping, larger intercoolers, larger driver or a foundation easily modified to accept a larger driver, gear that is oversized or provided with a high service factor. Compressors can also be purchased destaged or with less than maximum diameter impellers. If on-line washing with liquid to clean the compressor is anticipated, then tie-ins can be provided and compressor designed to accept on-line liquid washing.

35. Has the required nozzle position been specified? Would a barrel type compressor have any advantages?

_____

(Yes or No)

Many users prefer bottom-connected compressors to avoid removal of the major process piping when performing maintenance on horizontally split compressors. This minimizes the possibility of introducing pipe strain when major piping is reinstalled. It also reduces time to perform maintenance. Bottom-connected nozzles, however, require relatively costly mezzanine mounting structures and introduce the possibility that the structural resonances may cause vibration problems. Bottom-connected compressors often lead to a piping arrangement that creates a liquid trap between the suction knockout drum and the compressor inlet. If the process gas is near dewpoint, liquid may accumulate and eventually slug the compressor. Additionally, it is not uncommon for tools or parts to be dropped unnoticed into the compressor suction or discharge piping when the compressor top casing is removed; this of course could lead to a compressor failure.

One novel approach that is sometimes attractive is to use a barrel-type casing, even though pressure levels do not dictate this design. The barrel casing arrangement allows removal of all rotating and wearing compressor components without disturbing major process piping even if top connected process piping is provided. Assembly of many critical components can then be performed in a better controlled shop environment instead of in the field. The barrel-type casing virtually eliminates split line leaks sometimes experienced with horizontally split compressors. Use of a barrel-type casing has the advantages mentioned but is somewhat higher in initial cost.

36. If there are any special uprate or standardization considerations with regard to nozzle sizes, types, etc., have these been specified?

_____

(Yes or No)

It may be advantageous to provide flanges on compressor casings that are the same type or the same size as process piping. Special consideration may be given to the possibility of future throughput increases, or standardization requirements of a plant.

37. If this is a critical class of machinery, has a non-standard lube oil system been specified?

_____

(Yes or No)

All vendors will offer their standard lube oil system if equipment specifications allow for competitive reasons. Although the standard system will meet API 614 requirements and may be acceptable in some applications, it will not meet the minimum requirements of certain other applications. This is particularly true for larger unspared compressor trains. Typical potential constraints found in standard systems include:

- self-contained control valves instead of pneumatic control valves and controllers;
- inadequate oil capacity in overhead seal oil reference drums;
- two instead of three lube oil pumps;
- marginally sized oil pumps and/or drivers;
- seal oil and lube oil systems that will not allow a compressor coastdown without damaging bearings or gas leakage from seals in the event of a power failure;
- no facilities to coalesce oil mist emanating from oil reservoir vents;
- instrumentation and control devices that are not normally used in the plant;
- equipment hardware that is not normally used in the plant such as pipe fittings, valves, pumps, flexible couplings, filters, etc.;
- oil coolers that have little or no allowance for cooling water fouling and/or have low water velocities. Additionally, coolers tend to be vendor standard designs and not manufactured in accordance with TEMA (Tubular Exchangers Manufacturers Association).

38. Have all available data on the possible corrosiveness of the gas been specified?

_____

(Yes or No)

Obviously, selection of compressor materials of construction must consider corrosives in process gas. Typical offenders include H_2S, chlorides, and acids. Contaminants during abnormal operation such as start-up or regeneration must also be considered. The compressor will be more susceptible to corrosion than stationary equipment due to higher stresses, cyclic stresses, and higher gas velocities; hence, potential corrosive constituents must be identified prior to material selection.

Special coatings may be applied in some applications to standard compressor materials to achieve acceptable corrosion resistance at a reasonable cost. Operating procedures or conditions can be modified in some cases to minimize or eliminate compressor exposure to corrosives (for example, increasing operating temperature to provide a comfortable margin from the dewpoint of an acid known to be present).

39. Has an API data sheet been prepared?

_____

(Yes or No)

Preparation of the API data sheet will help highlight process and hardware areas that need definition or investigation.

40. Has an API data sheet been prepared for driver and gear (if required)?

_____
(Yes or No)

These data sheets are an excellent means to communicate plant preferences to the vendors. Additionally, these data sheets provide a reasonably comprehensive means to define construction and design details.

41. If the partial pressure of H_2 in the gas is 215 psia or more, has a vertically split case (barrel) been considered?

_____
(Yes or No)

Split line process gas leakage will be increasingly troublesome as the partial pressure of hydrogen increases above 215 psia when horizontally split casings are utilized. Barrel-type casing designs can virtually eliminate these potential leakage problems.

42. If flushing is required, has it been specified?

_____
(Yes or No)

Liquid injection into the compressor (flushing) may reduce wheel and diaphragm fouling in dirty services. The compressor vendor can provide special liquid injection nozzles located to maximize their effectiveness. If required these should be specified to capitalize on vendor expertise.

43. If a maximum casing temperature is required, has this been specified?

_____
(Yes or No)

Emergency, start-up, or other unusual operating conditions may expose compressor to abnormally high operation temperatures. The type of casing construction selected and the materials of construction may be affected by the maximum temperature anticipated. Vendor should be informed of such unusual conditions.

44. Has instrumentation gauge panel location and scope of supply been specified?

_____
(Yes or No)

Vendor's scope of supply for all instrumentation should be agreed upon early. Decisions concerning local versus control house instruments should be reviewed with operations and maintenance personnel. Design and fabrication of a control panel is an area of considerable vulnerability; compressor vendor expertise in this area is not always at the same level as the user's. Some users design and provide all protective and control instrumentation. The compressor vendor thus provides appropriately located connections for instrumentation. The compressor vendor would of course be consulted to ensure the control and protective instrumentation is adequate.

45. Has a suction strainer been included in compressor piping? (*Note:* Recommend removing as soon as possible after start-up.)

_____
(Yes or No)

A start-up strainer should be included to help to prevent construction or other debris from damaging compressor. The strainer mechanical design should be adequate to prevent collapse under maximum anticipated differential pressure. Piping should be arranged to facilitate strainer removal and cleaning. If possible the strainer should be installed in a manner that permits its removal and cleaning without disturbing major process piping; this minimizes the possibility of introducing pipe stress on the compressor casing during strainer maintenance.

46. Does the cost estimate and budget include major spare parts for the compressor?

_____
(Yes or No)

Cost engineers often overlook major compressor spare parts such as rotors, gears, and couplings. These items should be purchased at time of compressor order to obtain best pricing and to ensure critical spare parts will be available for start-up. Justification of these essential items later in the project will be difficult and time-consuming, especially if budget problems are being experienced.

47. Have failure modes for instrumentation and control devices been considered in the event that instrument air and control power is lost?

_____
(Yes or No)

The safest failure mode for all compressor instrumentation and control devices should be considered. The failure modes selected should be consistent with failure modes of the related process. For example, should a steam turbine-driven compressor shut down or continue operating in the event of loss of instrument air? This decision is dependent upon process impacts and must incorporate compressor protection impact (i.e. protective devices for compressor may be compromised if the compressor is allowed to operate).

Project Phase: Preorder Review with Vendor
Machine Category: Centrifugal Compressors

Designation: _____

Location: _____

Service: _____

1. Does the vendor have adequate experience?

_____
(Yes or No)

(a) Location of similar machines:

_____

_____

It is important that other users of similar machines be contacted to discuss their experience with vendors of interest. Vendors should be able to provide location and contacts. Information obtained from other users can be very useful in determining performance, maintenance, and vendor support histories. Vendor strengths and weaknesses may be revealed from discussions with other users. If possible a plant visit to discuss (with the user) the history of machines similar to that proposed is an excellent way to learn more about the equipment proposed.

(b) How closely do the "similar" machines conform to the proposal?

	Yours	Others
Suction pressure	_____	_____
Suction temperature	_____	_____
Discharge pressure	_____	_____
Discharge temperature	_____	_____
Molecular weight	_____	_____
Head	_____	_____
Tip speed	_____	_____
rpm	_____	_____
Number of stages	_____	_____
Seal types	_____	_____
Settling out pressure	_____	_____
Mach number	_____	_____
Separation of lube and seal oil	_____	_____

It is unlikely that all process conditions for the new compressor will match compressors in service. Comparisons will increase the user's confidence of vendor's capability if the machine proposed has operating conditions that fall between those of machines operating successfully. Operating conditions outside proven operating windows for a proposed compressor must be carefully considered. Use of proven impellers at similar pressure levels and similar molecular weight gases in the proposed compressor reduces the probability of performance deficiencies.

(c) Have any similar machines experienced difficulty either during testing or in the field?

_____
(Yes or No)

If so, what were the characteristics of the problem and what steps will be taken to avoid a repetition?

All vendors will have some problems. Significant unresolved deficiencies reported would of course render a vendor unacceptable. Unproven fixes for problems reported are suspect. A great deal can be learned about the vendor's overall capability from his past history in identifying problems and resolving them in a timely manner.

(d) Comments of users on these machines:

How long have machines been in service? _____

Operating experience that extends through one compressor overhaul is most valuable. Information concerning premature wear of components, case of maintenance, parts availability, price, and quality will likely be available after the first overhaul. One year of operating experience is the suggested minimum credible experience. It is possible that a vendor may propose a design that has overwhelming advantages but is not totally field-proven. Such advantages might be higher efficiency, lower initial cost, or mechanical simplicity. Alternatives may merit careful consideration if economic incentives are large.

What start-up difficulties were experienced? _____

Start-up difficulties are the result of problems in one of several areas. These are:

- basic design problems (such as rotor dynamics);
- quality control;
- instrumentation and controls (associated primarily with lube and seal oil system);
- vendor field support (quality of serviceman and vendor response to questions).

Once problem areas are identified, steps can be taken during procurement, engineering, or manufacturing to reduce vulnerabilities in any one of these areas. For example, independent verification of vendor rotor dynamics calculations is appropriate for a vendor with reported rotor dynamics problems.

If so, was the Service Department's response satisfactory?

_____

_____

Were problems resolved in a timely manner? Were solutions implemented after an adequate engineering analysis was made? Trial and error problem solving techniques are unacceptable. Were appropriate vendor personnel made available for problem solving? Did the vendor bear a fair portion of the cost to correct vendor caused problems? Were modified parts furnished quickly, if needed?

Did the machines meet the process specification?

_____

_____

The most reliable source of compressor performance information is the vendor performance test. Field testing is possible but difficult due to dependence on the accuracy of numerous instruments and possible variations in molecular weight. Vendors that are found to overstate efficiency consistently at the proposal stage should be appropriately penalized during evaluation.

What has been the maintenance experience? _____

_____

Are maintenance intervals reasonably spaced or too frequent? How does ease of maintenance compare with machines of similar size and type? Is the use of special tools reasonable? Can routine maintenance be performed by shops that the owner normally uses or must repairs be performed by the original equipment manufacturer? Will owner's craftsmen be able to attain a level of expertise that allows making in-house repairs? Are assembly tolerances reasonable or difficult to attain?

Name and position of contact: _____

_____

The rotating equipment engineer involved in the maintenance and troubleshooting of the compressor of interest is usually a reliable source of information. First line operating management is also a good source of information.

Name and position of contact: _____

_____

(e) Were the machines built in the shop where your machines are to be built?

_____
(Yes or No)

If a different shop is to be used to manufacture your machine, then one should question how the technology has been transferred to a new shop. Have some of the key personnel from the original shop been relocated to the new shop? Have other machines similar to the machine you are purchasing been fabricated in the new shop? Have any manufacturing techniques been altered in adapting to the new shop? If so, how and why, and are the new methods satisfactory? Problems are more likely if your machine is one of the first to be manufactured in a different facility.

(f) If your machine is to be built in a shop other than the vendor's main shop:
  • How is the parent company or licensor going to ensure that the machines will be built correctly?

_____

Will he have an inspector (or engineer) in the shop on a full- or part-time basis?

$$\overline{\hspace{3cm}}$$
(Yes or No)

How qualified will this person be?

The parent company should provide a qualified inspector and preferably an engineer to follow manufacture of the compressor in the outside shop. Inspection points the original manufacturer intends to enforce should be discussed and agreed upon early by all three parties involved (i.e. the purchaser, vendor, and manufacturing shop). If the particular type of machine is routinely subcontracted or manufactured by a licensee then the potential for problems is minimal.

- How many similar machines have been built in that shop?

Were any problems experienced with the machines built? Were schedules met? The purchaser should be reluctant to allow use of a shop that has little or no experience manufacturing the machine to be purchased.

- How similar were they? _____

Machines of the same frame size, similar molecular weight gas, number of stages, inlet conditions, and speed are ideally similar. Successful installations of such similar machines should increase user confidence in the shop's capability. Machines different by one frame size, especially larger, should be considered similar if all other conditions are approximately the same.

- Who will do the engineering and drafting for the machines?

The means by which technology will be shared should be discussed with the outside shop and original manufacturer. This should not be a major concern if the shop has previous experience with the machine being purchased; sharing of technology could then be handled as in the past. A review of engineering and drafting packages by the original manufacturer would be one means to minimize potential problems:

- Will the machines be an exact duplicate of the licensor's design?

$$\overline{\hspace{3cm}}$$
(Yes or No)

If not, what modifications will be incorporated and why?

The purchaser should understand all deviations. Some deviations will be obviously acceptable. Other deviations should be field proven. The purchaser should be assured that the latest improvements in equipment design by the parent company are incorporated by the outside shop. Design modifications not approved by the parent company should be rejected.

2. If machines are not identical, how much extrapolation has the bidder done on the established experience limits?

   (a) Are these extrapolations based on experience with other machines?

   —————————

   (Yes or No)

   (b) Do these extrapolations appear sound?   —————————

   (Yes or No)

   Interpolation of experience (for example, a vendor offering a 5-stage machine in a frame size that has been in operation in 4 and 6 stages) is preferred to extrapolation. Extrapolation of experience indicates that the offering is outside either the upper or lower limits of proven performance. Extrapolation must be made using sound engineering principles. Extrapolation by more than one standard frame size should be avoided, if possible. Machines that have near maximum number of impellers in a casing should be avoided if successful experience is not demonstrated. Rotor dynamics and aerodynamic performance are the areas of most vulnerability when exceeding the envelope of proven performance.

   (c) Is the vendor offering within the limits of any previous experience?

   —————————

   (Yes or No)

   (d) Is this experience applicable to the application being considered?

   —————————

   (Yes or No)

   Vendor experience in certain specific areas may be of particular value even if frame size or performance is not similar. These areas include such technology as low- and high-temperature applications, low molecular weight gases, dirty services, identical gases, and high-pressure compressors.

   Specifications: ———————————————————————————————

   (a) Does the offering completely meet your specification in all aspects?

   —————————

   (Yes or No)

   (b) If not, the vendor's responsibility is to list all deviations.

   Are there any deviations not listed in the proposal?   —————————

   (Yes or No)

Vendors attempt to standardize as much as possible and draw from their own operating experience. As a result they often propose deviations from user specifications. Deviations found during proposal review (including discussions with vendor's representatives) but not itemized in proposal indicate that the vendor is not fully aware of specified requirements. Deviations found that are not itemized indicate that others exist. Prior to order commitment the user should be certain that all major vendor exceptions (itemized or not) are understood and resolution reached.

Note that vendor execution in some instances may be acceptable or even superior to specified requirements. Rarely are machines purchased without accepting some vendor deviations from the inquiry specification. The final mechanical design of a machine is almost always negotiated so that it incorporates both the vendor standard execution and the inquiry specified requirements. Do not discount the experience of vendors without due consideration; they are in the best position to offer improvements in design. Vendors are also best able to offer design alternatives that may reduce cost.

3. What design and shop practice modifications have occurred since the similar machines were built?

   • Rotor, including shrink specifications _____

   • Casing   _____

   • Bearings   _____

   • Seals   _____

   • Clearances   _____

   • Wheel design and manufacture _____

   • Diffuser design   _____

   • Materials   _____

   • Balancing   _____

   • Tip speeds   _____

   • Overspeed testing   _____

   Which of these changes are definite design improvements and which are cost reduction items?

   _____

   _____

Any items that have had changes made for cost reduction or design improvement reasons should be tested to see that these changes do not lead to potential problems. Changes in several areas increase the possibility that unexpected problems may arise. It may be necessary to insist that previous practices be adhered to in some instances to increase the probability that any problems experienced will be minimal.

What types of seals are proposed? (Please refer to P&ID Specification Review, Item 18)

_____

_____

Does the vendor have experience with these seals at a higher pressure?

_____

(Yes or No)

Has the vendor operated on gas with similar contaminants?   _____

(Yes or No)

If so, what steps did he take to avoid lube oil contamination?

_____

_____

Examples of steps that might be taken include: buffer gas injection, double mechanical seals, clean-up systems to recover contaminated oil.

How many pressure reducing steps are there? _____

What is the maximum pressure per reducing stage and does prior experience exist?

_____

Are you inside this tolerance?   _____

(Yes or No)

What steps will the vendor take to prevent the bushings acting as bearings?

_____

_____

Seal bushings must be grooved or squeeze film dampers installed unless prior experience can be totally verified.

Has the vendor operated on gas with similar contaminants?_____

(Yes or No)

If so, what steps did he take to avoid seal oil contamination? _____

_____

How many pressure reducing stages are there?   _____

What is the design maximum pressure per reducing stage?_____

Are you inside this tolerance?   _____

(Yes or No)

What steps will the vendor take to prevent the bushings acting as bearings?

_____

_____

If none, what action will he take if this occurs? _____

_____

Why can't he do this before problem arises? _____

_____

If the seals are bushing type, will overhead tanks be provided? _____
(Yes or No)

If the differential is controlled by overhead tanks, a bladder must be inserted between the reference gas and the main seal system if gas in the seal zone will contaminate the seal oil.

4. How are machine expansions handled to minimize thrusts due to coupling slip forces and expansion (gear couplings)?

(a) Where is casing anchored axially? _____

Where is thrust bearing? _____

Are these on opposite ends of casing? _____

If not, there will be a large shaft movement into the coupling. Is the coupling adequare to take this movement? (Question wisdom of this approach.)

(b) In which direction will coupling slip force act relative to internal thrust forces? _____

(c) Assuming the coupling slip force to be

$$F = 0.3 \frac{T}{d} = \frac{18,900 \times P}{\text{rpm} \times d}$$

where T = torque in lb-in,
P = horsepower,
d = shaft diameter at coupling in inches,

are all thrust bearings designed for normal thrust forces plus or minus the coupling slip force? _____
(Yes or No)

(d) Can the coupling slip force completely cancel the normal thrust and position the shaft on the "inactive" side of the thrust bearing? _____
(Yes or No)

(e) Would a non-lubricated coupling be more appropriate? _____
(Yes or No)

If no, check again with designated machinery specialist.

5. How will the job be handled from proposal through design, drawings, fabrication, inspection, testing, and shipping?

   (a) Has a wheel layout already been established?   _____

                                                      (Yes or No)

   (b) If so, who did it and how? _____

      _____

   (c) Will the wheel layout be rechecked by others before drafting?

      Who? _____

      How? _____

   (d) If drafting selects the parts from a range of standard parts, who checks that these parts will meet the requirements?

                                       _____

      _____

   (e) What part does Design Engineering play in the machine selection?

      _____

      _____

   (f) If wheels must be cut who specifies the final diameter?

      _____

      _____

      Will this final diameter be rechecked?   _____

                                           (Yes or No)

      If so, who will do it? _____

      _____

   (g) If sidestreams are involved, who designs the sidestream inlets?

      _____

      Are these rechecked for Δp?   _____

                                           (Yes or No)

      If so, who rechecks? _____

   (h) Is there a contract engineer who has full responsibility for following the unit through all phases of design, manufacture, inspection and test?   _____

                                                        (Yes or No)

How much experience has the individual selected for our machines had? _____

How many other machines is this individual responsible for at this time? _____

(i) Who is responsible for inspection? _____

What function does inspection report to? _____

(j) What function is responsible for testing? _____

Will the vendor's shop loading affect delivery? _____
(Yes or No)

Ask to see the schedule on the compressor superimposed on a loading diagram for: Engineering, Drafting, Purchasing, Machining, Assembly, and Testing.

Are all of these functions without overload to the degree that they may delay delivery? _____
(Yes or No)

Potential problem areas.

What analysis work will be done to ensure no problems due to:

- Rotor flexibility and stability _____
- Torsional vibrations (if more than two elements tied together) _____

_____

- Lateral criticals _____
- Thrust loadings _____

Do these approaches seem satisfactory? _____

6. If more than 10 ppm of H_2S has been specified in the process gas, what steps will be taken to ensure the 90,000 psi max. yield or $22R_C$ hardness are not exceeded?

_____

_____

How will hardness in heat-affected zone of welded impellers be checked?

_____

_____

7. What overspeed tests will be conducted on the wheels? _____

_____

Will the wheels be spun to a given % above mechanical design or a given % above max. continuous for the machine? _____
(Yes or No)

What will be the percentage overspeed?   _____

Will the assembled rotor be tested at overspeed?   _____

(Yes or No)

Project Phase: Contractor's Drawing Review
Machine Category: Centrifugal Compressors

Designation: _____

Location: _____

Service: _____

P&ID's

1. Is there a KO drum on the suction and any cooled interstages?   _____

(Yes or No)

2. Do the recycle lines re-enter upstream of the KO drums?   _____

(Yes or No)

3. Are the KO drums equipped with a gauge glass, LHA, and shutdown switches?   _____

(Yes or No)

4. Have all the alarm and shutdown switches specified been provided?

Low lube oil pressure alarm and auxiliary pump start-up actuation _____

Low lube oil pressure alarm and trip   _____

Low seal oil level (or pressure) alarm and auxiliary pump start-up actuation   _____

Low seal oil level (or pressure) alarm and trip   _____

High seal oil level alarm   _____

High discharge temperature alarm   _____

Others (detail) _____

5. Are TI's and PI's specified for suction and discharge of each stage?_____

(Yes or No)

6. (a) Is there a check valve in the discharge downstream of the anti-surge recycle?

_____

(Yes or No)

   (b) If there are two machines in parallel, is there a check valve on the discharge of each?

_____

(Yes or No)

7. Is there a flow meter on each feed and discharge stream from the machine?

_____

(Yes or No)

   The mass flow on each stage is to be measured. More than one meter is only required if the mass flow changes from section to section.

8. Have pressure taps been provided up and downstream of the temporary suction strainers?

_____

(Yes or No)

9. Will all instruments be changeable on the run and are shutdown circuits testable on the run?

_____

(Yes or No)

10. Are all lines to remote pressure gauges valved at the tie-in to the main line?

_____

(Yes or No)

11. Is there an isolatable closed circuit including a cooler at the machine so that the pre-start-up run-in can be performed?

_____

(Yes or No)

12. If the machine is not in closed circuit and is not an atmospheric air machine, is there a suction flare release to dump the suction gas in the event of shutdown?

_____

(Yes or No)

13. If the process is "flow controlled", is the metering elements outside the recycle loop?

_____

(Yes or No)

14. Is the antisurge metering element inside the recycle loop?

_____

(Yes or No)

15. On refrigeration machines the TIC's must close on driver trip. Do they?

_____

(Yes or No)

16. On motor driven refrigeration machines the casing pressure must be reduced to about 40 psig before starting to prevent driver overload. Liquid in the suction drums impedes the pressure reduction. Can the liquid from the drums be pumped into the accumulator?

_____

(Yes or No)

17. Is there a safety valve on the discharge of the machine if the downstream equipment cannot stand the machine's discharge pressure under the combined conditions of:

Trip speed _____

High mol wt. _____

High suction pressure _____

Low temperature _____

Layouts

1. Are the main and interstage suction lines from the KO drums cleanable by the method proposed? _____

    (Yes or No)

2. By the time of this review the moment of inertia of compressor and driver should be available. Is the hydraulic energy inside the check valves less than 1.3 of the kinetic energy of the shafts? _____

    (Yes or No)

    If not, the check valves will either have to be relocated or additional ones installed.

3. Are there drain valves at the low points of the suction and discharge lines? _____

    (Yes or No)

4. Are all the check valves horizontal? _____

    (Yes or No)

5. Are all the check valves damped or equivalent? _____

    (Yes or No)

6. Will the recycle valves pass the compressor design flow? _____

    (Yes or No)

7. Are the control valve actuators adequate for contemplated operating conditions? _____

    (Yes or No)

    (Users have had trouble on both straight control valves and butterflies.)

8. If the machine is an atmospheric air compressor, does the antisurge vent have a silencer and is the intake of a sound attenuating type? _____

    (Yes or No)

9. Is the foundation separate from that of reciprocating machines? _____

    (Yes or No)

10. Can the temporary strainers be removed without disconnecting any piping?

_____
(Yes or No)

11. Are the suction, discharge, and compressor drains connected either to a blowdown system or vented to a safe place?

_____
(Yes or No)

12. Unless the compressor vendor has given dispensation, is there a straight section of at least three pipe diameters on the suction flanges?

_____
(Yes or No)

13. Are the pipe stresses and moments within the levels allowed by the vendor?

_____
(Yes or No)

14. Are all the piping supports and anchors as described in the piping stress calculation?

_____
(Yes or No)

15. Has sufficient allowance been made in the stress calculation for friction of supports?

_____
(Yes or No)

16. On refrigeration machines, are the liquid injection points at a sufficient distance from the drums to ensure vaporization of all the liquid?

_____
(Yes or No)

17. On refrigeration machines, do the liquid level control valves have blocks and bypasses?

_____
(Yes or No)

18. On refrigeration machines, it is necessary to adjust the TIC controllers during start-up. On motor-driven machines, is there a single switch to commission the TIC's immediately after start-up? Is it readily accessible from the platform?

_____
(Yes or No)

On turbine-driven machines are the TIC's readily accessible from the platform? (Adjustments to the set points are necessary during run-up.)

_____
(Yes or No)

19. Have ROV's been provided on all lines which can feed hydrocarbons to a fire at the machine and do not have block valves at least 25' horizontally from the machine?

_____
(Yes or No)

Has the electrical conduit and valve operator to such ROV's been fireproofed sufficient to permit operation of the valves after a 10-min fire?

_____
(Yes or No)

20. *Consider operability*

(a) Are all instruments clearly visible?

_____
(Yes or No)

(b) Has the operator safe and easy access to all bearings?

_____
(Yes or No)

(c) Has the operator safe and easy access to the handwheels on the ROV's, the flow control devices and the recycle valves?_____
(Yes or No)

(d) Run through a start-up sequence. Can all the operations required be done by one person?

_____
(Yes or No)

(e) Is there safe access to the suction and discharge line and all casing drains?

_____
(Yes or No)

(f) Can the oil drain sight glasses be readily seen?

_____
(Yes or No)

(g) If there are overhead seal tanks, has the operator a clear view of the level gauge?

_____
(Yes or No)

If the oil level control has to be put on hand control using the bypass, will the operator be able to see the level gauge from his position at the valve?

_____
(Yes or No)

(h) Can either seal trap be taken out of service for repairs with the other trap draining both seals?

_____
(Yes or No)

21. *Consider maintenance*

(a) Is there a suitable location where the casing top half can be put without interfering with maintenance?

_____
(Yes or No)

Note: If the machine has multiple casings or if it has a turbine driver, all top halves may be off at the same time.

(b) If the casing is a barrel type, is there room to pull the barrel internals *in situ*?

_____
(Yes or No)

(c) Is the crane big enough to carry the largest maintenance weight? Usually the top half of the largest casing.

_____
(Yes or No)

(d) Can the top halves be moved to the storage area without passing over operating machinery?

_____
(Yes or No)

(e) If the machine is motor-driven, is there access to that end so that the motor rotor can be pulled, if necessary, using portable equipment?

_____

(Yes or No)

(f) Are the motor cooling ducts so positioned that they do not unnecessarily interfere with the crane movement?

_____

(Yes or No)

(g) If the compressor is at grade with overhead piping, can the piping spools be readily removed and swung out of the way leaving vertical lift for casing?

_____

(Yes or No)

Have lifting provisions been made to facilitate this?

_____

(Yes or No)

(h) Can the rotors be removed to the maintenance shop without passing over operating machinery?

_____

(Yes or No)

(i) Piping should not run unnecessarily over parts which must be removed for maintenance. Has this been complied with?

_____

(Yes or No)

22. *Consider instrumentation*

Are the TI's installed in such a way that they will measure the correct temperature? If the lines are two phase, will they see the correct phase?

_____

(Yes or No)

Project Phase: Mechanical Run Tests
Machine Category: Centrifugal Compressors

Designation: _____

Location: _____

Service: _____

The mechanical run test for centrifugal compressors is basically a balance check. In some cases data on the vibration characteristics of a machine will also be disclosed.

The basic procedure should be to run up to 110% of max. continuous for turbine-driven machines, and run for a minimum of 15 min. Then drop back to max. continuous speed and make the overall test 4 h. For motor-driven machines, max. continuous speed is design speed.

	Conditions	
	Design	*Test*
(a) Speed rpm	_____	_____
(b) L.O. inlet pressure	_____	_____
(c) L.O. inlet temperature	_____	_____
(d) Max. vibration	_____	_____
(e) Calc. critical	_____	_____
(f) Max. noise level	_____	_____

1. Do the first four test conditions match the design conditions to your satisfaction?   _____
   (Yes or No)

2. Is the actual critical speed within 5% of the calculated value?   _____
   (Yes or No)

3. Is the temperature rise across each bearing less than 60°F?   _____
   (Yes or No)

4. *Vibration*

   (a) Frequency survey when running at max. continuous speed. Note vibration at running speed and other frequencies.

Probe location	*Magnitude (mils)*	*Frequency (cpm)*
_____	_____	_____
_____	_____	_____
_____	_____	_____
_____	_____	_____
_____	_____	_____
_____	_____	_____

   (b) Is there undue vibration at critical frequency, also at frequencies between 35 and 50% of running speed?

   _____
   (Yes or No)

   (c) Shaft and bearing vibration attenuation:

	A, shaft	*B, Housing*	*A/B, attenuation*
I.B. bearing	_____	_____	_____
O.B. bearing	_____	_____	_____

   Is the attenuation less than 4? (If no, mention in report.)   _____
   (Yes or No)

(d) Vibration readings (mils):

	110% design speed	*Max. continuous*	*Difference*
I.B. bearing	_____	_____	_____
O.B. bearing	_____	_____	_____

Is the difference in vibration levels at these speeds less than 20%? _____

(Yes or No)

5. Is a thorough check being made for oil leaks? _____

(Yes or No)

6. *Seals*

   (a) Is the oil collected from each seal drain less than 5 gal/day? _____

   (Yes or No)

   (This is significant only on carbon seals, or bushing seals with normal differential pressure.)

   (b) Is the seal oil outlet temperature less than 180°F? _____

   (Yes or No)

   If not, insist that it be lowered. However, operation of normal running seals at low pressures may make the outer bushing run hot. An assurance that this is the cause should be accepted.

7. Bearing inspection. Are all bearing surfaces showing normal running pattern? _____

(Yes or No)

Demand replacement otherwise.

8. Seals to be inspected only if it is the vendor's standard practice.

9. Internal inspection to be carried out if a spare rotor is to be fitted and run (normally specified).

   Is there an absence of rubbing? _____

   If rubbed, demand a clearance check.

10. Check the internal alignment and clearance data from final assembly.

    (a) Is alignment good, clearances within tolerances? _____

    (b) Likewise for a spare rotor if it has been fitted. _____

11. Were copies of vendor's test log sheet and final internal clearance diagrams obtained? _____

(Yes or No)

12. When witnessing a test, always try to find out if any difficulties occurred in preparing for it. Such problems could be repeaters.

13. Gas Pressure Test following mechanical run.

    (a) Is the test being carried out with visible soap bubbles?   _____
                                                                        (Yes or No)

    (b) Is the shaft being rotated to check for freedom of seals?   _____
                                                                         (Yes or No)

Project Phase: P&ID and Design Specification Review
Machine Category: Reciprocating Compressors

                         Designation: _____

                             Location: _____

                              Service: _____

P&ID's

1. Based on project philosophy (availability required) are two machines called for?   _____
                                                                              (Yes or No)

2. Have KO facilities been specified on all suction and interstages? Are you aware of King-type coalescer-separators?   _____
                                                                             (Yes or No)

3. Do recycle lines re-enter upstream of the KO facilities?   _____
                                                                             (Yes or No)

4. Are KO facilities equipped with a gauge glass, LHA, and shutdown switches?   _____
                                                                             (Yes or No)

5. Are the suction lines traced between the KO drums and the machine flanges, including pulsation bottles?   _____
                                                                             (Yes or No)

6. Do KO facilities have automatic drains and bypasses to allow checking?   _____
                                                                             (Yes or No)

7. Is there a safety valve on each compression stage and are there both pressure and thermal relief valves on the coolant header?   _____
                                                                             (Yes or No)

8. Has a remote shutdown switch been provided?   _____
                                                                             (Yes or No)

9. Have all the alarm and shutdown devices been specified?

    (a) Low lube pressure alarm   _____

    (b) Low lube pressure trip   _____

(c) High temperature alarm on each cylinder discharge _____

(d) Low cylinder lube flow _____

(Yes or No)

Others (detail): _____

_____

10. Are TI's and PI's specified for suction and discharge of each stage and TI's on discharge of each cylinder and on the coolant outlets from each cylinder? _____

(Yes or No)

11. On machines which are specified to be reaccelerated, have controls been provided to unload the cylinders during reacceleration? _____

(Yes or No)

12. If these controls are of the bypass type, is there a check valve on the compressor discharge? _____

(Yes or No)

13. Have automatic unloading facilities been specified for start-up? _____

(Yes or No)

14. Are there coolant block valves on each cylinder? _____

(Yes or No)

15. Does each machine have double block valves and a vent to avoid blinding for valve repairs? _____

(Yes or No)

16. Have pressure taps been provided around the spool which will contain the temporary suction screen? _____

(Yes or No)

17. Do the oil coolers have provision for back-flushing the water side? _____

(Yes or No)

18. Will all instruments be changeable on the run? _____

(Yes or No)

19. Are all lines to remote pressure gauges valved at the tie into the main lines? _____

(Yes or No)

20. Are individual packing vents used and is each packing vent discharging to a safe location?

_____

_____

If to atmosphere, will this cause a pollution or safety problem?

_____

21. Is there an isolatable closed circuit at the machine so that the pre-start-up run-in can be performed?

                                                                _____

(Yes or No)

22. (a) Is the suction drum adequate?

_____

(Yes or No)

(b) Has a variable or high density crinkled wire mesh been specified?

_____

(Yes or No)

(c) Has a coalescing section been specified?

_____

(Yes or No)

23. How is the suction pressure controlled?

(a) On underload (less gas than design) _____

(b) On overload (more gas than design) _____

24. (a) If the compressor is handling a flammable gas and is not suction pressure controlled or if so controlled and the control fails, will the suction remain above atmospheric?

_____

(Yes or No)

(b) With variations in the suction pressure will rod failure be avoided?

_____

(Yes or No)

(c) With variations in the suction pressure will excessive compression ratios and temperatures be avoided?

_____

(Yes or No)

(d) Based on (a), (b), and (c) above should the machine have a low suction pressure shutdown?

_____

(Yes or No)

25. Has the possible need for superior filtration been thoroughly considered?

_____

(Yes or No)

If the KO drum is a long distance from the machine it may be cheaper to install a filter than clean a long suction line.

26. Can the suction line be thoroughly cleaned by the method proposed? Should it be coated?

_____

(Yes or No)

Describe the method proposed _____

_____

27. (a) Has an API data sheet been prepared for the compressor?

_____

(Yes or No)

(b) Have all applicable specifications been indicated? Normally there should be several in-house specifications.

_____

(Yes or No)

(c) Have all required accessories been specified? _____

(Yes or No)

Always specify pulsation bottles. Specify interstage piping on two-stage machines whenever this pressure level is not used for process. Specify frame intercoolers and moisture separators where practical for multistage compression. Consider use of frame aftercoolers on air machines. Specify cooling water manifolding. Specify instrument panel. Specify sight flow indicators on all systems.

(d) Have capacity control requirements been specified? _____

(Yes or No)

Don't use suction valve lifters. Suction valve unloaders must be air-operated if control is manual. If this is what you want, indicate suction valve unloading; indicate manual and note in remarks. Unloaders to be air-operated.

Note in remarks if reacceleration is required.

(e) Has distance piece requirement been specified? _____

(Yes or No)

Normally standard with solid cover. Abnormal conditions would be over 0.1% H_2S or some other contaminant which could degrade the lube oil. Extra long distance pieces are required for non-lube machines.

(f) Have site data been completed? _____

(Yes or No)

In remarks, indicate cooling water temperature range, especially on open systems.

(g) Has a helium test been called up if partial pressure of H_2 at the discharge is 100 psia? _____

(Yes or No)

(h) Has the gas analysis data sheet been completed? _____

(Yes or No)

Does this fully describe the full range of foreseeable operation? _____

(Yes or No)

Do the sheets of operating conditions also fully describe this range? _____

(Yes or No)

(i) Has the driver data sheet been prepared? _____

(Yes or No)

(j) If TEMA inter- and/or aftercoolers are required, has this been noted? _____

(Yes or No)

Project Phase: Preorder Review with Vendor
Machine Category: Reciprocating Compressors

Designation: _____

Location: _____

Service: _____

1. *Vendor experience*

    (a) Location of similar machines _____

    _____

    (b) How closely do the machines resemble yours?

	Yours	*Others*
Suction (psia)	_____	_____
Discharge (psia)	_____	_____
Suction temperature (°F)	_____	_____
k (C_P/C_V)	_____	_____
Bore and stroke	_____	_____
rpm (max. 600)	_____	_____
Discharge temperature (°F)	_____	_____
Piston speed (ft/min) (max. lubricated, 850; unlubricated, with piston rings, 650)	_____	_____
Valve velocity (ft/min)	_____	_____
Piston rod loading (pounds at design conditions)	_____	_____

    The last five are significant. If data for the other machine are lower than yours, it is not a suitable machine for comparison.

    (c) Have any similar machines experienced difficulty in the field? _____
    (Yes or No)

    If so, what were the characteristics of the problem and what steps were taken to avoid a repetition?

    _____

    _____

    (d) Comments of users on their machines.

    How long have machines been in service (min. 1 year)? _____

    What start-up difficulties were experienced? _____

If so, was the Service Department's response satisfactory? _____
(Yes or No)

Did the machines meet the process specifications? _____
(Yes or No)

What has been the maintenance experience? _____

_____

Name and position of contact _____

(e) Were the similar machines built in the same shop where your machines are to be built? _____
(Yes or No)

(f) If the machine is to be built in a shop other than the vendor's main shop:

- How is the parent company or licensor going to ensure that the machines will be built correctly?

_____

_____

Will he have an inspector (or engineer) in the shop on a full- or part-time basis? _____
(Yes or No)

How qualified is this person? _____

_____

- How many similar machines have been built in that shop? _____

_____

- How similar are they? _____

_____

- Who will do engineering and drafting for the machines? _____

_____

- Will the machines be an exact duplicate of the licensor's design? _____
(Yes or No)

If not, what modifications will be incorporated and why? _____

_____

2. If the comparison machines are not identical, how much extrapolation has the bidder done on the established experience limits?

_____

_____

   (a) Are these extrapolations based on experience with other machines?

                                                  (Yes or No)

   (b) Do these extrapolations appear sound?

                                                  (Yes or No)

   (c) Are any extrapolations within any previous limits or experience?_____

                                                  (Yes or No)

   (d) Do these extrapolations appear sound?

                                                  (Yes or No)

In all cases of extrapolation, consult with your responsible Machinery Specialist.

3. What design and shop practice modifications have occurred since the comparison machines were built?

   (a) Bearings _____

   (b) Valves _____

   (c) Piston to rod attachment _____

   (d) Rod to crosshead attachment _____

   (e) Liner location device _____

      Which of these are definite design improvements and which are cost reduction items?

      _____

      Consult with your responsible Machinery Specialist on all cost reduction items.

4. Will the vendor conduct a torsional analysis? _____

                                                  (Yes or No)

If not, what evidence is he providing to comply with applicable API specification?

_____

5. Do the cylinder materials meet experience requirements and conservative industry guidelines, i.e.

Relief valve setting	Cylinder material
0–1000 psig	CI, cast or forged steel
1000–2500 psig	Cast or forged steel
2500 psig	Forged steel

6. Does the vendor agree to run rod load reversal checks at all expected loadings? _____

(Yes or No)

7. Will the motor be adequate to run the machine at all expected loading conditions? _____

(Yes or No)

8. If the machine is for vacuum service, is the motor big enough for the drawdown peak hp? _____

(Yes or No)

If not, how will the machine be started and will unloading be acceptable to process?

_____

_____

Also, is it possible to trip one unit without causing an overload trip on the other? _____

(Yes or No)

9. Does the vendor take any exceptions to the specifications? _____

_____

If so, consult with your responsible Machinery Specialist on all exceptions. Check data sheet if specifying against no negative tolerance.

10. Does the expected performance meet with that specified? _____

_____

Note: Many vendors quote flow as 3% higher than specified to get around API's no negative tolerance.

11. What is rod stress at area of rod under thread root?

$$\text{Stress} = \frac{0.785[\text{cylinder bore}^2(P_d - P_s) + \text{rod diameter}^2 P_s]}{\text{area at root (in square in)}}$$ _____

Does this comply with your specification, i.e. maximum stress? _____

(Yes or No)

8000 psi for rolled threads
7500 psi for ground threads
7000 psi for cut threads

On multistage machines, check rod loadings at all unloading conditions to ensure that the above values are not exceeded.

12. The requirements of 11 are for the worst design conditions. Will the rod loadings at SV setting be within the vendor's rod loading limits? _____

(Yes or No)

13. If the piston has a tail rod, is it retained by a substantial steel cover? _____
    (Yes or No)

    (A broken rod outboard of the piston can cause the tail rod to be driven out like a projectile.)

14. Are the main bearings either babbitted or aluminum? _____
    (Yes or No)

15. Are the cross-head rubbing surfaces either babbitted or aluminum? _____
    (Yes or No)

    Are these surfaces replaceable? _____
    (Yes or No)

16. Will the cylinder lube connections in the water jacket run through solid metal? _____
    (Yes or No)

    (Piping through the water jacket is not permitted.)

17. Are there external connections for the coolant between the cylinder jackets and heads? _____
    (Yes or No)

    (Internal cooling connections are not permitted except on air compressors.)

18. Does the machine have vented packings? _____
    (Yes or No)

    If the gas will degrade the lube oil (H_2S or chlorine, etc.), is a double distance piece required? _____

    Accept vendor's recommendation for this.

19. Will the valve gaskets be proven execution? _____
    (Yes or No)

20. Is there a full flow lube oil filter? _____
    (Yes or No)

    On engines, is the filter twinned? _____
    (Yes or No)

    Is the filter of the type which can be cleaned on the run? _____
    (Yes or No)

    If not, how long is it expected to last? (6 months minimum) _____

    Will there be pressure gauges on both inlet and outlet of the filter? _____
    (Yes or No)

21. If needed, where will piping pulsation study be performed? _____

22. Will the cylinders have outboard supports not attached to the heads? _____
    (Yes or No)

23. Will the suction and discharge valves be non-interchangeable?   _____
    (Yes or No)

24. Will the machines be suitable for rail mounting?   _____
    (Yes or No)

25. Will the compression cylinders be equipped with dry type liners?   _____
    (Yes or No)

26. Will hollow pistons be easily ventable for disassembly or are safer,
    continuously vented types furnished?   _____
    (Yes or No)

27. If the specification calls for reacceleration on power failure, will the
    starting unloaders automatically operate?   _____
    (Yes or No)

28. If strainers are required on packing coolers, will they be twinned?   _____
    (Yes or No)

29. If intercoolers and/or aftercoolers are to be supplied for flammable
    or toxic gas, will they be to TEMA R?   _____
    (Yes or No)

Gas engines

This section is only to be completed if an integral or separate gas engine is involved.

30. If the engine is to burn refinery gas (i.e. fuel gas, hydrogen, propane,
    etc.), will the engine compression ratio be less than 7:1?   _____
    (Yes or No)

31. Will the ignition system be the solid state type (i.e. no magneto)?   _____
    (Yes or No)

32. On integral engines will there be anti-blowback type explosion doors
    on the crankcase?   _____
    (Yes or No)

33. On integral engines compressing hydrocarbons will the distance
    pieces be of the two-compartment type?   _____
    (Yes or No)

34. Will the engine have a proven electronic governor? State make. _____
    (Yes or No)

35. Will an electrical tachometer be provided?   _____
    (Yes or No)

Project Phase: Contractor's Drawing Review
Machine Category: Reciprocating Compressors

Designation: _____

Location: _____

Service: _____

P&ID's

1. Have KO facilities been installed on all suction and interstages? _____
   (Yes or No)

2. Do recycle lines re-enter upstream of the KO facilities? _____
   (Yes or No)

3. Are KO facilities equipped with a gauge glass, LHA and shutdown
   switches? _____
   (Yes or No)

4. Are the suction lines traced between the KO drums and the machine
   flanges, including pulsation bottles? Alternatively, is cylinder cooling
   medium (preferably water/glycol mixture) warmer than incoming
   gas? _____
   (Yes or No)

5. Do KO facilities have automatic drains? _____
   (Yes or No)

6. Is there a safety valve on each compression stage and on the coolant
   header? _____
   (Yes or No)

7. Has a remote shutdown switch been provided? _____
   (Yes or No)

8. Have all the alarm and shutdown devices specified been provided? _____
   (Yes or No)

   (a) Low lube pressure alarm _____

   (b) Low lube pressure trip _____

   (c) High-temperature alarm on each cylinder discharge _____

   (d) Low cylinder lube flow alarm _____

   (e) Others (detail) _____

   _____

9. Are TI's and PI's specified for suction and discharge of each stage
   and TI's on discharge and coolant outlet of each cylinder? _____
   (Yes or No)

10. On machines which are specified to be reaccelerated, have controls been provided to unload the cylinders during reacceleration? _____
    (Yes or No)

11. If these controls are of the bypass type, is there a check valve on the compressor discharge? _____
    (Yes or No)

12. Have unloading facilities been provided for start-up? _____
    (Yes or No)

13. Is there a coolant block valve on each cylinder with a low point drain and a high point vent? _____
    (Yes or No)

14. Does each machine have double block valves and a vent to avoid blinding for valve repairs? _____
    (Yes or No)

15. Have pressure taps been provided around the spool which will contain the temporary suction screen? _____
    (Yes or No)

16. Do the oil coolers have provision for back-flushing the water side? _____
    (Yes or No)

17. Will all instruments be changeable on the run? _____
    (Yes or No)

18. Are all lines to remote pressure gauges valved at the tie-in to the main line? _____
    (Yes or No)

19. Is there an isolatable closed circuit at the machine so that the pre-start-up run-in can be performed? _____
    (Yes or No)

20. Are there purge connections which allow gas-freeing the compressor in preparation for maintenance? _____
    (Yes or No)

Layout, etc.

1. Are the main and interstage suction lines from the KO drums or filters cleanable by the method proposed? _____
    (Yes or No)

    If the lines are becoming quite long it may be cheaper to put filters adjacent to the machine to reduce cleaning cost.

2. Are all the cylinder, snubber, and gas cooler supports from the machine foundation? _____
    (Yes or No)

    Note: This requirement does not apply to remote coolers.

3. Are all the piping supports either on the machine foundation or on separate footings going down below the frost line? _____
(Yes or No)

4. The pulsation study will indicate which gas lines have high shaking forces and an estimated magnitude of these forces. Are these lines suitably supported and clamped? _____
(Yes or No)

5. Is there sufficient clearance to pull all pistons and cooler bundles? _____
(Yes or No)

6. Is all contractor's piping such that it does not interfere with access to any valve, distance piece, crosshead, or crankcase cover? _____
(Yes or No)

7. Can all piping to each cylinder be removed to permit cylinder removal? _____
(Yes or No)

8. Where the contractor is doing some of the oil piping, is the piping between the filters and the machine stainless steel? _____
(Yes or No)

9. Are packing vents run separately to a safe location? _____
(Yes or No)

If these vents run into a disposal header under pressure, is there a block valve and check valve at the connection to the header? _____
(Yes or No)

Has each packing vent line a means of individually monitoring for leakage? _____
(Yes or No)

10. Have distance piece vents and drains been provided? _____
(Yes or No)

11. Have all small connections been gussetted? _____
(Yes or No)

12. Has piping been modified as required per acoustic study results? _____
(Yes or No)

Process piping layout: suction piping

13. Are the main and interstage suctions steam traced and insulated? _____
(Yes or No)

14. Are the machine laterals taken off the top of the header? _____
(Yes or No)

15. Is there a manual drain on the header? _____
(Yes or No)

16. If the compressor is handling a flammable gas, are the isolation valves 25 ft from the machine? _____
(Yes or No)

Is the spool piece for the temporary strainer readily accessible for both inspection and removal? _____
(Yes or No)

17. Normally, it should be mounted immediately adjacent to the suction bottle. A strainer is required at each stage unless frame type intercoolers are used.

18. Are blind flanges available at each end of the suction headers to facilitate cleaning, inspection, etc.? _____
(Yes or No)

19. If the contractor's piping ties directly onto CI cylinders, are the flanges flat faced? _____
(Yes or No)

20. Are all valves supported? _____
(Yes or No)

21. Vertical unbraced lines can result in excessive vibration. Has unnecessary flexibility been avoided wherever possible? _____
(Yes or No)

22. Is the suction KO drum within 50 ft of the machine, or if not, has a proven separator been provided within this distance? _____
(Yes or No)

Discharge piping system

23. Are all fittings such as oil separators adequately supported? _____
(Yes or No)

24. Is the piping system flexible enough to keep thermal load stresses on cylinders within acceptable limits? _____
(Yes or No)

25. The liquid condensate on intercoolers and aftercoolers will not run uphill into KO facilities. Assume that it will mist. Therefore, on intercoolers and on aftercoolers where condensate removal is desired, have liquid separation facilities been provided at the low point in the line between the cooler and the KO drum? _____
(Yes or No)

26. On machines mounted above grade with mezzanine floor, has grating been provided for access to all machines valves, distance pieces, and block and bypass valves? _____
(Yes or No)

27. Cylinder ends sometimes blow out. Is all equipment requiring operator attention such as instrument panels, instruments, and block and bypass valves out of direct line with the cylinder ends? _____
(Yes or No)

28. Consider maintenance on the machines. Is there a way of removing all sections of the machines for maintenance? Suitable laydown available for cylinders? If on mezzanine is deck strong enough?

    _____
    (Yes or No)

29. *Consider operability*

    (a) Are all instruments clearly visible?

    _____
    (Yes or No)

    (b) An operator's primary sense is touch. Can he feel all valve covers? _____
    (Yes or No)

    (c) Run through the starting sequence. Can all the operations required be done by one person?

    _____
    (Yes or No)

30. For Engine Compressors consider regrouting. During the life of the machine it will likely have to be regrouted. Is the building design such that heavy lifting equipment can be brought in to lift and move the whole unit?

    _____
    (Yes or No)

Project Phase: Mechanical Run Test in Shop
Machine Category: Reciprocating Compressors

Designation: _____

Location: _____

Service: _____

The purpose of an in-shop running test on reciprocating compressors is to determine and correct any flaws or errors in the machinery manufacture which would delay commissioning of the machine on site. Test should consist of a flush check of frame lubrication cleanliness before run, with a 100-mesh or smaller screen before filter; a run of approximately 8 h; and a physical examination of machine internals after the run to ascertain that all working parts will operate satisfactorily in the field. Usually, because of horsepower limits of shop driver, the test will be run with the machine unloaded. Screens should be inserted in suction and discharge, but valves should be left in. The surge bottles need not be mounted. The manufacturer must have taken all necessary steps to satisfy himself that the machine is ready to run before the witness test is begun. Special note: The review engineer may wish to include certain of the following checklist items in his pre-purchase review with vendors.

1. Is lube screen free of foundry sand and weld slag? If not, crankcase and oil cooler should be reopened to find where it came from.

    _____
    (Yes or No)

2. Does the lube screen show sufficiently clean that the machine may be safely run? Minor amounts of matter on the screen that will be removed by frame filter are acceptable.

_____
(Yes or No)

3. Is auxiliary lube pump capable of supplying pressure?

_____
(Yes or No)

4. Is the pressure drop across frame oil filter normal?

_____
(Yes or No)

5. During the run is there sufficient oil on rod to indicate satisfactory operation of the lubricator?

_____
(Yes or No)

6. Are oil drops showing in all lubricator pumps and is lubricator developing full pressure?

_____
(Yes or No)

7. Are all valve covers cool and do all valves appear to be operating properly?

_____
(Yes or No)

8. Is the machine free of knocks or undue vibrations? If not, machine must be stopped immediately and problems corrected.

_____
(Yes or No)

On completion of run request a contact thermometer, dial indicators, suitable micrometers, and feeler gauges. Have the valve covers, valves, cylinder heads, and crankcase covers been removed and have all packing cases been pulled out of bores? Record measurements.

9. Have all measurements been recorded?

_____
(Yes or No)

10. Are rod clearances in head and partition bores large enough to prevent scraping of rod after normal piston and crosshead wear?

_____
(Yes or No)

11. Are all gasket seats free of paint, casting defects, or tool chatter marks? (Use a flashlight oriented on its side and look for radial shadows.)

_____
(Yes or No)

12. Are valve gaskets solid metal type?

_____
(Yes or No)

13. Is the bore for stuffing box in frame end head free of casting defects?_____
(Yes or No)

14. Is the rod securely fastened to piston and crosshead such that it cannot back off?

_____
(Yes or No)

15. Is there sufficient space in distance piece to readily change packing in cylinder, intermediate diaphragm, and oil scraper rings?

_____
(Yes or No)

16. Are separate covers supplied for each space in distance pieces for packing access? _____
                                                                    (Yes or No)

17. Are vent and drain connections from packing and distance pieces securely piped and labeled? _____
                                                                    (Yes or No)

18. Is the cylinder head gasket face free of any openings to cored water passages in either cylinder or head? _____
                                                                    (Yes or No)

19. Are the gas passages cored such that there are no cavities or depressions for a liquid trap? _____
                                                                    (Yes or No)

20. Are the cylinders firmly supported on the distance piece or cylinder body and not on the head? _____
                                                                    (Yes or No)

21. Is the frame oil circulation connected such that the filter is last in the stream before injection to bearings? _____
                                                                    (Yes or No)

22. Is there an easily removable plug in the piston? _____
                                                                    (Yes or No)

23. Is there any indication of penetration of stud drill or tap in the drilling for cover studs? _____
                                                                    (Yes or No)

24. Are the cylinder analyzer holes drilled and located properly? They should be open to bore through liners and not covered by rings when piston at end of stroke. _____
                                                                    (Yes or No)

25. Are the suction and discharge valves truly non-reversible? _____
                                                                    (Yes or No)

26. Are the supports for the lubricator and piping, and frame oil cooler and piping firm enough to prevent vibration in operation and sturdy enough to withstand shipping without damage? _____
                                                                    (Yes or No)

27. Are the liners, pistons, rods, and crossheads free of any score marks deep enough to catch a finger nail? If not, item must be washed clean and score must be smoothed off. _____
                                                                    (Yes or No)

28. Is the crankcase nameplate securely fastened and does it indicate machine description and serial number and will it be sufficiently viewable when on-site? _____
                                                                    (Yes or No)

29. Does each cylinder carry a nameplate securely fastened and does it indicate cylinder description and serial number and piston end clearances?

    _____
    (Yes or No)

30. Are the rails or sole plates for the machine precoated with epoxy paint for proper adherence to epoxy grout?

    _____
    (Yes or No)

31. Are laminated shim packs of stainless steel available for shipment with machine?

    _____
    (Yes or No)

32. Is rust protection being applied to machine before shipment and is it equal to that specified?

    _____
    (Yes or No)

33. Is the shipping crate study enough to protect the machine, especially the external piping and gauges? All integral pipes and lines must remain installed and connected for shipment.

    _____
    (Yes or No)

34. Question the supplier if the shipper has adequate experience and if shipping method is completely satisfactory to prevent any damage to the machinery in transit.

    _____
    (Yes or No)

Measurements taken immediately after run test (*dimensions in thousandths of an inch*)

Name of Inspector:＿＿＿＿＿＿＿＿＿＿＿＿＿＿＿＿＿＿＿＿＿＿＿＿＿

Date: ＿＿＿＿＿＿＿＿＿＿＿＿＿＿＿＿   Manufacturer: ＿＿＿＿＿＿＿＿＿＿＿＿

Compressor No.: ＿＿＿＿＿＿＿＿＿＿   Model: ＿＿＿＿＿＿＿＿＿＿＿＿＿＿

Serial No.: ＿＿＿＿＿＿＿＿＿＿＿＿＿   Size: ＿＿＿＿＿＿＿＿＿＿＿＿＿＿＿

	Cylinder numbers							
Temperatures	*1*	*2*	*3*	*4*	*5*	*6*	*7*	*8*
Oil at pump $<180°F$								
Main bearing lube end $<225°F$								
Main bearing behind throw to cylinder number $<225°F$								
Conrod shell to cylinder number $<225°F$								
If temperature excessive, rerun until satisfactory and then examine part for wipes								
Rod drop to cylinder (piston high (H) low (L) with dial indicator) <5 mils								
Rod centerline runs below head bore centerline. Micrometer in stuffing box bore top and bottom of rod								
Rod clearance on bottom-cylinder head bore. Feeler gauge reading								
Clearance at crosshead shoe-top. Long feeler gauge. Ample clearance required, but must be uniform across the whole shoe								
Piston clearance in liner with feeler gauge. (If rider rings supplied must be ample to allow for wear)								

Frame oil pressure before filter/after filter (<5 psi across filter) ＿＿＿＿＿＿＿＿＿＿

Standby lube pump pressure ＿＿＿＿＿＿＿＿＿＿＿＿＿＿＿＿＿＿＿＿＿＿＿＿＿＿

Cylinder lubricator pressure if available ＿＿＿＿＿＿＿＿＿＿＿＿＿＿＿＿＿＿＿＿

Project Phase: P&ID and Design Specification Review
Machine Category: Special-purpose Steam Turbines

Designation: _____

Location: _____

Service: _____

Note: The following sections are in two parts. The questions raised under "General" apply to all turbines. Those under "Condensing Turbines" are additional for those machines.

P&ID's

General

1. Is there a warm up vent (at least $1\frac{1}{2}$ in) on the inlet line? _____
   (Yes or No)

2. Does the inlet block have a 1-in bypass for line warm up? _____
   (Yes or No)

3. Does the exhaust valve have a 1-in bypass for warm up? (Back pressure turbines only.) _____
   (Yes or No)

4. Is there a trap and bypass upstream of the trip and throttle valves? _____
   (Yes or No)

5. Is there a trap and bypass on the steam chest of single valve turbines? _____
   (Yes or No)

6. Is there a trap and bypass on the low point of the exhaust casing? _____
   (Yes or No)

7. Is there a low pressure seal vent line on both seals? _____
   (Yes or No)

8. (a) What devices cause a trip of the turbine other than the built-in ones?

   _____

   (b) Have these been specified? Have you considered 2-out-of-3 trip logic? _____
   (Yes or No)

   (c) All special-purpose turbines have a separate trip and throttle valve which is a shutdown device and can be actuated by an electrical signal. Do the process safety shutdowns utilize this device? _____
   (Yes or No)

   (d) Is this provision called out on the turbine data sheet? _____
   (Yes or No)

9. What does the overspeed trip pressure switch actuate? _____

_____

Have these been specified?   _____
                                                              (Yes or No)

10. (a) Has an exhaust line safety valve been provided between the
       turbine and the exhaust block valve?   _____
                                                              (Yes or No)

   (b) Is the safety valve setting at or below the maximum design
       exhaust system pressure?   _____
                                                              (Yes or No)

11. All special-purpose turbines should have either an approved/
    redundant electronic or a hydromechanical/electric governor. If
    automatic control of the turbine speed is desired, has provision of
    an air head been called out on the data sheet?   _____
                                                              (Yes or No)

   If an increase in signal is not to increase speed, has this requirement
   been specified?   _____
                                                              (Yes or No)

12. Has the following instrumentation been specified?

   (a) Inlet and exhaust TI's   _____

   (b) Inlet and exhaust PI's   _____

   (c) Steam chest PI (only on single valve units)   _____

   (d) 1st stage pressure   _____

   (e) Other stage pressures   _____

   (f) Steam flow   _____
                                                              (Yes or No)

13. Have the requirements for turbine washing been specified? _____
                                                              (Yes or No)

   Based on experience, multistage turbines may often require washing
   during first year of operation. Consider full wash facilities. For some
   locations only tie-ins may have to be considered. All large (over
   2000-hp) turbines will probably be multistage on 600 psi to 125 psi
   steam.

14. (a) Has the API data sheet been prepared?   _____
                                                              (Yes or No)

   (b) Has location been specified?   _____
                                                              (Yes or No)

   (c) Have applicable specifications been detailed?   _____
                                                              (Yes or No)

(d) Have the instruments required on the local panel been specified? _____
(Yes or No)

(e) Has the lube oil specification sheet been completed if this is not to be from the driven equipment?
_____
(Yes or No)

Condensing turbines

1. Has a suitable shaft seal system been provided?
_____
(Yes or No)

Usually, the sealing steam will be taken from a pressure tap on the casing but for start-up a pressure-controlled live steam supply must be provided. Adequate traps must be installed to keep water out of the seals.

2. Is there an exhaust pressure safety device to relieve pressure in the event of cooling water failure?
_____
(Yes or No)

3. Is this safety device water sealed?
_____
(Yes or No)

4. Is there a minimum flow recycle arrangement on the condensate pumps?
_____
(Yes or No)

5. Is the condenser level control and bypass (if valved) arranged for max. pumpout on air failure?
_____
(Yes or No)

6. Are the pump glands sealed by discharge pressure?
_____
(Yes or No)

7. Is the pump suction chamber vented back to the condenser steam space?
_____
(Yes or No)

Project Phase: Preorder Review with Vendor
Machine Category: Special-purpose Turbines

Designation: _____

Location: _____

Service: _____

1. Are there any deviations to your specification other than those previously disclosed?

_____

If yes, list and discuss with your Machinery Specialist.

2. Does the vendor have adequate experience? _____

                                              (Yes or No)

(a) Location of similar machine _____

(b) How closely does it conform to the proposal? _____

	Yours	*Others*
Blading	_____	_____
Inlet pressure	_____	_____
Inlet temperature	_____	_____
Exhaust pressure	_____	_____
Exhaust temperature	_____	_____
rpm	_____	_____
Tip speed	_____	_____
Blade passing frequency	_____	_____
Number of stages	_____	_____
Type of seals	_____	_____
Governor type	_____	_____
Reviewed by Machinery Specialist	_____	_____

(c) Have any similar machines experienced difficulty either on test
or in the field? _____

                                              (Yes or No)

If so, what were the characteristics of the problem and what steps were taken
to correct it and to avoid repetition?

(d) Comments on users and their machines.

How long have machines been in service (min. 1 year) _____

What start-up difficulties were experienced? _____

Is so, was the Service Department's response satisfactory? _____

                                              (Yes or No)

Did the machines meet the specified duty and efficiency? _____

                                              (Yes or No)

What has been the maintenance experience? _____

_____

Name and position of contact: _____

(e) Were the comparison machines built in the shop where your
machines are to be built?   _____
                                                          (Yes or No)

(f) If your machine is to be built in a shop other than the vendor's main shop:

- How is the parent company or licensor going to ensure that the machines
  will be built correctly?

  _____

  _____

- How many similar machines have been built in that shop? _____

  _____

- How similar are they? _____

  _____

- Who will do the engineering and drafting for the machines?

  _____

- Will the machines be an exact duplicate of the licensor's
  design?   _____
                                                          (Yes or No)

- If not, what modifications will be incorporated and why?

  _____

  _____

(g) *Tip speed*

- Is the tip speed above 825 ft/sec at maximum design speed? _____
                                                          (Yes or No)

- If so, API-612 says an integrally forged shaft and wheel
  arrangement is preferred. Is this what you are getting?   _____
                                                          (Yes or No)

If not, consider following up on the vendor's experience in this area. Look
especially at the rpm of "similar" machines because centrifugal forces
increase as the square of rpm while tip speed is a linear function. Also
check on steam temperatures.

3. If the comparison machines are not identical, how much extrapolation has the bidder done on the established experience limits?

   (a) Are the extrapolations based on experience with other machines? _____

   _____

   (b) Do these extrapolations appear sound? _____

                                                  (Yes or No)

   (c) Describe extrapolations beyond previous limits of experience _____

   _____

   (d) Do these extrapolations appear sound? _____

                                                  (Yes or No)

4. What design and shop practice modifications have occurred since the similar machines were built?

   (a) Rotor including shrink specifications _____

   (b) Casing _____

   (c) Bearings _____

   (d) Shaft seals _____

   (e) Running clearances _____

   (f) Wheel design and manufacture _____

   (g) Diaphragms and fixed nozzles _____

   (h) Materials _____

   (i) Balancing _____

   (j) Tip speeds _____

   (k) Overspeed trip design _____

   (l) Overspeed testing of rotor _____

      Which of these changes are definite design improvements and which are cost reduction items?

      _____

      _____

      Consult with your designated Machinery Specialist on all items which are cost reduction items.

5. *Shaft sealing*

   (a) What type of external shaft seals are proposed? _____

   _____

(b) If carbon ring:

- Is the rubbing speed below a conservative maximum of 160 ft/sec?

  _____

  (Yes or No)

- The shaft should be hard chrome or ceramic coated in the seal zone. Is it?

  _____

  (Yes or No)

- How many rings are there? _____

  Maximum pressure per ring is 35 psi and minimum number of rings is 4.

- Is there a satisfactory vent to atmosphere part way down the seal to prevent lube oil contamination if the seal partially fails?

  _____

  (Yes or No)

(c) If labyrinth packing:

- Has a vacuum vent system been provided?

  _____

  (Yes or No)

- Is there an inter-packing vent which could be connected to the 15-psig steam system?

  _____

  (Yes or No)

  (Manufacturers often underestimate seal leakage and vent condenser is therefore undersized.)

  Is the labyrinth compatible with the shaft bearing surface?_____

  (Yes or No)

6. How are the machine expansions handled to minimize thrusts due to coupling slip forces and expansion?

   (a) Where is the casing anchored axially? _____

   Where is the thrust bearing? _____

   If these are not at opposite ends of the casing, there will be a large shaft movement into the coupling. Is the coupling adequate to take this movement? (Question the acceptability of this approach.)

   (b) In which direction will the coupling slip force act relative to the internal thrust forces?

   _____

   (c) Assuming the coupling slip force to be

   $$F = 0.3\,\frac{T}{d} = \frac{18{,}900 \times P}{\text{rpm} \times d} \quad \text{lb}$$

   where T = torque (lb-in),
   P = horsepower,
   d = shaft diameter at the coupling (in).

Are all thrust bearings designed for normal thrust forces plus and minus the coupling slip force? _____

(Yes or No)

If not, why not? _____

(d) What residual thrust capacity is available in the bearings to overcome the thrust due to blade deposits?

_____

_____

Is this capacity sufficient to cover fouling which would reduce the hp output by 20%? _____

(Yes or No)

Are the wheel discs provided with balance holes to reduce thrust increase due to fouling? _____

(Yes or No)

7. *Machine integrity*

(a) As mounted on the baseplate, will the machine retain its internal alignment during shipping? _____

(Yes or No)

(b) Will the machine have to be opened up on site for final adjustments before starting? _____

(Yes or No)

(This is a contractor's problem, but allowance must be made in the schedule for it. Also, the turbine service persons will be required.)

8. *Governing*

(a) Is the governor to be a hydromechanical/electric type or an electronic model? _____

(Yes or No)

(b) How stable will it be at minimum specified temperature? _____

_____

(c) If the governor oil system is separate, has an oil heater been provided? _____

(Yes or No)

(d) How is the maximum speed stop override operated in testing the overspeed trip? Does this seem a controlled operation?

_____

_____

(e) Is a minimum speed stop provided? _____

(Yes or No)

9. *Critical speeds*

   (a) Have the critical speeds been determined and do these meet your requirement of a 20% speed margin from all running speeds? _____

      (Yes or No)

   (b) How were the criticals calculated?

      Rigid supports _____

      Flexible supports _____

      Only flexible support calculations are acceptable.

   (c) What bearing stiffness was used in the critical calculation? _____

      _____

      This stiffness should be between 1 and 10×10^6 lb/in.

   (d) How closely have actual criticals established on test agreed with calculations?

      _____

      If there is any doubt about the turbine vendor's ability to calculate criticals, insist that the vendor with train responsibility check the calculations.

10. *Blade vibrations*

   (a) Has the vendor submitted Campbell diagrams for all turbine blades over 5 in long? _____

      (Yes or No)

      These diagrams must show all three fundamentals (tangential, axial, and torsional) and their harmonics up to the maximum frequency of excitation (usually blade passing frequency). Margins of 10% must be maintained. Excitation due to harmonics above passing frequency can be tolerated because of low energy levels.

   (b) How does the vendor propose to meet nozzle and blade experience criteria?

      By test _____

      By demonstration _____

      Be very cautious about demonstration. Are there the same number of nozzles and blades? Is the speed range the same (5% may be critical).

11. *Journal bearings*

   (a) What is the journal speed? _____ ft/min

   (b) What is the bearing loading? _____ psi

   (c) If over, what type of bearing does the vendor propose to use? _____

      _____

(d) Can he demonstrate that he has used exactly the same bearing in a similar condition of load and speed and with a critical speed as low or lower in percent of running speed than our application? _____

_____

This bearing design problem is critical. We would like to have tilting pad bearings but many turbine manufacturers have no experience with them or will not use them. If the investigation leaves one in doubt about the bearing experience, try to insist on tilting pad journal bearings.

12. *Combined T&T valve or separate T&T valve*

(a) Manufacturer's name and model number _____

(b) Does the valve have a built-in 5-mesh monel strainer?   _____

(Yes or No)

Note: An additional separate external strainer would be advantageous.

(c) Does the valve have a pilot arrangement to assist opening?   _____

(Yes or No)

If not, you should either get another type of valve or put a small bypass on the isolation valve for run-up.

(d) Does the valve have the partial stroke feature as specified?   _____

(Yes or No)

13. *Starting*

If the turbine is to operate at an exhaust pressure of 30 psi or above:

(a) Will heating the casing from the exhaust header with the turbine stationary be acceptable?

_____

(Yes or No)

(b) If not, how must the turbine be heated to prevent shaft distortion?

_____

Dependent on the answers to 13(a) and (b), check the flow plan to ensure that provision has been made to run up in the correct manner.

(c) Has the entire train been checked out to see that turbine warm-up procedure will not cause lubrication problems in any bearing in the train?

_____

(Yes or No)

(d) What procedure will be followed in design to ensure that the vibration when passing through the critical will not damage the labyrinths? _____

_____

Most manufacturers have programs which will predict shaft movement but may not use them unless you ask.

14. *Shaft thermal stability*

   (a) What is the chance of the shaft developing a thermal bow if the machine is
   tripped with the exhaust left open?

   _____

   _____

   (b) Is there some time limit to get the machine rolling again?   _____

   (Yes or No)

   (c) If there is a strong possibility of a thermal bow, should a turning
   gear be provided?   _____

   (Yes or No)

   If the vendor thinks a thermal bow is likely, then turning gear
   should be provided.

   (d) If turning gear is provided, will the speed be sufficient to provide
   hydrodynamic lubrication to all the bearings in the train?   _____

   (Yes or No)

   If not, what is proposed to ensure adequate lubrication? _____

   _____

15. *Lube viscosity*

   (a) Is there any incompatibility between the lube viscosity requirement of the turbine and of the driven equipment?   _____

   (Yes or No)

   There should be no problem with this but the question should be
   asked.

16. *Allowable piping forces*

   (a) Will the allowable piping forces be to NEMA SM-21?   _____

   (Yes or No)

   (b) What evidence can the vendor supply to indicate that these forces will not
   cause excessive casing strains?

   _____

   _____

   There has been feedback from a number of sources which indicates that many
   turbines will not tolerate the forces allowable under SM-21 as calculated by
   piping stress analysis.

   (c) Does the vendor have sufficient knowledge of piping arrange-
   ments that he could analyze the contractor's proposed
   arrangement and suggest necessary modifications?   _____

   (Yes or No)

   (d) Would the contractor accept this arrangement?   _____

   (Yes or No)

17. *Tachometer*

   (a) Whose tachometer will be provided? _____

   API-612 says the tachometer will be the pulse counter type or equal. The vibrating reed unit is not equal.

   (b) Will the tachometer be mounted so that it can be seen by an operator at the T&T valve?   _____

                                                                        (Yes or No)

18. *Shaft access for vibration*

   (a) How does the vendor propose to provide access to the shaft at both bearings to permit shaft vibration readings with a hand-held pickup?

   _____

   _____

19. *Nozzles*

   (a) Are all bladed nozzles replaceable in the field? They must be.   _____

                                                                        (Yes or No)

   (b) Will all the nozzle block bolts be wired to prevent unscrewing?_____

                                                                        (Yes or No)

   If not, how will they be fixed? _____

   _____

   If in doubt, request wiring.

   (c) Are any of the trailing edges in the blades less than 0.02 in thick?   _____

                                                                        (Yes or No)

   If they are, insist they be beefed up to 0.02 in.

20. *Washing*

   Will turbine washing produce any problems with:

   (a) Thermal expansion? _____

   (b) If the water were to fail during the wash when the steam inlet was at saturation, would this result in insufficient differential expansion to cause failure?

   _____

   (c) Thrust bearing load? _____

   (d) What load could be expected from the turbine with the inlet saturated?

   _____

Project Phase: Contractor Drawing Review
Machine Category: Special-purpose Steam Turbines

Designation: _____

Location: _____

Service: _____

Note: The following sections are in two parts. The questions raised under "General" apply to all turbines. Those under "Condensing Turbines" are additional for these machines.

P&ID's

General

1. Is there a warm up vent (at least $1\frac{1}{2}$ in) on the inlet? _____
   (Yes or No)

2. Does the inlet block have a 1-in bypass for line warm up? _____
   (Yes or No)

3. Does the exhaust valve have a 1-in bypass for warm up? (Back pressure turbines only.) _____
   (Yes or No)

4. Is there a trap and bypass upstream of the trip and throttle valve? _____
   (Yes or No)

5. Is there a trap and bypass on the steam chest of single valve turbines? _____
   (Yes or No)

6. Is there a trap and bypass on the low point of the exhaust casing? _____
   (Yes or No)

7. Is there a low pressure seal vent line on both seals? _____
   (Yes or No)

8. If the vendor has specified a pressure for this vent, has satisfactory control been provided? _____
   (Yes or No)

9. What devices cause a trip of the turbine other than the built-in ones? _____

   _____

   Is this as specified? _____
   (Yes or No)

10. What does the overspeed trip pressure switch actuate? _____

    _____

    Is this as specified? _____
    (Yes or No)

11. If there is no built-in strainer in the trip and throttle valve, is there a Y-type strainer in the inlet line? (Prefer to have both!) _____

(Yes or No)

12. Has an exhaust line safety valve been provided between the turbine and the exhaust block valve if the exhaust pressure is above 75 psig? _____

(Yes or No)

Is the safety valve setting at or below the max. design exhaust system pressure (including exhaust casing)? _____

(Yes or No)

13. Back pressure turbines with labyrinth seals must have an eductor and condenser. Are these shown? _____

(Yes or No)

14. Has the following instrumentation been provided?

Inlet and exhaust TI's _____

Inlet and exhaust PI's _____

Steam chest PI (only on single valve units) _____

1st stage pressure _____

Steam flow _____

15. Have turbine washing facilities been provided? _____

(Yes or No)

Condensing turbines

1. Has a suitable shaft seal system been provided? _____

(Yes or No)

Usually the sealing steam will be taken from a pressure tap on the casing, but for start-up a pressure-controlled live steam supply must be provided to keep water out of the seals.

2. Is there an exhaust pressure safety device to relieve pressure in the event of cooling water failure? _____

(Yes or No)

Is this safety device water sealed? _____

(Yes or No)

3. Is there a minimum flow recycle arrangement on the condensate pumps? _____

(Yes or No)

4. Is the condenser level control and bypass (if valved) arranged for max. pumpout in case of air failure? _____

(Yes or No)

5. Are the pump glands sealed by discharge pressure? _____

(Yes or No)

6. Is the pump suction chamber vented back to the condenser steam space? _____

(Yes or No)

7. Has the following additional instrumentation been provided?

   (a) Vacuum gauge on inlet and interstage of ejector _____

   (b) Seal steam pressure gauge _____

Piping layouts

1. Is the inlet steam taken off the top of the main header and is there a trapped dead leg on the header downstream of the turbine take-off? _____

(Yes or No)

2. Does the inlet slope continuously between the washing desuperheater connection and the machine flange with no pockets of any kind to trap water? _____

(Yes or No)

3. Can the inlet pipe be readily diverted outside for initial blowout? _____

(Yes or No)

   The piping not to be blown must be capable of being thoroughly inspected internally.

4. Can the trip and throttle valve be manipulated easily from the main platform? _____

(Yes or No)

   This valve is used to start up the turbine and control its speed when out of governor range.

5. The turbine exhaust safety valve must be removable for testing with the machine in service. Is it located so that if it is dropped it will not cause damage to other equipment? _____

(Yes or No)

6. On air blower drivers, are all steam vents on both inlet and exhaust lines away from and above the air intake hood? _____

(Yes or No)

7. Has the inlet and exhaust pipe been provided with sufficient direction anchors so that all piping growth will be away from the turbine? _____

(Yes or No)

8. Are the inlet and exhaust pipe supports, guides, anchors, etc., as described in the piping stress calculation? _____

(Yes or No)

   Has sufficient allowance for friction been made in the stress calculation? _____

(Yes or No)

9. Before start-up the operator must blow all steam line and casing drains. Are all these valves accessible for him to do this? _____

(Yes or No)

These drains are normally taken to a funnel. Is the location of the funnel such that the steam venting from it will not interfere with other equipment nearby or cause a hazard? _____

(Yes or No)

10. Piping must not run unnecessarily over parts which must be removed for maintenance, i.e. bearing covers, top half casing, governor, and trip and throttle valves. Has this problem been avoided and have crane capacity and lifting height been checked? _____

(Yes or No)

11. Is there a place where the casing top half can be set down during maintenance without interfering with maintenance? _____

(Yes or No)

Can it be moved to this location without passing over other equipment which might be running? _____

(Yes or No)

Can the rotor be removed to the maintenance shop without passing over running equipment? _____

(Yes or No)

12. Is the exhaust steam trap adequate to dispose of all the water required to make the inlet steam 1% wet? _____

(Yes or No)

Is it located on the low point of the exhaust? _____

(Yes or No)

(If the casing will readily drain into the line the trap should be on the line. Otherwise, it should be on the casing.)

13. Can the operator safely open the inlet and exhaust block valves? _____

(Yes or No)

14. Will the operators be able to manipulate the turbine washing system safely and in a controlled manner? _____

(Yes or No)

Consider operability

1. Are all instruments clearly visible? _____

(Yes or No)

2. Does the operator have safe and easy access to all the bearings? _____

(Yes or No)

3. Has he clear access to the manual trip? _____

(Yes or No)

4. Can he see clearly the tachometer from the governor overspeed device? _____

(Yes or No)

5. Can he see clearly the tachometer from the trip and throttle valve?_____

(Yes or No)

6. Run through a starting sequence. Can all the operations required be done by one man? _____

(Yes or No)

Project Phase: Mechanical Run Tests
Machine Category: Special-purpose Steam Turbines

Designation: _____

Location: _____

Service: '_____

The mechanical run test for steam turbines is basically a means of checking rotor balance, controls, safety trips, and checking for leaks.

The basic procedure should be to run up to speed with steam conditions as close to design as possible. When conditions stabilize, including bearing and lube oil temperatures, the turbine should be operated for a period of 1 h, with no further rise in bearing and lube oil temperatures.

(On condensing turbines, steam inlet temperature or test period may be reduced to prevent excessive temperature in the turbine casing.)

	Conditions	
	Design	*Test*
1. Steam inlet pressure	_____	_____
2. Steam inlet temperature	_____	_____
3. Steam exhaust pressure	_____	_____
4. Steam exhaust temperature	_____	_____
5. Lube oil pressure	_____	_____
6. Lube oil inlet temperature	_____	_____
7. Speed (a) Max. continuous	_____	_____
(b) Normal operating	_____	_____
(c) Trip	_____	_____
(d) Calculated critical	_____	_____
8. Max. vibration	_____	_____

1. Do conditions 1, 2, 3, 5, 6, 7, and 8 on test match the design conditions to your satisfaction?

_____
(Yes or No)

2. Is the turbine half coupling (with adaptor, if necessary) fitted for the test?

_____
(Yes or No)

3. Is there a steam strainer on the inlet?

_____
(Yes or No)

4. Is the actual critical speed within 5% of the calculated value?

_____
(Yes or No)

5. Bearing temperature: Is the temperature rise across each bearing less than 60°F?

_____
(Yes or No)

6. *Vibration*

(a) Frequency survey when running at max. continuous speed. Note vibrations at running speed and other frequencies.

Probe location	Magnitude (mils)	Frequency (cpm)
_____	_____	_____
_____	_____	_____
_____	_____	_____
_____	_____	_____
_____	_____	_____
_____	_____	_____

Is there an absence of vibration at critical frequency, also at frequencies between 35% and 50% of running speed?

_____
(Yes or No)

(b) Shaft and bearing housing vibration attenuation:

	A, shaft	B, housing	A/B, attenuation
I.B. bearing	_____	_____	_____
O.B. bearing	_____	_____	_____

Is the attenuation less than 4? (If not, state actual figures in the report.)

_____
(Yes or No)

(c) Vibration readings (mils):

	Just below trip speed	*Max. continuous speed*	*Difference*
J.B. bearing	_____	_____	_____
O.B. bearing	_____	_____	_____

Is the difference in vibration levels at these speeds less than 20% _____

(Yes or No)

7. *Overspeed trip*

(a) The overspeed trip must be actuated at least three times. Is the difference between the highest and the lowest trip speeds less than 0.5% of the highest? _____

(Yes or No)

(Note any problems with adjustment of trip, and operation of the trip valves.)

Note the actual final setting of the trip. _____ rpm

Is this approximately 110% of max. continuous speed? _____

(Yes or No)

8. Are all auxiliary trips being tested – low lube oil pressure, etc.? _____

(Yes or No)

9. Is a thorough check being made for leaks, steam, oil, air to governor, etc.? Finally satisfactory? _____

(Yes or No)

10. Is the machine stable at all speeds? (Hunting within 0.5%.) _____

(Yes or No)

11. Is the speed control operating satisfactorily? _____

(Yes or No)

12. (a) Are the data being taken, including the linearity of speed versus control signal? _____

(Yes or No)

Record	Signal											
	Speed											

(b) On loss of control air, is the result as specified? _____

(Yes or No)

13. Bearing inspection: Do bearing surfaces show normal wear patterns? _____

(Yes or No)

14. If there is a spare rotor to be run, request an internal inspection (normally specified).

 (a) Is there an absence of rubbing?   _____

    (Yes or No)

    If rubbed, demand a clearance check.

15. Check the internal alignment and clearance data from final assembly drawing (if available).

 (a) Is alignment good; clearances within tolerances?   _____

    (Yes or No)

 (b) Likewise for a spare rotor if it has been fitted.   _____

    (Yes or No)

16. Have copies of vendor's log sheets, and final internal clearance diagrams been obtained?   _____

    (Yes or No)

17. When witnessing a test, always try to find out if any difficulties occurred in preparing for it. Such problems could be repeaters.

Project Phase: Mechanical Run Tests
Machine Category: Lube and Seal Oil Consoles

Designation: _____

Location: _____

Service: _____

Lube and seal oil consoles are inspected during fabrication, erection, and the flushing operation carried out in the manufacturer's shop.

On consoles and seal oil drainer packages, the following procedure should be followed by the inspector:

1. Is the oil piping at the pumps rigidly supported?   _____

    (Yes or No)

2. With pump piping flanges unbolted, is the piping alignment satisfactory?   _____

    (Yes or No)

3. Are all valves and strainers accessible?   _____

    (Yes or No)

4. Have all piping and valves been inspected internally, probing with a magnet? Finally, no machining chips, welding spatter, machining burrs, weld dross, burn through, flux, and other contaminants?   _____

    (Yes or No)

5. Has all lacquer been removed from bends and fittings?

    _____
    (Yes or No)

6. Have cooler bundles been pulled and checked for cleanliness?

    _____
    (Yes or No)

7. (a) Reservoir interior clean?

    _____
    (Yes or No)

   (b) Paint work satisfactory?

    _____
    (Yes or No)

8. *Pumps*

   (a) Alignment to drivers satisfactory?

    _____
    (Yes or No)

   (b) With dial indicators on pumps, is alignment satisfactory while
       jumping on console base?

    _____
    (Yes or No)

       (If not, check whether console is to be grouted, or is supported
       as level in shop as expected in the field.)

9. Whole system successfully hydrostatically tested?

    _____
    (Yes or No)

10. *Flushing*

   (a) Are all console discharges connected to the tank by temporary
       bypasses?

    _____
    (Yes or No)

   (b) Is the flushing oil temperature at 180°F?

    _____
    (Yes or No)

   (c) Are vibrators being used to shake pipework?

    _____
    (Yes or No)

   (d) Are all control valve bypasses, four-way valves on filters and
       coolers, being swung periodically?

    _____
    (Yes or No)

   (e) Is this initial flushing carried out for 8 h uninterrupted?

    _____
    (Yes or No)

   (f) Control function tested successfully?

    _____
    (Yes or No)

   (g) After initial flush, system checked by installing felt pads, backed
       up with SS mesh in the temporary bypasses and filter outlets.
       System re-flushed for 2 h, oil at 180°F, flow as high as possible,
       pipework vibrated?

    _____
    (Yes or No)

   (h) Bypass pads clean?

    _____
    (Yes or No)

If no, flushing must continue.

(i) Filter outlet pads clean? _____

(Yes or No)

If no, filters to be overhauled to determine the cause of the filter leakage.

11. Will the console be shipped with the temporary bypasses installed? _____

(Yes or No)

(Wanted so that flushing can start in the field as soon as the console is set on its foundation.)

12. *Drainer packages*

(a) Are all valves accessible? _____

(Yes or No)

(b) Have all volume chambers, sight glasses, traps, pipework, and valves been inspected internally, probing with a magnet? _____

(Yes or No)

(c) Check made for no evidence of lacquer? _____

(Yes or No)

Project Phase: Field Handling, Storage and Installation
Machine Category: All

Designation: _____

Location: _____

Service: _____

1. Are machinists available to assist in checking for damage, etc.? _____

(Yes or No)

2. Has machine/unit been checked for transit damage? _____

(Yes or No)

3. Are blinds on flanged openings still tight? (If not, retighten or renew.) _____

(Yes or No)

4. Are all other openings plugged or blinded? _____

(Yes or No)

5. Is the paint covering on machine/unit still good? No signs of rust? (Rust should be removed and area repainted.) _____

(Yes or No)

6. Check all items against packing list.

    (a) Anything short?

    _____
    (Yes or No)

    (b) Anything damaged?

    _____
    (Yes or No)

    (c) Has this been reported?

    _____
    (Yes or No)

7. Have all loose items been restored in closed boxes?

    _____
    (Yes or No)

    (a) Have these been stored in a limited access area?

    _____
    (Yes or No)

    (b) Has a record been made of where these items are stored?

    _____
    (Yes or No)

8. How much time is expected between receipt and start of installation? Give strong consideration to using oil mist as a "preservative blanket" for all machine internals!

    _____

9. Have specific instructions regarding rotation of rotors, crankshafts, etc., been included in the vendor's service manual?

    _____
    (Yes or No)

10. Have oil reservoirs been checked for presence of water and drained if necessary?

    _____
    (Yes or No)

11. Have oil reservoirs been topped up with lube oil or rust preventative? _____

    (Yes or No)

12. Following (11) above, has the rotor or crankshaft been turned two complete revolutions? (This includes small pumps as part of a package.)

    _____
    (Yes or No)

    (a) On reciprocating compressors, operate the hand pump if available, and crank the cylinder lubricator if the machine has the cylinders installed.

13. Has a program been established for regular rotation of shafts – two turns at 2-week intervals? Include draining water from oil.

    _____
    (Yes or No)

14. If time in (8) above is over 1 month, have blinds been removed, and machine internals inspected to determine condition of protective coatings? Renew if necessary. This should be repeated at 2-month intervals.

    _____
    (Yes or No)

15. Has plastic sheeting been placed over casings?   _____

    (Yes or No)

    Has a breathing space been left open?   _____

    (Yes or No)

16. Are all exposed machined surfaces coated with rust preventative?   _____

    (Yes or No)

17. Have reciprocating compressor valves been stored in a container of light oil?   _____

    (Yes or No)

18. Has it been arranged that lube and seal units will be installed as soon as possible in order to put them into operation?   _____

    (Yes or No)

    (These can be flushed by discharging directly back to the tank while waiting for hook up to machinery piping.)

19. Has all major equipment not stored within warehouse, etc., been stored in a place where damage from construction activities and traffic is least likely? Again, has oil mist preservation been considered?   _____

    (Yes or No)

20. Gear units – preserved in vendor's shop for extended storage – should be stored such that unit will not be turned. No oil to be added until finally installed. Consider use of appropriate diester or polyalphaolefin synthetic lubricant.

Glossary

Availability: The probability that a system or piece of equipment will, when used under specified conditions, operate satisfactorily and effectively. Also, the percentage of time or number of occurrences for which a product will operate properly when called upon.

Concept: Basic idea or generalization.

Confidence limit: An indication of the degree of confidence which one can place in an estimate based on statistical data. Confidence limits are set by confidence coefficients. A confidence coefficient of 0.95, for instance, means that a given statement derived from statistical data will be right 95% of the time on the average.

Configuration: The arrangement and contour of the physical and functional characteristics of systems, equipment, and related items of hardware or software; the shape of a thing at a given time. The specific parts used to construct a machine.

Corrective maintenance: Unscheduled maintenance or repair actions, performed as a result of failures or deficiencies, to restore items to a specific condition. *See also Unscheduled maintenance* and *Repair.*

Cost-effectiveness: A measure of system effectiveness versus life-cycle cost.

Critical: Describes items especially important to product performance and more vital to operation than noncritical items.

Discounted cash flow analysis: A method of making investment decisions using the time value of money.

Distributions: *See Probability distribution.*

Downtime: That portion of calendar time during which an item or piece of equipment is not able to perform its intended function fully.

Emergency maintenance: Corrective, unscheduled repairs.

Engineering: The profession in which knowledge of the mathematical and natural sciences is applied with judgment to develop ways to utilize economically the materials and forces of nature.

Environment: The aggregate of all conditions influencing a product or service, or nearby equipment, actions of people, conditions of temperature, humidity, salt spray, acceleration, shock, vibration, radiation, and contaminants in the surrounding area.

Equipment: All items of a durable nature capable of continuing or repetitive utilization by an individual or organization.

Exponential distribution: A statistical distribution in logarithmic form that often describes the pattern of events over time.

Failure: Inability to perform the basic function, or to perform it within specified limits; malfunction.

Failure analysis: The logical, systematic examination of an item or its design, to identify and analyze the probability, causes, and consequences of real or potential malfunction.

Failure Mode Effect Analysis (FMEA): Identification and evaluation of what items are expected to fail and the resulting consequences of failure.

Failure rate: The number of failures per unit measure of life (cycles, time, miles, events, and the like) as applicable for the item.

Function: A separate and distinct action required to achieve a given objective, to be accomplished by the use of hardware, computer programs, personnel, facilities, procedural data, or a combination thereof; or an operation a system must perform to fulfill its mission or reach its objective.

Hardware: A physical object or physical objects, as distinguished from capability or function. A generic term dealing with physical items of equipment – tools, instruments, components, parts – as opposed to funds, personnel, services, programs, and plans, which are termed "software".

Item: A generic term used to identify a specific entity under consideration. Items may be parts, components, assemblies, subassemblies, accessories, groups, equipment, or attachments.

Life-cycle: The series of phases or events that constitute the total existence of anything. The entire "womb to grave" scenario of a product from the time concept planning is started until the product is finally discarded.

Life-cycle cost: All costs associated with the system life-cycle, including research and development, production, operation, support, and termination.

Maintainability: The inherent characteristics of a design or installation that determine the ease, economy, safety, and accuracy with which maintenance actions can be performed. Also, the ability to restore a product to service or to perform preventive maintenance within required limits.

Management: The effective, efficient, economical leadership of people and use of money, materials, time, and space to achieve predetermined objectives. It is a process of establishing and attaining objectives and carrying out responsibilities that include planning, organizing, directing, staffing, controlling, and evaluating.

Material: All items used or needed in any business, industry, or operation as distinguished from personnel.

Mean Time Between Failure (MTBF): The average time/distance/events a product delivers between breakdowns.

Mean Time Between Maintenance (MTBM): The average time between both corrective and preventive actions.

Mean Time Between Replacement (MTBR): Average use of an item between replacements due to malfunction or any other reason.

Mean Time To Repair (MTTR): The average time it takes to fix a failed item.

Median: The quantity or value of an item in a series of quantities or values, so positioned in the series, that when arranged in order of numerical quantity or value, there are an equal number of values of greater magnitude and of lesser magnitude.

Model: Simulation of an event, process, or product physically, verbally, or mathematically.

Modification: Change in configuration.

Normal: Statistical distribution commonly described as a "bell curve". Mean, mode, and median are the same in the normal distribution.

On-condition maintenance: Inspection of characteristics which will warn of pending failure, and performance of preventive maintenance after the warning threshold but before total failure.

Operating time: Time during which equipment is performing in a manner acceptable to the operator.

Predictive maintenance: Predictive maintenance is a maintenance method that involves a minimum of intervention. In its simplest form it is based on the old adage "don't touch, just look". In the context of process machinery, predictive maintenance is practiced through machinery health monitoring methods such as vibration and performance analysis.

Preventive Maintenance (PM): Actions performed in an attempt to keep an item in a specified operating condition by means of systematic inspection, detection, and prevention of incipient failure. *See also Scheduled maintenance.*

Probability distribution: Whenever there is an event E which may have outcomes $E_1, E_2, \ldots, E_n$, whose probabilities of occurrence are $p_1, p_2, \ldots, p_n$, one speaks of the set of probability numbers as the p.d. associated with the various ways in which the event may occur. The word probability distribution refers therefore to the way in which the available supply of probability, i.e. unity, is "distributed" over the various things that may happen.

Production: A term used to designate manufacturing or fabrication in an organized enterprise.

Random: Any change whose occurrence is not predictable with respect to time or events.

Re-rating: Alteration of a machine, a system or a function by redesign or review for change in performance; mostly, but not always, for increased capacity, etc.

Rebuild/recondition: Total teardown and reassembly of a product, usually to the latest configuration.

Redundance: Two or more parts, components, or systems joined functionally so that if one fails, some or all of the remaining components are capable of continuing with function accomplishment; fail-safe; backup.

Refurbish: Clean and replace worn parts on a selective basis to make the product usable to a customer. Less involved than rebuild.

Reliability (R): The probability that an item will perform its intended function without failure for a specified time period under specified conditions.

Repair: The restoration or replacement of components of facilities or equipment as necessitated by wear, tear, damage, or failure. To return the facility or equipment to efficient operating condition.

Repair parts: Individual parts or assemblies required for the maintenance or repair of equipment, systems, or spares. Such repair parts may be repairable or nonrepairable assemblies or one-piece items. Consumable supplies used in maintenance, such as wiping rags, solvent, and lubricants, are not considered repair parts.

Repairable item: Durable item determined by application of engineering, economic, and other factors to be restorable to serviceable condition through regular repair procedures.

Replaceable item: Hardware that is functionally interchangeable with another item but differs physically from the original part to the extent that installation of the replacement requires such operations as drilling, reaming, cutting, filing, or shimming in addition to normal attachment or installation operations.

Safety: Elimination of hazardous conditions that could cause injury. Protection against failure, breakage, and accident.

Scheduled maintenance: Preplanned actions performed to keep an item in specified operating condition by means of systematic inspection, detection, and prevention of incipient failure. Sometimes called preventive maintenance, but actually a subset of PM.

Spares: Components, assemblies, and equipment that are completely interchangeable with like items and can be used to replace items removed during maintenance.

Specifications: Documents that clearly and accurately describe the essential technical requirements for materials, items, equipment, systems, or services; including the procedures by which it will be determined that the requirements have been met. Such documents may include performance, support, preservation, packaging, packing, and marking requirements.

Standards: Established or accepted rules, models, or criteria by which the degree of user satisfaction of a product or an act is determined, or against which comparisons are made.

Standard deviation: A measure of average dispersion or departure from the mean of numbers, computed as the square root of the average of the squares of the differences between the numbers and their arithmetic mean. It is also a measure of uncertainty when applied to probability density distribution.

Standard item: An item for common use described accurately by a standard document or drawing.

Surveillability: A qualitative factor influencing reliability. It contains such considerations as accessibility for surveillance and monitoring of a machine or its function(s), etc.

System: Assembly of correlated hardware, software, methods, procedures, and people, or any combination of these, all arranged or ordered toward a common objective.

Training: The pragmatic approach to supplementing education with particular knowledge and assistance in developing special skills. Helping people to learn to practice an art, science, trade, profession, or related activity. Basically more specialized than education and involves learning what to do rather than why it is done.

Troubleshooting: Locating or isolating and identifying discrepancies or malfunctions of equipment and determining the corrective action required.

Unscheduled Maintenance (UM): Emergency maintenance (EM) or corrective maintenance (CM) to restore a failed item to usable condition. Often referred to as breakdown maintenance.

Warranty: Guarantee that an item will perform as specified for at least a specified time.

Index

The *WEIRD* Disappearance of Jordan Hall

A Richard Jackson
Book

The WEIRD Disappearance of Jordan Hall

○

○

●

JUDIE ANGELL

Orchard Books

New York and London
A division of Franklin Watts, Inc.

Orchard Books
387 Park Avenue South, New York, New York 10016
Orchard Books Great Britain
10 Golden Square, London W1R 3AF England
Orchard Books Australia
14 Mars Road, Lane Cove, New South Wales 2066
Orchard Books Canada
20 Torbay Road, Markham, Ontario 23P 1G6

Orchard Books is a division of Franklin Watts, Inc.

Manufactured in the United States of America
Book design by Mina Greenstein
The text of this book is set in 12 pt. Galliard.
9 7 5 3 1 2 4 6 8 10

Library of Congress Cataloging-in-Publication Data
Angell, Judie.
The weird disappearance of Jordan Hall.
Summary: When Emma's sixteen-year-old boyfriend steps
   an old box in her father's magic shop and becomes
   they become involved in bringing him back to vis-
   bility. [1. Magic—Fiction] I. Title.
   .A5824We 1987 [Fic] 87-7781
ISBN 0-531-05727-5
0-531-08327-6 (lib. bdg.)

For the kids at
Windward High School

C. C. Major cleared his throat and raised his glass.

"All right, everyone," he said, "it's time for the toast!"

Emma Major, fifteen, lifted her glass of punch in a serious tribute, but her little sister, Polly, sang out with glee:

> *"Hail to the victors valiant,*
> *Hail to the conquering heroes,*
> *Hail, hail to Mich-i-gan—"*

"Wait a minute, *wait* a minute!" Harvey Leff cried. "Hold it, Polly—I'm going to Syracuse University, not Michigan!"

"I know," Polly said with a shrug, "but 'Hail to the Victors' is the only college song I know. And you *are* going to college. . . ."

"Please." C.C.'s commanding voice took charge

again and everyone quieted down. "I will make the toast," he said, "and Polly, no sipping the punch until I've finished. Harvey—" he began and cleared his throat again. "Harvey, we're going to miss you here at Major Magic. Your help in the shop has always been above and beyond the call of duty—"

"Like when he let Artemis's doves loose in the stockroom?" Polly interrupted. Emma nudged her.

"You've been our stockboy, salesperson, demonstrator, and all-round factotum for—what is it, two years now?"

Harvey nodded.

"Two years," C.C. continued, "after school, weekends and summers. And here you are, off to college!"

"What C.C. means, Harvey"—Deirdre Major smiled at the boy—"is we'll miss you!" She drank from her plastic champagne glass.

"We'll miss you, Harvey. And good luck!" C.C., Polly, and Emma chimed in and drank.

"Well, thanks," Harvey said with a shy smile. "Thanks for everything. I'll sure miss you all and" —he looked around —"everything here at Major Magic. It's a wonderful store . . . a great place to work. . . ."

"Cut your cake, Harv!" Polly coaxed, tugging at his sleeve.

•2

Emma sighed. She took her plate of ice cream cake to the counter near the cash register and leaned over it dreamily, her two hands cradling her chin.

Harvey came to stand near her. "What's wrong, Emma?" he asked. "I'll be back every vacation."

Emma laughed and patted his shoulder. "I was thinking how much harder I'll have to work now that you're going, Harv. And especially with school starting. It'll just be the three S's—school, store, and sleep!"

"But your dad'll be hiring a replacement for me."

"Whenever he finds one," Emma complained. "He's been interviewing for over a week. And then we'll have to train him—"

"Is that you groaning, Emma?" Deirdre called from the other end of the store.

"I'm not groaning, Mother. . . ."

"Yes, you are. I can tell."

"It's just her age, Mrs. Major," Harvey called. "Fifteen is a difficult time. I remember."

In spite of herself, Emma laughed. "Harvey, can you really remember back that far?"

"There are people looking through the window," Polly said. "We'd better turn the sign around and open up again. They don't know why we're closed in the middle of the day."

But C.C. held up his hand. "We're closed for

an hour," he said, "for a double celebration. First and foremost, of course, was to bid farewell and Godspeed to Harvey Leff, the best assistant Major Magic's ever had!"

"Hear, hear!" Deirdre cried.

"And the second celebration is for an announcement!"

"An announcement?" Emma looked questioningly at her father. Then her face lit up. "Oh," she cried, "you've found a replacement!"

"Not yet, Emma. That's not it."

C.C. smiled at his wife. "Emma, Polly, Harvey—and Deirdre, who already knows—I'd like to announce Major Magic's acquisition of the Langhorn estate!"

"What?" Emma blinked.

"Our store has acquired, through auction, the entire Louis Langhorn estate! Isn't that cause for celebration?"

"It certainly is, C.C.," his wife said. "We'll have a grand sale! It'll be a real windfall."

"Louis Langhorn!" Harvey exclaimed. "The famous magician?"

"The same!" C.C. pointed to a picture on the wall. "I hung that portrait of him for good luck sixteen years ago, Harvey. Before any of the stock went on the shelves. He was the best of the best, all right. Here's to you, Langhorn!" C.C. saluted the man in the frame.

"What'd you get, Daddy?" Polly asked. "What'd you buy?"

"Oh, lots of things, lots of wonderful things. You'll see when the crates arrive. We'll open and sort every one."

Emma rolled her eyes.

"Emma, you're groaning again," Deirdre said.

"I'm not! I haven't said a word!"

"Yes, you have. I can hear you. 'Harvey's leaving, just when all this stuff arrives, I'll never get out of the stockroom, I'll be a slave to the store all my life. . . .' "

"Mom, I didn't say that!"

"Didn't you?"

"No."

"Emma?"

Emma laughed. "Well, not out loud, anyway," she said.

"But I can always hear you," her mother said, grinning. "After all, this is a magic store."

●

The four Majors lived on the West Side of New York City in a large loft above their store: a magician's supply shop that C.C. and Deirdre had owned and operated since before Emma was born.

The loft consisted of a huge living and dining area with a small kitchen alcove off to one side and two bedrooms, each with its own bathroom, at the far end. The bedrooms were divided by a long hall at the end of which was the back door of the loft and a staircase leading down into the stockroom of the shop. The front of the loft was entered by a door and stairway from the street. The Majors had no neighbors in the building.

"We can make as much noise as we want," Polly often said, "because there's no one to hear us downstairs. All my friends who live in apartments have to be quiet after seven o'clock."

And tonight Polly was taking advantage of that freedom by jumping rope in the bedroom she shared with Emma.

"*E*, my name is Ellen and my husband's name is Ed; we come from East Brunswick and we sell eggs! *F*, my name is Frances and my husband's name is Frank; we come from Florida and we sell frankfurters! *G*, my name is Gloria and my husband's name is—"

"Polly!" Emma cried.

"No. George. My husband's name is George."

"No, I mean you have to stop now. You're making the floor shake, and I have to color my signs."

Emma had three large poster boards spread out before her on the floor, and she had lettered lightly on each one with pencil. Polly put down her jump-rope and bent over her sister's shoulder.

"*Help Wanted*," she read out loud.

"They're signs for a new shop assistant," Emma said. "I'm going to put them up at school. And I'm going to make little cards to stick on the bottom with our phone number and address on them—so the kids can just peel the cards off and call us when they get home or from a phone booth or something."

"That's a good idea," Polly said.

"It was Harvey's."

"It's still a good idea. Let's see . . . Hold up the sign."

"I haven't finished coloring it."

"Hold it up anyway. Hmmmm . . . *Assistant in a Magic Store. We need combination stockperson-salesperson. Will train. Lots of Fun—Meet Famous People.* What famous people?" Polly asked.

"Magicians. People like that. Uncle Meissen, for example."

"Oh. Yeah. I keep forgetting—Uncle Meissen's pretty famous."

"And The Great Gilhooley, Diavolo, Mr. Barrymore—"

"Yeah, I see, I see. But mostly, the ones in the shop are amateurs . . . people off the street. And kids . . ."

"Well, I'm not going to say *mostly amateurs* on my sign," Emma said. "I want to attract kids at school. I don't want to say it's just like working in any store. Which it really is . . ."

"No, it isn't, Emma." Polly sounded so serious that Emma looked up from her sign painting. "Working in a magic store is special. A magic store's a special place."

Emma smiled. I guess, she thought, a magic store *is* a special place if you're nine. But when you're fifteen . . . well, it just takes a lot of time. And when it's your parents' store, you don't even get paid! "Say," she said out loud, "I think I'll do the 'Meet Famous People' part in yellow and outline it in black. What do you think, Pol?"

But Polly was looking toward their bedroom window. "Emma . . ." she whispered. "Look at the window. On the fire escape."

Emma glanced over. "Awwww," she said. "A kitty. How'd you get all the way up here, kitty?"

Polly made a face at her sister. "How do you think—wings? Fire escapes are for climbing, Emma. I'm letting it in."

"You better not. Mom wouldn't like it."

"But it's a stray. And it found its way to *our* window, which means it's our job to take it in. Besides, you don't want it to freeze out there."

"In September? Come on!"

•8

"I'm letting it in!" Polly declared and walked to the window. She looked at Emma out of the corner of her eye, and when her sister went back to coloring, she opened the window slowly. The cat stalked through as if it had always belonged and began to wash itself on top of the girls' radiator.

"Look, Emma . . . He's all black except for one white eye and ear. Isn't that strange?" Polly began to stroke the cat's fur.

Suddenly, without warning, the cat leaped off the radiator and walked deliberately toward the spot where Emma was sitting on the floor.

"Kitty, kitty, *pssssss*," Polly hissed, but the cat didn't even hesitate.

"Hey!" Emma cried as the cat put one sooty paw down on the word "Magic" on her clean white poster. As Emma jumped to her feet, the cat scooted back to the radiator and out the window to the fire escape.

"Hey!" Polly cried, but the cat was gone.

"Oh, great," Emma muttered. "Close the window before you get the rest of New York's animals in here. Look what he did! A paw print, right on my sign . . ."

Polly was giggling. "I think it looks good," she said. "It'll attract people."

Emma sighed. "It'll probably attract other cats."

○

● Emma looked up when the bell rang, and blinked. She and one boy were the only students left in the classroom. She hadn't seen anyone leave: How could the time have gotten away from her like that?

"May I have your papers, please?" the teacher asked, and Emma and the boy brought them to her desk. The boy stretched his arms.

"Where is everyone?" he asked, just then noticing the empty room.

"They all left," Emma answered. "Well, it is just a placement test. I guess they didn't care where they got placed if it meant they could have a period free. So they finished the test in ten minutes."

The boy shrugged. "Guess so," he said. "It's important to me, though. I need honors math. It'll look good on my applications for college. I need a scholarship."

•10

They began to walk toward the door together.

"Are you a senior?" Emma asked. "This test was for eleventh-grade math."

"Oh no, I'm a junior. I just"—he grinned—"like to plan ahead."

Emma smiled at him as she pulled a piece of paper from her unzipped purse. She stopped walking to study it.

"Good. I thought so," she said.

"What's that?"

"I have lunch this period. I'm not late for some class and I don't have to rush." She tucked the schedule back into her purse. "It takes me forever to memorize my schedule. . . ."

"I have lunch this period, too," the boy said. "Want to, uh, eat together?"

Emma raised her eyebrows. "Sure. Why not? Are you new here?"

"You mean at Franklin Pierce High? No . . . but I was going to ask you the same thing. You're a junior?"

"Uh-huh."

"I guess we just haven't had the same classes or knocked each other down in the halls."

Emma smiled. "Well," she said, "it is a huge school. Anyone could get lost in it. I mean, this year I only have one class with any of my friends. French, first period."

•11

"Well, if we both get into honors math, you'll have a class with another friend," the boy said. "Me."

They had reached the door of the cafeteria, and he held it open for her.

"What's my friend's name? Besides 'me', I mean," Emma asked as they walked toward the stack of trays.

He laughed. "Jeez, I'm sorry. It's Jordan. What's yours?"

"Emma."

"Well, great," he said, grinning at her. "Great."

●

They had lunch. Or almost had lunch. They both talked so much they barely ate. They found out they both liked movies, hated heavy metal, read murder mysteries, were on time with assignments, and sometimes played the same record over and over and over for hours, driving everyone else crazy.

"This was a great lunch," Jordan said when the bell rang.

Emma said, "Yeah. . . ."

"I'd really like—" Jordan began.

"What?"

"I was going to say I'd really like to meet you after school, but . . . I have some stuff to do. . . ."

Emma nodded. "Me too."

•12

"Another time, though."

"Oh, definitely."

"Great." Jordan grinned.

They went to their classes.

●

Emma whispered "Jordan" to herself as she walked down West 38th Street.

He was really nice, she thought.

Two years at Franklin Pierce and she'd never even run into him. Well, maybe she had and never noticed. But she was noticing now!

He was really nice!

She stopped outside a boutique and stared at the clothes in the window. Black leather skirt, yellow cotton sweater, a plaid scarf you could wind around your whole body three times . . .

Emma wrinkled her nose. She liked long skirts and high-buttoned collars for dressing up. Victorian.

Jordan.

He was *really* nice.

Emma wondered what it was he had to do after school. Maybe he had a job, as she did. If he had to work, then that meant he'd be busy *every* day after school.

Well, but so was she, so it didn't matter. There were still weekends . . . Friday nights . . . Saturday nights . . .

But he hadn't asked her out yet. He'd only said "another time. . . ."

No, they had had too good a time at lunch. He'd find her again. And they could probably have lunch again together. Tomorrow! And the day after!

Emma nearly stumbled over a Chihuahua at one end of a long leash. " 'Scuse me," she said to the woman at the other end.

He said he needed a scholarship for college, Emma remembered. So he probably *is* working after school. He needs the money.

He really was *nice*. . . .

I daydream too much, Emma decided.

She entered the loft from the street entrance. She wanted to put off checking in at the store until she'd arranged her new books and changed out of her skirt and blouse.

"Hi, dear," her mother called from the kitchen alcove. "Better hurry. Dad's going crazy down there with all the kids."

"What kids?"

"The ones applying for the job, of course. You should know; you put up the signs. Here. Have a cookie."

Emma bent over, took the cookie in her mouth, and put her books on the counter.

"You mean the kids are from Franklin Pierce? They came because of my sign?"

"Uh-huh. They're all carrying those little cards you stuck to the bottom. Have a glass of punch. I have gallons left. Don't know why I made so much. . . ."

"No, thanks. Did you pick anyone yet?"

"I don't think so. They're still coming in. Listen, honey, I really just came up for a quick snack. I'm going down again now, and you should, too. Dad's getting overwhelmed, what with the customers and the interviewees—"

"Okay, okay, I'm going." Emma finished another cookie.

"How was the first day?"

Emma beamed at her mother. "Wonderful!" she cried.

"Really? Well, that's some reaction to the start of school. Polly just said 'okay.' "

"Well, Polly didn't meet Jordan."

"Ohh," her mother said and nodded.

"I'll tell you about it downstairs," Emma promised.

"You'd better. I'll be in the stockroom." And Deirdre swept down the hall to the back of the apartment in one of the three long flower-print skirts she liked to wear around the store.

Emma came down through the back door in her usual costume: jeans. Her mother looked up from her bookkeeping niche—a desk crammed into a corner surrounded by cartons and crates.

"I'll have to hear about Jordan later," she told Emma. "Just go on into the store. Dad's about to go bonkers."

"It's high time, Emma!" C.C. said, waving a hand at her. The other hand was holding the phone. "I think you'd better come down," he was saying. "I can't hire over the phone." He signaled Emma to the counter where a customer was waiting. "Yes, we're open until nine. That's right, nine. Thank you, 'bye." He hung up. "Emma, what did you put on those signs, anyway? 'Free food?' 'Free money?' I think your whole high school's been in or called."

"Oh, good! Did you hire anyone?"

"I haven't had time to consider. Take the cash register, will you? Polly!"

Polly appeared from between two aisles. "Yes?"

"Honey, go in the back and ask Mom if Artemis's scarves arrived. And if they did, make sure they're either pale-blue-and-white or pale-pink-and-white or a combination of both but nothing else. Got that?"

"Got it."

"Thank you." He turned to a teenage girl with spiky blond hair who had come into the shop. "And how may I help you, young lady?" he asked.

"I came about the after-school job?" she said. "The poster said you were looking for someone?" She spoke in questions. "It said you would train?"

"Well, we will, of course, but it would help if you've had some experience. . . ."

"Gee, I never worked in a store before, but it looked like fun. I mean, y'know, magic and all. I like to watch magic tricks."

C.C. smiled. "We don't see a lot of performances here in the store. Do you know my daughter, Emma, from school? She's the one who put the posters up."

The girl looked up and for the first time noticed Emma at the far end of the counter.

"Oh, hi," she said and waggled her fingers.

Emma said, "Hi."

"We had a class together last year," the girl said. "Gym."

"Uh-huh," Emma concurred. The girl turned away, and Emma held up her hand and made the thumbs-down sign at her father.

"Well, look, uh—"

"Marjorie. Marjorie Malone."

"Well, Marjorie, why don't you write your name and phone number down on that pad over there—

it's a list of applicants. We'll get back if we need you. Okay?"

"Okay."

When she had gone, C.C. said, "Emma, take a look at that list and put a star next to anyone who might work out, okay?"

The phone rang and C.C. took care of another applicant.

Polly came out of the stockroom.

"The scarves are purple," she announced.

"*Purple!*" C.C. bellowed.

"Dark purple."

"Tell your mother to call Arnold Nagle. Tell him Artemis is sending a messenger to pick up her pale-blue-and-white or pale-pink-and-white or a combination-of-both scarves and they'd better be here or I start doing business again with his brother!"

Polly scooted back.

The door opened, and a tall boy entered the shop.

"Billy!" Emma cried. "Hi! Are you here about the job?"

"Yep."

"Dad, this is Billy Dempsey. He was my bio partner last year."

Her father and Billy shook hands.

"Guess everybody's been coming down, huh, Emma?" Billy said.

"Lotta kids . . ."

"Yeah, we were talking at lunch. Nice to have a job where you can meet some famous people. Most jobs are boring, you know? We'd, like, haul crates off the trucks, or maybe run errands around town, or, you know, sometimes check the customers out. What would I do here?"

C.C. scratched his head. "Oh, probably haul crates off the trucks, run some errands, help out at the cash register, stuff like that. . . ."

"Oh," Billy Dempsey said.

"Emma, what *did* you put on those signs?" C.C. asked again, and Emma replied, "I *wrote* what the job was, honest, Dad! I guess the magic store was what got them."

●

It was nearly six and nearly quiet when the door opened and Jordan walked in. He was wearing a suit and tie, and he had put something on his hair to make it sit flat on his head. Emma didn't recognize him.

"May I help you?" she asked, barely looking up as she made notes on a pad.

"Emma?" Jordan's eyes were wide.

"J-Jordan? Is that *you?* I mean, under that hair?"

Embarrassed, Jordan smoothed a cowlick.

"I'm sorry. It looks . . . nice, except I didn't recognize you," Emma stammered. "What are you doing here?"

Jordan reached into his pocket and pulled out one of Emma's neatly lettered cards. "Gee . . . if you're working here, then that means the job's already taken, huh? I *thought* about coming here right from school, but then I figured maybe if I changed into something nice I might stand a better chance of—"

"Wait, Jordan, wait," Emma said, holding up her hand. "I do work here, but it's my parents' store. I made the sign you saw and that little card with our phone number. Didn't I tell you my last name at lunch today?"

"I— You might have, but maybe I wasn't paying attention. Well, I mean, there was so much to pay attention to— I mean— Listen, no, I guess I didn't get your last name. Or if I did, I didn't associate it. I'm really sorry. I didn't put you and the job together."

"It's okay." Emma was laughing. "But it really wasn't necessary to get dressed up."

"I know it's a stockroom job, but for the first meeting I just wanted to look my best."

C.C. clapped Jordan on the shoulder from behind, where he'd been listening.

"Frankly, I think that's an impressive way to apply for a job. Don't you, Emma?"

"Oh, definitely," Emma replied.

"And you didn't come just to meet famous people?" C.C. asked.

"No, sir. I came for the job. I worked all summer at the aquarium, but I wouldn't have the time to commute all the way out there with school starting and everything. And I do need a job." His back was turned to Emma, and above his head she enthusiastically gave her father a thumbs-up sign.

C.C. held out his hand. "I'm C. C. Major," he said. "Your new boss."

"Great!" Jordan cried. "That's great!" He shook the hand.

"Well?" C.C. asked.

"Well . . . what?"

"What's *your* name, son? I need that, you see, so I know what to yell when I want you. Also your address, phone number, and social security number."

Jordan turned red, which charmed Emma. "I'm really sorry, Mr. Major. I was just excited about the job. Do you want references, too?" He reached into his pocket. "I have a letter from the man I worked for at the aquarium and one from the director—"

"It's all right, son, I'm trusting my own instincts—and my daughter's—on this one. But I still need your name."

"Jordan. Hall. Jordan Hall."

"Jordan . . . Hall," Emma said softly. "That's a nice name."

●

After they had discussed hours and salary and Jordan had left, C.C. gave his daughter a hug. "Thanks, honey. I think you picked a good one. And saved me the trouble of finding someone myself, too."

"So that's *the* Jordan," Deirdre said from the stockroom doorway. "I can see why you're impressed with him, Emma."

Emma smiled at her mother.

"Well, since he's so impressive, why didn't you mention him before?" C.C. wanted to know. "We could have gone straight to the source instead of dealing with all those glamour-seeking kids."

"I didn't know him before, Dad. I just met him today."

"You just met him *today*? I hired someone you only met *today*?"

"Uh-huh," Emma said and went back to filing her receipts.

"I like him," Deirdre said.

"Well, I do, too. . . ." C.C. rubbed his chin. "I

have a feeling about him. He strikes me as one of those all-together people, you know? Really *there*. I think he'll be just fine."

"A good choice," Deirdre put in.

"I have the same feeling," Emma said.

○

● Jordan had put the chicken in the oven to roast before his mother came home from work, so it was almost ready as she walked through the door.

If Jordan and Emma found many things in common, their mothers were certainly not one of them. Deirdre Major was tall, with black hair swept into a tight chignon. She wore bright colors and lots of jewelry and red polish on her fingernails, and sometimes big fake flowers in her hair. She looked like an off-duty opera star.

Olivia Hall was small, with rounded shoulders. She wore tailored suits and flat shoes. Her pale brown hair was brushed back from her face and then hung straight to her shoulders. She was a harassed executive secretary working for the vice president of a toy manufacturer, and she looked it.

•24

"Hi, Jordy," she said, hanging her shoulder bag on the hook near the door. "You remembered to put the chicken in. . . . Smells good. . . ."

"It'll be ready in a minute. Guess what? You know how you always said to look presentable for any job interview? Well, I took your advice and I got a job today!"

"Oh, honey . . ."

"What's the matter?"

"Well, I didn't mind your working in the summer. It kept you busy and the money came in handy, but with school starting . . . I hate to see you spend your off-time—"

"Don't worry about it. We need the money. And besides, it has a nice benefit."

Olivia took off her shoes and sat on the couch. "What kind of benefit?" she asked as she rubbed her tired feet.

"The benefit is fifteen years old, a junior at F.P.H. She's about five-two with short black hair that's kind of curly, and grayish eyes. Her name's Emma Major."

"Does she go with the job?"

"Yes, she does. Her parents own the store where I'm working. I met her today."

"Well, Jordy."

" 'Well, Jordy' what?"

"Oh . . . nothing . . . It's just that you're growing up so fast. I can't believe you're a junior in high school already."

"You've been saying that all summer, Mom."

"Have I?"

"Yeah. And before that, you said, 'I can't believe you're a sophomore.' And before that—"

"All right, you can stop. I get the picture."

"No, no, not yet. I can remember when you said, 'I can't believe you're in kindergarten already!' " He gave her a wink. "Get ready, Mom, because the time is coming when—"

"I know, I know, don't say it." She got up from the couch and went into the kitchen to unload the dishwasher, which she started every other morning as she left for work. Since there were only the two of them, there weren't that many dishes used, and Olivia liked to think she was saving energy.

She looked up and smiled at her canary, whose cage hung from a ceiling bracket over the tiny kitchen table.

"Hel-lo, Henry," she sang at the bird, but received only a slight ruffle of feathers in response. "Henry's quiet today!" she called to Jordan.

"Henry's always quiet!" Jordan called back.

It was true. Henry sang beautifully, but only twice a year, usually in the spring. Olivia had acquired him after Jordan's father left, but Henry wasn't much of a companion, remaining hunched up in his feathers most of the time. Still, she'd heard Henry sing more times than she'd heard from her wandering husband—who now lived somewhere in Arizona. Or Nevada. Or Wyoming. Or all of the above—Olivia and Jordan were never sure.

But she counted her blessings. She had Jordan.

"Tell me about the first day of school!" she called.

"I had lunch with Emma Major!" Jordan called back.

Olivia smiled to herself. "What *else*? Surely there must have been more than lunch . . . Subjects? Teachers?"

There was a pause. Then Jordan yelled, "I don't remember anything else!"

●

Neither Jordan nor Emma remembered much about that first day of school except each other.

Jordan was supposed to work at Major Magic on Tuesdays, Fridays, and Saturdays, but he showed up on the other days, too, and helped out, though without pay. He and Emma just wanted to be together.

"But without customers!" Emma sighed at Jordan Friday evening as they were closing up. "We haven't had a real date since we met!"

"But that's only been nine days," Jordan said.

"And besides, we did have a date. We went to the movies last Saturday night."

"Mmmmm," Emma grumbled, "but we had to take Polly!"

"Yeah, I know. . . ."

"And you were the one who offered to take her!"

"I know, but I felt bad for the kid. She doesn't really do anything but hang around the store. We're all too busy to do much with her."

Emma nodded. "You're right. And it really was sweet of you," she told him. "Polly didn't realize there were other kinds of shows besides magic shows. And you made a friend for life. She thinks you're wonderful."

"Do *you* think I'm wonderful?" Jordan touched her arm.

"Wonderful," Emma said.

"I wish we could go out tomorrow night. By ourselves," Jordan said, looking into her eyes. "But I did promise my mother I'd go with her to Aunt Elizabeth's—"

"I know. It's always something," Emma said. "I just wish—"

The phone shrilled, piercing the air and making them jump.

"Oooooh," Emma said through clenched teeth.

"I'll get it," Jordan said. He squeezed her fingers

and raced to the counter. "Good evening," he announced. "You've reached Major Magic, but we're closed— Oh. Hi, Mom."

Exasperated, Emma shook her head and moved off to the stockroom.

Her parents, she noticed, had left the light on so that she could find her way upstairs and Jordan could see his way out the back door into the alley. Emma looked around at the shelves, at all kinds of magicians' paraphernalia: brightly colored fans and fake flowers and birds, glittering golden boxes and ice-blue satin top hats. She smiled to herself. Other girls and their boyfriends sit in their living rooms watching TV, she thought. We get to stand in this little room and look at each other through feather boas! She picked up a pink feather from the floor, held it for a moment in her palm, then blew it away. That's how our time goes. . . .

Suddenly she shrieked. Something had rubbed against her bare leg.

"You again!" Emma cried as she spotted an all-black cat with one white eye and ear. "You scared me to death! How did you get in here, anyway?" As she bent to stroke its back, Jordan hurried into the little stockroom.

"Did I hear you scream?" he asked anxiously. "Are you okay?"

Emma looked up at him. "Sorry," she said. "It

was this cat. He scared me." She stood up and faced him.

"That was my mom," he said.

"I know."

"She wanted to know if I'd missed a dentist appointment today. She said I hardly seem to remember these important things since I started 'hanging out' here."

"Did you?"

"Did I what?"

"Miss your appointment."

"Oh. No, it's not until next week."

"Well, what did she say when you told her that?"

Jordan smiled. "She said she thinks I'm bothering you and your family, being here all the time. Even on days I'm not supposed to be working."

Emma smiled back. "Bothering us, huh?" she said.

"Yup. I said I didn't think I was bothering you."

"Uh-uh," Emma said. She sat down on a three-legged stool, and Jordan pulled up a crate next to her.

"I told her I'd be home a little later tonight," he said. "You're right, Emma. We hardly have any time together."

"It's not that," Emma said. "We have lots of time together. Except we never have any peace and quiet just to talk and think and be alone. A whole world of parents and teachers and magicians and delivery people and customers and salespeople, and *everybody* is always right there with us!"

Jordan nodded. "That's true," he said.

"I wish we could disappear together, go to never-never land or someplace. Just the two of us."

"Mmmmm. I think the M-104 bus goes there. Hey, look at that. . . ." He pointed to the cat, who was poised and ready to jump to a top shelf.

"Oh, get him, Jordan!" Emma cried. "He's going to hurt himself or knock something down! Send him outside. I wonder how he got in here in the first place. . . ."

Jordan opened the outside door and gently nudged the cat out. "Why, on the M-104 bus, of course," he said.

"From never-never land," Emma said with a smile.

"*Emma!*" A yell from the top of the stairs.

"Yes, Dad?"

"*Check the lock on the back door, will you?*"

"I did!"

"*What?*"

"I said, I *did!*"

"*Oh! And is it okay?*"

"Yes! It's okay!"

"*Because if it isn't, have Jordan pick up a good one on his way to work tomorrow! The estate stuff's coming in the morning, and I want everything secure!*"

"*It's okay!*" Emma shouted up the stairs.

"*Oh. Good. Thanks. And come on up. You've got homework.*"

Emma turned back to Jordan. "You see?" she said. "You see what I mean?"

○

● On Tuesday, Emma and Jordan sat across from each other at a table in the school cafeteria.

"Is this the tuna?" Emma asked, staring at a mound on her paper plate. "Or is this the cole slaw? I can't tell the difference—it all tastes like mayo with sticks in it."

Jordan laughed. "I see what you mean," he said. "Remember yesterday? The 'Monday Special' was cottage cheese with fruit, but it still looked exactly like what you've got there. Here." He pushed a wrapped sandwich toward her. "Have this," he said. "My mother insists on making my lunch. She's done it since I was in kindergarten, and she says she'll do it for my honeymoon. Anyway, you can easily tell what it is. Roast beef."

"Thanks." Emma took the sandwich and waved to a friend, who smiled back. "See that girl?" she said to Jordan. "I had lunch with her every day last year. And this year I've been eating only with you since the first day."

"Do you feel bad about that?"

"Oh, no! But maybe I feel bad that I don't feel bad. Or something. Anyway, I hope she's not mad at me."

"She didn't look too mad when she smiled and waved at you. Listen, I used to have lunch with Avery Weisfeld last year. And when we finished eating, we'd go shoot baskets in the gym or out in the yard."

"Do you miss that?"

"Not at all. Avery's in my French class, and he must've grown about three inches during the summer. I'd never be able to sink anything with him in the way now."

"Very funny."

"Say, how about *you* going one on one with me? At least I'd have a fighting chance with you. You're a lot shorter than I am."

"Yeah, but a lot better! Come on."

•

"You let me win!" Emma said, as they walked home that afternoon.

Jordan looked at her with wide eyes. "I did not," he said.

"You did. I could tell."

"No, I didn't," he said, smiling, "but if you want

to think so, it's okay with me. And I wish you'd tell it to Avery Weisfeld next time you see him. His laughter was what really ruined my game!"

When they arrived at the store, a surprise greeted them. Boxes were piled near the door on the sidewalk. C.C. and Polly were carrying them in, and someone special was helping them.

"Uncle Meissen!" Emma cried. "I didn't know you'd be here today!"

"Ah! Give your Uncle Meissen a kiss, sweet girl. It's been a while since I last saw that pretty face."

Smiling, Emma gave their old family friend a hug. The Illustrious Meissen, as he was billed in the business, wasn't really a relative, but he had known C.C. even before the opening of Major Magic and the girls loved him.

"Well, now!" Meissen boomed, his arm around Emma's shoulder, "Who's *this?*"

"*This* is Emma's boyfriend," Polly volunteered before anyone could speak.

"He has a *name*," Emma said tartly to her sister.

"His *name*," Polly repeated in the same tone, "is Jordan Hall."

Jordan extended his hand, which Meissen took and pumped up and down.

"Pleased to meet you, sir," Jordan said. "I've certainly admired your work. I loved your last TV special. . . ."

•35

"Thank you, Jordan. You like Emma Major and you liked my TV show . . . I must say, you certainly are a man of taste."

"Let's not stand around," C.C. said. "Emma, your mother's in the store so that we all can bring in and sort these things, so let's get a move on. Jordan, why don't you grab that big box over there? I'll help you, and we'll stand it up against the wall inside."

"What is this, Mr. Major?"

"This? It's all from the Langhorn estate."

"I mean this box."

"*That?*" Meissen interrupted. "*That* box? Be careful how you hoist it, now. . . ."

"Why?" Jordan bent to lift the lower end.

"That was Louis Langhorn's disappearing box. C.C. thinks it should bring the biggest price of anything in the collection, don't you, C.C.? I do too. It should. Hmmm," he said, as Jordan and C.C. carried it past him. "Needs a good polishing, I should say."

With five of them working, the collection was soon stacked neatly in the stockroom, and C.C. cleared an area for unpacking and sorting.

"Geeee," Polly breathed. "Look at all this stuff. . . ."

There were long boxes. Short cartons. Paper-wrapped parcels. Loosely packed shopping bags.

There was even a huge red-framed poster with a picture of Louis Langhorn—an announcement of an appearance by "The Greatest Magician the World Has Ever Known."

"Fascinating," C.C. said, holding up a brightly colored stuffed parrot. "Wouldn't you swear this was real?"

"Yeah, I guess so," Jordan said.

"I wonder how he used this thing?"

"Maybe it was a pet once," Jordan offered.

"And will you look at these!" C.C. ran his fingers through an open box of gold bangles, bracelets, and hoops of different sizes. Some were encrusted with chips of colored stones and beads.

"He did love flash and dash," Meissen said with a sigh. "I was just a boy when Langhorn was the Man of the Hour, but I saw him. And I still remember some of the wonderful things he did. All modern magicians owe him a debt."

Polly reached behind the knot of Meissen's tie and pulled out a quarter.

"This is for my part of the debt!" she announced, and as Meissen laughed and mussed her hair, she told him, "I want to be the Woman of the Hour, the best magician there is someday."

"Careful, Meissen, she's after your job," C.C. warned.

Meissen laughed. "She can have it. At least for

a while. Seems like I've been on the road forever! And tonight I'm off to Atlantic City for a week."

"Atlantic City sounds like fun," Emma said.

"Oh, it can be, unless you're working two shows a night. Then you don't see much of it because you use your days for rest. At least," he added, "you do if you're *my* age."

Deirdre poked her head into the stockroom. "C.C.? Madame Nadia's on the phone."

"Speaking of *age!*" Meissen chortled.

"Now, Meissen," C.C. said as he went into the store, "she's not *that* old."

"Not *that* old!" the magician roared. "Eighty-four if she's a day and still on the road! I want to speak to her, C.C.!"

Polly watched the men leave and then turned to Jordan. "Madame Nadia mostly works the southern states," she explained. "She loves Florida. She says the climate's easy on her bones, and the retirees are crazy about her act. They say she gives them hope for the future."

Emma smiled at Jordan. "I have hope for the future too," she said. "Now you'll really have to be here all the time. Your mother can't keep saying you're hanging around and bothering us."

"Yeah . . ." Jordan sighed and they sat and stared at each other.

"Mph," Polly sniffed. "Guess you want me to leave."

Jordan and Emma didn't answer.

"Guess you want me to go into the store now and help Mom."

Jordan and Emma continued to gaze at each other.

Polly got up. "Guess maybe I will," she said and headed for the door. Just once, she looked back. "Oh boy," she muttered.

●

"Does everybody look that goony when they're in love?" Polly asked Emma that night as they sat on their beds doing homework.

Emma looked up. *"Goony?"*

"You and Jordan looked really icky today in the stockroom. Will I look like that someday?"

"No," Emma said and went back to her book.

"No?"

"No! Not if you're still a pest when you get older. No one will want to look goony with you."

"Awwwww . . ."

Emma smiled. "I was just teasing," she said. "Someday you'll like a boy and he'll like you back and it'll be very nice. Okay?"

"Okay," Polly said. "What's nice about it?"

"It makes you feel nice. Makes you feel pretty.

And terrific. Makes you feel . . . pretty terrific."

"Thanks, Emma, thanks a lot."

Emma threw up her hands. "I can't explain it," she said. "Jordan's just wonderful, that's all!"

"I gathered. . . ."

"We finally have a kind of date together—we've been trying to be alone together for such a long time—it's not easy!"

"Where are you going?"

"Well, nowhere, really. I haven't met his mother yet, so tomorrow after school we're going shopping for groceries, and then we'll go back to Jordan's and make dinner. And then we'll eat when she comes home from work. And then Jordan and I will go out together for a while."

"But it's a school night," Polly reminded her. "Tomorrow's Wednesday."

"I know, but Mom and Dad said it would be all right if we get home early."

Polly sighed. "I'll miss hearing that phone ring and one of us yelling, 'It's your moth-er, Jordan!' "

Emma giggled in spite of herself. "Maybe she'll call anyway," she said. "Out of habit."

5

○

● "Jordan," Emma said as they put their bags of groceries on the kitchen table. "Do you know what you're doing?"

"About what?"

"Dinner, dodo," she laughed. "Can you cook? Because *I* can't. I don't understand half this stuff we brought home!"

"Sure, I can cook! My mom and I have been alone all these years, and since I always get home before she does—or used to, anyway—I usually get dinner started."

Emma looked around. "It's a nice apartment," she said. "Nice kitchen."

Emma smoothed a sleeve of her shirt. "Do I look all right?" she asked.

"Sure!"

"Don't you think I should have worn a skirt instead of—"

"Honest, Emma—You look . . . terrific."

They smiled at each other.

"It's okay for the two of us," Jordan said finally. "I mean, it's fine. Guess I never thought about it."

"Oh, Jordan! A birdcage! What is that, a canary?"

"Yeah. Henry. He only sings in the spring. Don't ask me why or how he even knows what month it is. Hand me the head of lettuce, okay? You *do* know what that is, don't you?"

Emma threw it at him.

They made beef bourguignon. Or rather, Jordan did, with Emma acting as assistant, passing him mushrooms, onions, and marjoram leaves. But she fixed the salad by herself.

When Jordan put the lid on the casserole, Emma asked, "What happens now?"

"It simmers."

"That's all? It just simmers?"

"That's all." He brushed some flour from her shoulder.

Emma smiled. "That's great," she said, "because now we really do have some time to be all by ourselves!"

"I know," Jordan said and smiled.

"I keep expecting the phone to ring, or a little bell over the door. Or my father. Or mother. Or Polly to—"

"Shh—" Jordan put a finger over her lips. "No one will bother us here. No one *ever* bothers anyone here. Even Henry doesn't sing. Let's go sit in the living room."

"Should I be nervous about meeting your mother?" Emma asked as they sat together on the couch.

"No. Why should you? I wasn't nervous about meeting your parents."

"That was different," Emma said. "You met them before you knew they were my parents. Well, almost . . ."

Jordan looked at her face and swallowed. It felt as if something were caught in his throat. Emma looked so pretty, sitting there under the Currier & Ives snowstorm print. Her cheeks were all flushed and her eyes were so large and—and—

Jordan reached for her hand and squeezed it tightly. He leaned toward her. Emma blinked, took a breath, and leaned toward him. They both closed their eyes at the same time.

Suddenly, the door burst open.

"Hel-lo there!" Olivia called.

Jordan and Emma pulled apart. "Hi, Mom," he said.

Emma, remembering her manners, got to her feet.

"Hi, Mrs. Hall," she said softly. "I'm so pleased to meet you."

"Thank you, dear," Olivia said, and they both sat down, Emma on the couch again and Olivia in a chair facing them.

"So," Olivia said, as she crossed her ankles and folded her hands, "your parents own a magic store."

"That's right," Emma said and folded her hands, too.

"You sell . . . magic tricks? I guess I haven't talked to Jordan too much about what the store actually—"

"Well, we supply professional magicians with things they need for their acts, mostly. Including costumes because lots of times they need special things built into what they wear. But we do have 'tricks' and things in the shop for the general public. I hope you'll come by sometime and see for yourself."

Jordan coughed into his sleeve.

"It must be fun," Olivia said.

"It's really like most other businesses," Emma said. "It gets hectic sometimes. Some magicians have artistic temperaments—that's what my mother calls it—but most of them are very nice. And they're friends, too."

"Mmmm," Olivia said. "People always think the

toy business is lots of fun, but it's really . . . just another business too."

"Uh-huh," Emma said and threw Jordan a glance.

"Emma's in three honors classes!" Jordan said quickly, and Olivia said, "Isn't that nice."

Emma sighed.

But things warmed up when the meal began. Olivia loved the beef bourguignon, and Jordan said Emma fixed most of it, which made Emma blush and Olivia smile. Soon they were chatting together like old friends.

"I really love this place," Emma said. "It's so warm. And quiet, too. You have those really thick walls that nobody puts up anymore."

"Yes, well, it's an old building," Olivia said. "And I've been here since I was first married. They're talking about going co-op now . . . I don't know what we'll do if that happens. I certainly can't afford to buy the place. . . ."

"That's why I needed a job," Jordan said. "Anything at all that I can bring in helps."

Emma looked at them—mother and son—Jordan sitting straight and tall and smiling at her, Olivia a little hunched in her chair like the quiet canary in his cage above their heads.

"I think it's nice," Emma said, "how supportive you both are of each other."

"Well, you know, we're all we have," Olivia said.

•

"I liked watching you . . . in your own house . . . with your mom," Emma told Jordan as they rode down in the elevator on the way out. "You're so . . . considerate. And helpful."

"That's me," Jordan said. "C and H."

"I mean it! You're so good."

Jordan laughed.

"I bet you were always 'the good boy' in school. I bet you always did the right thing."

Jordan opened the elevator door for Emma and then moved her toward the lobby wall, where he leaned over her.

"When I was six," he said, "I caught a mouse in the basement of this building, and I brought it to school in a shoebox with holes punched in the sides."

"Really?"

"Really. And I opened the box during the pledge of allegiance and dropped the mouse by its tail into Amanda Peebles' King Kong lunch box."

"I don't believe it." Emma moved away from the wall, but Jordan gently pushed her back with his free hand.

"When I was eight, I skipped school for a day to watch them set up the Ringling Brothers Circus at Madison Square Garden. No one even showed me how to get there—I did it all myself. And when I got home, my mother was waiting—hysterical—with two policemen."

"Gee, I'll bet!"

"And when I was ten, I bought a joint from a sixth-grader and smoked it in the boys' room. I thought my throat would burn out, and I was sick all night."

"Jordan—"

"And when I was twelve—"

"Okay! Enough! You weren't such a good boy. I was a much better 'good girl' than you were a 'good boy.' But I did notice one thing, while you were going through your confessions. . . ."

"What?"

"I noticed that you only did things in your even-numbered years. How old are you now?"

He grinned. "Sixteen."

"That's right. So what is it you're going to do? After all, this is your year."

"You bet it is!" Jordan took her hand, and they walked through the lobby to the big front doors.

"Your mom's nice," Emma said. "Even if she is a little—well, uh—"

"Overprotective? Yeah. She really is terrific, though. She had to raise me by herself, and it wasn't easy."

•47

"Especially during the even years," Emma laughed.

"Especially! Now: it's only eight o'clock. What would you like to do for the next hour and a half, Miss Emma Major of Franklin Pierce High School and West Thirty-eighth Street in Manhattan?"

They ended up sitting under the statue of George M. Cohan in the middle of Times Square where Broadway and Seventh Avenue cross each other. They made their own little world while it seemed all of humanity walked, ran, drove, yelled, honked, laughed, cried, argued, and made up around them.

●

They got back to Major Magic at 9:35, only five minutes later than Emma had promised her parents.

"They'll be upstairs by now," Emma said. "Let's go through the stockroom."

They walked around to the side of the building. Jordan took out his own keys to the outside door and opened it for Emma. As she started through, her foot touched something soft that moved. She uttered a little cry and stumbled against Jordan. And whatever she had stepped on screamed too.

"I'll get the light," Jordan said and quickly turned it on. They both saw it—the black cat with one white eye and ear.

"That's the second time he's done that to me!"

Emma said angrily. "I can feel my heart pounding!"

"I bet he can, too," Jordan told her. "You probably scared him just as much. Say, look at that."

The cat was nosing the tall box that Jordan and C.C. had stood against the wall the day before. He sniffed all along the bottom edge and along both sides. Then he stood on his hind legs with his front legs pushing against the door of the box and sniffed the door, too.

"Funny, huh?" Jordan asked Emma as they both watched.

"What's going on down there?" C.C. stood at the top of the stairs with a lit flashlight. He was wearing a brown silk robe with a gold ascot at his throat. "Emma? *Is that you?*"

"Yes, Daddy! And Jordan. We just got back!"

"Made enough noise! We thought someone was breaking in. Your mother's behind me here with Polly's lacrosse stick!" He turned and spoke over his shoulder. "Put it away, Deirdre; it's just the kids!"

"I'll be up soon, Daddy. . . ." Emma called.

"Yes, all right. Soon, Emma. Don't dawdle, now. Big day tomorrow. You too, Jordan. Big day. Start cataloging."

"Yes sir."

The upstairs door clicked shut.

"Peace and quiet time is over," Emma sighed.

"I'd better get home before my mother calls," Jordan said, but still they stood and looked into each other's eyes until a scratching sound interrupted them.

"Look at that, Emma. The cat's clawing at the box."

"Oh gosh, he'll ruin the finish. Take him out with you, Jordan—"

"Uh-oh, he's got the door open. He's going in—I'll get him—" Jordan made a dive for the cat, whose tail was about to disappear into the black box. There was a thump and a piercing little cry as Jordan tumbled in and the box door slammed shut behind him. Emma stared with wide eyes at the closed box and listened to the thumps and bumps inside it.

"Emma! Open the door, will you?" Jordan's voice was muffled, and Emma moved quickly to help him. "It's dark in here! I didn't think the box was that deep, but—"

Emma pulled the door wide.

"Wow! Thanks," Jordan said. "I couldn't seem to push it open from the inside. Weird, huh? There wasn't any trouble opening it from the outside. A cat could do it. And did. And here's the really weird thing: What happened to the cat? Look at that. Door's wide open and he's not in there!"

Emma blinked. "Come on, Jordan. . . ." she said hesitantly. "Where are you?"

"Huh?"

"Where are you?"

"What—uh—what do you mean, where am I?" Emma looked to her right, then to her left. The stockroom was small. Jordan might be hiding behind some piled crates, but then his voice would be coming from there. And why would he hide, anyway?

"I'm not kidding, Jordan," Emma said. "I can't see you." She stared straight ahead of her and caught her lower lip between her teeth.

"Emma. Hey! Look. Look! Blue shirt. Gray cords. White running shoes. Right here! Emma, *here!* Jeez, you look like you're looking right *through* me!"

Emma now began to gnaw her lower lip, and her eyes seemed to grow wider.

"Emma," Jordan said, "I'm going to reach out and touch your hand. Hold it out for me like this, okay?"

"Like what?"

"Never mind. Just hold it straight out. Okay. Now I'm reaching for it. I'm *touching* it, okay?"

Emma gasped. "Jordan, I can feel you touching my hand," she said, "but I can't see you at all. This is spooky, Jordan, and I don't—"

They were interrupted by the sound of the ringing phone behind the upstairs door, the store phone extension.

"Guess who," Jordan whispered as they heard the door open.

"Emma, is Jordan still here?" C.C. yelled down. "His mother's on the phone. She expected him about twenty minutes ago. And you said you were on your way up! Is he still here? Emma?"

Emma pulled her shoulders back and marched to the foot of the stairs. She looked up at her father's silhouetted figure.

"Uh . . ." she managed, "not exactly."

○

● "What do you mean, 'not exactly'?" C.C. roared. "Emma, a simple question! Either he's here or he's not here!"

"Well, he's . . . uh . . . not here."

"Okay, great. I'll tell his mother he's on his way home. And you come on up!" The door slammed.

"I didn't say he was on his way home, either," Emma said softly to the closed door. "Jordan," she said, turning to the empty room, "what are we going to do?"

"Okay, okay. There's some kind of answer to this. It's probably a trick of the light or something."

"A trick," Emma said. "A trick!"

"What?"

"It's a trick, Jordan, that's what it is! It's a magician's trick! You got into a magician's box, and this is a magician's trick. That's all it is."

"You're right! It's the box! I'll get back into it again and I'll reappear. I mean, it figures, doesn't it? That's where I'll *re*appear!"

"Try it, Jordan, try it!"

"Okay!"

"Okay."

"Okay . . ."

"Okay, do it, Jordan. Stop saying 'okay'!"

"I said 'okay' because I *did* it, Emma. Couldn't you tell by my voice? Where it's coming from? Anyway, I guess it didn't work because I can tell from your face you still can't see me."

Emma took a deep breath. "No, I can't," she said, "but now I know there has to be *some* rational explanation. This is just magicians' hocus-pocus and there's a way to figure this out like any other trick."

"Polly told me magicians don't tell their tricks to anyone."

"They don't, but my father will know." She turned to the stairs again. "Dad-dy!" she called. "Daddy, you'd better come down here . . . We've got a problem!"

●

"Okay, Jordan, it's cute. It's cute. I like it. But it's late. Your mother is frantic and I told her you were on your way."

"Mr. Major—sir—this isn't really what I had

planned for this evening—making you mad, that is. Or my mother. I'm not kidding. Emma's not kidding. We called you down here because we thought you'd understand what happened—" Jordan heard his own voice crack and hoped he was convincing. He understood how C.C. felt. He wouldn't have believed himself either.

"How'd you do it, Jordan? Throw your voice like that? I'd swear you were standing right in front of me here! A professional ventriloquist couldn't do a better job. I mean that, son, but enough's enough!"

Jordan put his hand on C.C.'s shoulder, and Emma's father leaped into the air.

"I'm sorry!" Jordan cried, "but I just had to show you I wasn't throwing my voice! Emma can't see me and you can't see me, which means. . . ."

"Which means he can't be seen, Daddy. That's why we called you. Jordan's disappeared."

C.C. frowned. "This isn't a put-on," he said.

"No, sir!" Jordan and Emma said together.

"And it's no trick. . . ."

"If it is, we don't know how to finish it," Emma said.

"All right . . . He got into this box—" C.C. pointed to it—the tall, black, harmless-looking prop once used by the famous Louis Langhorn— "and when he got out, he was—he was—I can't say it."

"Invisible." Jordan said it for him.

"That's ridiculous!" C.C. bellowed.

"I know."

"He was chasing this cat—it's been hanging around the building. Anyway, you know how curious cats are, and it pushed open the box and sneaked into it, and Jordan went in after it and—" She took a deep breath. "And when he came out—I never saw him come out."

"Did you try getting back in the box?"

"Sure. That was the first thing we did."

C.C. scratched his head. "There has to be some reasonable explanation. Mirrors, maybe . . ."

"Mirrors?"

"I don't know. . . . Emma, go upstairs and get your mother. Jordan, hold out your hand again. I want to make sure this is really happening." He groped gingerly in the air until Jordan touched him; then he flinched and drew back as if he'd been burned.

"This is spooky, kid," he said.

Deirdre swept down the back stairs from the loft in a flowing red-and-gold Chinese robe. She had already creamed her face for the night, and it glowed under the harsh stockroom light.

"Now what's going on here?" she asked.

"Didn't Emma tell you?" C.C. asked.

"Tell me what?"

"Where's Jordan?" her husband asked.

"He went home, didn't he? He'd better have or his mother will just— He *did* go home, didn't he?"

"No. He's here."

"Where?" Deirdre lifted the edge of the framed Langhorn poster and peered behind it. "Yoo-hoo, Jor-dan!" she called. "I give up," she said. "Where is he?"

"Right here, Mrs. Major," Jordan said.

Deirdre cried, "Whoops!" and then burst out laughing. "Why, that's wonderful! How did you do that, Jordan?"

Emma sighed.

Polly clattered down the stairs wearing Emma's clogs. "Why is everyone down here?" she asked. "Is this a family meeting?"

"Sort of . . ." Emma said.

"If this is a family meeting, why wasn't I invited?" Polly demanded.

"You explain, Jordan," Emma said.

●

Deirdre sat quietly on a crate as she thought. Every now and then she would mutter something out loud, shake her head, and begin to think again, as she twisted the braided cord on her robe.

Polly walked back and forth in front of Langhorn's box, leaning forward, staring at it, studying it.

•57

"Don't you *dare* get too close to that door, Polly Major," her mother warned.

"I'm not, I'm just inspecting," Polly answered. She picked up a broom and poked its handle into the blackness.

"Polly!" her mother cried.

"I'm just poking," Polly said. "Gee, this could be a neat sales promotion."

"Polly . . ." Emma said.

"I mean it! Think how things would sell if Jordan stood there—or *there*—well, anyway, wherever he is—and talked about all the merchandise!"

"Polly . . ." Emma said again, and Polly sighed, "I'm just *kid*ding."

"Well, it's not funny!"

"She's right, though," Jordan piped up. His voice, coming from empty space, made them all jump.

"Stop it, Jordan," Emma said, "and I mean it."

"Just sounded like fun. . . ."

"Some fun."

"Speaking of fun," Deirdre said, "I just this minute thought of something that won't be."

"What's that, Mrs. Major?"

"Telling your mother, Jordan. You're supposed to be home by now as it is."

"Oh no," Jordan moaned.

"Oh yes. We'll certainly have to tell her. You

can't just walk into your apartment and not be there. I mean, of course, you'll *be* there—"

"She's right," Jordan said, grabbing Emma's hand. "I'll be there, but I won't be there. I don't think my mother will make it through this. First she lost her husband—"

"Jordan, she hasn't *lost* you!" Emma said. She held up the hand Jordan was holding, and it looked as though she were gripping air tightly.

"Maybe we're all going crazy for nothing," Polly said. "I mean, maybe it'll just . . . wear off in a little while. . . ."

"Well, you must call her, Jordan," Deirdre said. "Tell her something, uh, came up, and you'll be later than you thought."

"Louis Langhorn's disappearing box . . ." C.C. mused, stroking his chin. He raised his head and hands to the ceiling. "*What's the gimmick, Langhorn?*" he yelled. "Give us a sign how this trick works and I'll—I'll sponsor you for a national monument!" He waited. Nothing happened. "Hey, Langhorn! Think of it! A statue of *you* in Central Park! Taller than Christopher Columbus over at Columbus Circle!" Nothing happened. He smiled sheepishly. "Doesn't hurt to try, right?" he asked.

●

"Hi, Mom? Yeah, it's me, Jordan. I know I should have been home by now, but— Yeah, I

know that and I'm sorry, honest, but . . . well
. . . see, something's happened. No, no! No accident, Mom, nothing like that! Although . . .
actually, I called to prepare you for a slight . . .
change." He covered the mouthpiece and whispered to Emma, "What kind of change, she wants
to know."

"Of course she does, Jordan. Want *me* to tell
her?"

"Yes, but only after I wipe out. Mom? Still there?
No! No, Mom, I'm not married! Yes, I *swear* I'm
not married! Mom, I'm only sixteen! What *is* the
change? Well . . . the thing is, I'm invisible. Invisible: I-N-V-I-S-I-B-L-E. Right. Right. Want to
talk to Emma?" The phone looked as if it were
floating in the air toward Emma's hand.

"Mrs. Hall? It's Emma, and no, he's not kidding, and it's not a joke. Jordan must have trig-
gered something in a piece of equipment that once
belonged to this old magician and— What? *Radiation?* Oh no, Mrs. Hall, nothing like that,
honest. He's just—well, I mean, you can touch
him and you can hear him . . . The thing is, you
just can't *see* him." She reached out to touch Jordan's shoulder for comfort and accidentally slapped
his nose. "Yes," she said into the phone, "I don't
blame you. Sure, by all means. Use the door on

Thirty-ninth Street. Okay." She hung up. "Your mother's coming right over," she announced to the air.

●

They ushered Olivia Hall into the big living room, and Deirdre put an arm around her shoulders to steady her.

"Now, now, Mrs. Hall, it's all right," she said soothingly. "We'll get to the bottom of this, and everything will be just fine."

Olivia sat on the couch and groped in the air with her fingers until Jordan grabbed her hands again and held them.

"I'm here, Mom, I'm here," he murmured.

"*Where?*"

"It's just some special gimmick that we haven't figured out yet, that's all," C.C. said. "Believe me, Olivia—may I call you Olivia?—we'll lick this silly thing."

Olivia gasped softly. "Can't you just ask the magician who owned the—the box or whatever it was? Can't you ask the man who devised the trick? I simply don't—"

"We can't ask him, Mrs. Hall," Emma said. "He's been dead for over forty years."

"Oh, good Lord!" Olivia wrung her hands.

"I'm really fine, though, Mom," Jordan

quickly assured her, "I mean, other than the obvious . . ."

His mother burst into tears.

●

They took Olivia down to the stockroom. Emma and Jordan recreated the entire incident with the box, both as a demonstration and as an experiment to see if it would produce a visible Jordan again. It did not.

"Maybe Meissen will know," C.C. suggested.

"I thought he was out of town," Deirdre said.

"Atlantic City. Caesar's, I think. The number's upstairs. Maybe there's a simple answer he can give us over the phone!"

"Well, you can't call now. It's just eleven. He'll be on stage again. Let's all go up and have something to eat and drink, and we'll try him later."

●

The four Majors and Olivia Hall sat at the big oak dining table in the loft and stared. Down at the end of the table a glass of milk rose into the air, tipped, and the milk disappeared. A fork ascended from a plate, turned on its side, and cut a bite of cake. The fork took aim, struck, and the cake disappeared.

"Fascinating," Deirdre whispered.

The milk glass rose again, stopped in midair, and came back down to the table.

"I just remembered something," Jordan said. "There's a big history test tomorrow."

"You're not going to school looking like that!" his mother cried.

"Like what?" Polly giggled.

"I mean, in that state. You're coming home with me, and that's where you'll stay until you—until you're—until I can see you again." Suddenly Olivia felt Jordan's fingers on her arm.

"Mom, I can't do that. I mean, I'm here! I feel fine! I can't stay cooped up in the apartment just because I'm, uh . . ."

"I'm calling Meissen now," C.C. said. He walked to the little telephone stand near the kitchen entrance, pulled open a drawer, and began to rummage through papers. "Here it is," he said. "Here's the card he left me. He must be back in his room by now."

"Face this way, C.C.," Deirdre said, "so we can all listen."

Everyone remained at the table and watched C.C. dial, mop his face with his ascot, and sit heavily on the couch next to the phone stand.

"Let me have The Illustrious Meissen's room, please," he said. He frowned and covered the mouthpiece. "The clerk wants to know if he's listed under *I* or *M!*" he groaned. Then back to the phone: "*M!* It's *M!*" he roared, then muttered, "Good grief . . ."

"Hello? Is that you, Jack? It's C. C. Major in New York. Is Meissen there? He is? Wonderful. Put him on." He smiled at the group. "That was his manager," he explained. "Meissen's taking off his makeup. He'll be right with us."

They heard fingers drumming on the tabletop. All those visible had their hands in their laps.

"Meissen? It's C.C. Sorry to bother you at this hour, but it's rather urgent. How'd your show go?" He nodded to Deirdre. "It went fine."

"Good." Deirdre nodded.

"Anyway, Meissen . . . remember that box of Langhorn's that came with the estate? You know, the big black one, kind of coffin-shaped—the one he used for his vanishing act. Yes! That's it. Well, it seems that our young helper, Jordan Hall— that's right, the one our Emma fancies—"

Emma glanced at where Jordan was sitting and smiled. She knew he smiled back.

"It seems," C.C. went on, "that the boy got into the box and he, uh, disappeared. I said, he *disappeared*. Vanished! Poof! Gone! Is this connection all right, Meissen? . . . Well, no, we're not

•64

worried about *where* he disappeared *to*, Meissen. He's *here*. I said, he's *here*, but no one can see him. No, look, let me explain. . . ."

He did, and very well too, according to his family. He took Meissen through the scene with as much detail as he could, and when he was finished, he asked if anyone had anything to add. Then he listened for a few moments and hung up.

"What happened?" Deirdre said.

"Did he tell you how the trick works?" Polly asked eagerly.

Only Emma and Jordan, holding hands at the table, were silent as they waited.

"He's going to check it out," C.C. said.

"He's going to *check it out?*" his wife cried. *"How?"*

"Easy, take it easy. He's going to try some old contacts he has who either knew Langhorn or watched him work or worked with him—or *something*. Anyway, he has more access to that kind of secret than we do, so he's going to make some phone calls first thing in the morning and get back to us."

Emma's face fell. "Daddy, why 'first thing in the morning'? Why not now?"

"It's late, Emma," her father said. "When I said 'old contacts,' I meant '*old* contacts'! These folks are tucked in at eight-thirty! Meissen will see what he can do tomorrow, and we've done all *we* can. Besides, as Polly said, it all might wear off by tomorrow."

"Let's hope so," Emma sighed. "This wasn't the kind of disappearance I had in mind." She gazed up at where Jordan's face should have been.

"Disappearance . . ." Olivia moaned. "Just like his father . . ."

●

Jordan and Olivia sat in the taxi on their way home, with Olivia firmly gripping her son's sleeve.

"How do you feel?" she asked him.

"Fine," Jordan answered.

The taxi driver glanced into his rear view mirror. One woman had gotten into his taxi; he was sure of it. One woman. And now she was talking, and a male voice was answering. In the rear view mirror he saw—one woman.

"Let me feel your head," Olivia said. "Do you have a fever?"

"Mom, of course I don't have a fever." But he let her rest her palm on his forehead.

The taxi driver raised his eyebrow.

"I still think you should stay home tomorrow. I simply cannot grasp this entire *awful* situation! Emma seemed like such a sweet girl—"

"Emma *is* a sweet girl. This isn't Emma's fault."

"Well, but really—"

"Lady," the driver said, turning in his seat as they pulled up at a red light, "you okay back there?"

"Of course I am, young man!"

The driver faced front again and clicked his tongue against the roof of his mouth. "Sure you are, lady," he muttered. "Sure you are. . . ."

7

○

● The phone rang at seven-thirty in the Majors' loft. Emma picked it up in the kitchen.

"*Jordan?*" she squeaked.

"Yeah, it's me."

"Still the *same?*"

"Yeah. The same."

"Oh, Jordan, what are we going to *do?*"

"Did Meissen call yet?"

"No. It's too early for him. He's in show business. People in show business get up late. 'First thing in the morning' usually means noon."

"Great . . ."

"Do you want me to come over?"

"No. You have school. *I* have school, too. And I'm going."

"Come on, Jordan, you *can't.*"

"Yes, I can. I'm not sick. I have a history test I

really studied hard for, and I have this after-school job . . ."

"Don't I know . . ."

"Besides . . . I wouldn't miss lunch with you, Emma."

"Oh, Jordan . . ."

"Hey, don't cry! Come on. I'm going. I'm going to get on with my life. I mean, if this thing doesn't go away, I still have to live, right? I'm still a person, right?"

"Jordan, don't even *say* that!"

"Say what?"

" 'If this doesn't go away.' "

"Okay. It *will* go away. Like a virus. But I'm still going to school. My mother insists on taking me, and I've decided not to argue with her. It really would be a little awkward without someone to explain."

"*I* could explain for you."

"No, it has to be an adult, Emma. You know that. Kids have no credibility."

She sighed. "You're right. Okay. I'll see you at school. I mean—"

"I know what you mean."

●

Olivia Hall marched into the principal's office holding onto her son's arm and asked the secretary for an immediate appointment.

"He *has* immediate appointments," the secretary told her.

"Mine has to be more immediate," Olivia said. "It's urgent."

"Urgent?"

"Quite urgent."

"Well, I'll see what I can do, Mrs. Hall." She went into the inner office and returned a moment later.

"He can squeeze you in," she told Olivia.

"Thank you."

The principal of Franklin Pierce High School was a lanky man with thinning red hair. Olivia thought he was a little young-looking for a principal, but decided that these days you never could tell about a person's age.

"How do you do, Mrs. Hall," he said, rising and extending his hand.

"How do you do, Mr. Beaumont. This is my son, Jordan."

Mr. Beaumont looked from side to side, and then his face went blank. He looked at his hand as if he might have caught something from contact with Olivia. "Your son?" he managed.

"Yes, we've met, Mr. Beaumont. Jordan Hall." Jordan stuck out his hand too, but withdrew it quickly. He and his mother watched the blood drain from the principal's face.

"Here's our problem, Mr. Beaumont," Olivia began.

•

The principal and his mother decided that Jordan should not carry a large wooden pass through the halls. It might frighten unsuspecting students and faculty. Instead, he could move from class to class and listen to the lectures, though not participate.

But Jordan had questions. "How am I supposed to take notes? And people have to know I'm present in the classroom or someone might try to sit at my desk! And I have a history test today!" he complained.

Mr. Beaumont had no answers. They left him fanning himself with a sheaf of menus from the cafeteria.

"I'll stay with you, Jordy," his mother said.

"No, you won't. Go to work, Mom. I'll be fine."

"I can't leave you here . . . like this. . . ."

"Sure you can. It'll be okay. Really."

"Are you okay, lady?" A security guard appeared at Olivia's elbow.

Olivia glared at him. "Certainly I am, young man. Why does everybody keep asking me that?"

"Sorry, but you looked like you were, uh, talking to yourself."

He gaped as Olivia seemed to be thrust forward and propelled down the hall.

"Go on, Mom," Jordan whispered. "Thanks for coming with me, but it's time to leave me on my own. See you tonight."

"I think I should stay," his mother said. "I really think it's a mistake to leave you, Jordan . . . Jordan?" He didn't answer.

Of course, he could be anywhere. . . .

●

Emma leaned against her locker, hugging her books and waiting. Suddenly, she felt a light touch on her shoulder, and she smiled.

"There you are," she said.

"Told you I wouldn't miss lunch." Jordan took her hand.

"I kept staring at your empty seat in math."

"I was in it."

"I know. Did you take the history test?"

He sighed. "No. I chickened out. After the principal's reaction, I decided I wasn't ready for the whole school. I thought maybe I could take the test into the closet, but everyone still would have seen it floating across the room."

"Well, now everyone will see a sandwich disappearing into the air if we eat in the cafeteria."

"We won't eat there. You buy something for

me and meet me outside. We'll eat behind the gym."

"This is *terrible*, Jordan. . . ."

"The southeast corner of the gym building. Go and get us some food. Even if it looks like mayo with sticks in it."

●

"Jordan?" Emma whispered, as she glanced around. "Are you there?"

There wasn't any answer. Emma put her wrapped sandwiches and milk cartons on the ground next to the wall and called softly again. "Jordan? Where *are* you?"

No answer.

Oh no, she thought. It's like *The Incredible Shrinking Man*. It's getting worse, and he'll be lost forever!

She was almost in tears until she detected running footsteps approaching.

"Emma! It's me! You'll never guess what just happened!"

Emma felt herself enveloped in a bear hug. She couldn't see Jordan, but she didn't care. She could feel his arms around her, so she just closed her eyes and hugged him back.

For his part, Jordan thought for only a moment about how Emma must look, standing there hugging air with her eyes closed. He was happy just being so close to her. But then they both remembered where they were and stepped apart.

"I got carried away," Jordan whispered.

"Me too."

"Guess I've got more courage now that I'm invisible."

"I don't mind," Emma whispered, looking up at where his face should be. "Anyway, tell me what happened."

"Oh! Right! Well. I was on my way over here when I saw a kid sawing away at a bicycle chain."

"What?"

"This kid! He was stealing a bike from the rack near the side door. *You* know, the rack where kids who ride to school chain up their bikes!"

"Oh."

"Anyway, the kid was stealing one! So—" Jordan was laughing now. "I went up to the kid—face to face, only of course he didn't see mine—and I grabbed his hands and knocked the handsaw out of them!"

Emma's mouth dropped open.

"Emma, you should have seen his expression! He freaked completely! He didn't even scream or anything. Just had this crazy look in his eyes while he started backing away, backing away. Then he tripped in a pothole, got up, and took off like somebody set fire to his pants! It was incredible!"

"Good for you, Jordan."

"Emma, it was wonderful, I mean it! And as I was running over here to meet you, I started thinking all these nutty thoughts!"

"What kind of nutty thoughts?"

"Well . . . don't laugh."

"Believe me, Jordan, I'm not laughing."

"I could really *use* this thing! I could really— you know—be of use to people. This is a kind of power, this being invisible, and I could use it to— do some good! Catch thieves, things like that. You know, like—"

"Superman?"

"Yeah!"

"Jordan, I'd make a lousy Lois Lane."

"Aw—"

"And besides, you're *not* Superman. *He* couldn't be hurt, and you can. And once this thing gets out, everyone will know that all you are is invisible. *Not* superpowerful. And besides, what about school? And college? And your mother? And *me*? And *us*?"

He sighed. "It sounded good," he said.

"Noble," Emma agreed. "But only when you can travel faster than a speeding bullet. Eat your sandwich. I think the guard patrols here in a few minutes."

●

After school they walked to the store, holding hands all the way.

"I think," Jordan said, swinging Emma's hand up and down, "that New York City is probably the only place where you can walk down a street, swinging one hand back and forth and talking out loud to yourself, and no one will even turn around to look at you."

Emma giggled. "You're right," she said. "No one's looking at me right now. There's so much that happens all the time in New York that nothing surprises anyone. Anyway . . ." She smiled. "I wouldn't care if it did."

"You might," Jordan said. "Can you imagine us on the eleven o'clock news? 'What's it like to be invisible, Mr. Hall? Could you lean a little closer to the microphone, please?' 'And you, Ms. Major—what's it like to have an invisible boyfriend?' "

"Stop!" Emma cried. "Oh, it would be awful. Jordan, I really wish none of this was happening!"

●

Deirdre was on the phone near the cash register when Emma and Jordan walked in.

"Is that Uncle Meissen?" Emma asked, but her mother shook her head and Emma sighed heavily. Jordan squeezed Emma's hand, and Deirdre hung up the phone.

"That was Artemis. She says the pale-blue-and-white scarves arrived and they're fine. She likes the color scheme—they go with her blue sequined costume and her blue-ribboned doves, she says. But now she wants more colors because she's decided to dye her doves." Deirdre began to write on an order pad.

"She can't dye them," Jordan said. "It's against the law."

"No, she said she's already talked to the Humane Society, and they said it was all right provided she didn't use any harmful chemicals. She was so indignant, you should have heard her. 'As if *I* would ever do *anything* harmful to my *babies!*'"

C.C. walked over to them, and Emma gave him a kiss on the cheek. Jordan took his hand and shook it. C.C. was only momentarily startled.

"How was school?" he asked.

"It was boring. I couldn't do anything," Jordan complained.

"And I couldn't do anything because I was worrying about Jordan," Emma added.

Her father nodded. "It will all work out, kids. I just know it will. Meissen will call soon, and we'll have our answers."

"Uncle Meissen hasn't called *yet*," Emma wailed, "and it's almost four!"

"Calm down, dear, he did call," her mother told

her. "He said he was on the trail of something and he'd call back as soon as he got to the bottom of it."

"On the trail of what?" Jordan asked.

"Now, you know magicians. They'll keep you in suspense up until the last moment."

"That's right." C.C. groped for Jordan's shoulder and squeezed it. "Meanwhile, why don't you go into the stockroom and—" He stopped talking because two customers had come into the store. Unobtrusively, he took his hand from Jordan's unseen shoulder and slid it into his pocket.

Emma went to them—a pleasant-looking graying couple—and said, "Hi, may I help you?"

"We're just browsing, thank you, dear," the woman said. "We're visiting from West Virginia, and we have a grandson who's simply crazy about magic! We thought we'd bring him some presents from a real New York magic shop. They surely don't sell too many of these things in our small town." She looked around the shelves.

"Well, help yourself," Emma said professionally and returned to the counter. The bell over the door jingled, and two more people came in.

Suddenly, there was a cry from somewhere among the shelves. Emma craned her neck, but she couldn't see anything. All she heard were shrieks and giggles.

"Roger! Isn't that just *darling?*" declared the pleasant grandmother from West Virginia. In front of her nose dangled two bright red juggling balls. They flew into the air and crossed each other in an arc as they dropped, only to be thrust into the air again.

"Why, that's just the darlingest thing I've ever seen!" the woman squealed and the other customers in the shop ooh'd, ah'd, and nudged each other.

Emma had come out from behind the counter and now observed the "magic" from the end of the aisle. She cleared her throat loudly.

"Uh, Jordan," she said, while pretending to examine something on the shelf, "didn't Daddy ask you to go into the stockroom? The estate stuff needs to be sorted. . . ."

"How do you *do* that?" The grandfather had stopped Deirdre in the aisle. "Can we buy that wonderful juggling trick for our Timmy?"

"I'm afraid not," Deirdre answered with a knowing smile. "It's a house secret."

"Now, Roger, look at *that!*"

A shiny lasso with gold tassels on its ends began to twirl by itself in front of the small group of customers.

"Jordan!" Emma said, her voice a little shriller, "I think you'd better start work in the—"

"Now, that's—*that's* a trick I've got to have!" the man stammered.

The customers at the front of the store were staring at the dancing golden lasso as it bobbed and twirled. People on the street caught sight of it and began to pile into the store, or just stand and gape through the plate glass window. Jordan had broadened his show to include a magic wand conducting a silent symphony, water being poured from pitcher to glass all by itself, and flowers blooming suddenly from thin air—until the phone rang and Deirdre cried, "It's Meissen! It's Meissen!" She waved at Emma, who now cried, *"Jordan, stop!"* at the top of her lungs. A small glass of water that had been hovering near the face of a young girl suddenly dropped and smashed.

Emma, embarrassed, regained her composure and forced a smile for the customers. "Uh, the show's over," she said. "Please feel free to browse." She raced for the counter.

"Let's have the phone, Deirdre, quick!" C.C. cried, as he hurried over, but his wife was already holding the receiver out to him. Emma, nervously hopping from one foot to the other, felt Jordan's arm steal around her shoulder. Polly appeared from the stockroom and stood in the doorway, staring anxiously.

"Meissen? Hello?" C.C. said. He glanced briefly

at the collection of customers, who were beginning to disperse, saying, ". . . terrific, terrific." "Great show!" "How did they do that?" "Gee, we've got to come back here. . . ."

"Meissen? Forget everything! This is good for business!" C.C. said into the mouthpiece.

Jordan laughed, and Emma looked at him with sudden realization. She felt tears spring to her eyes. He's enjoying this, she thought. He doesn't want to come back. And I'm—I'm just—

"Ha ha, just kidding, Meissen," C.C. said. "Now tell me slowly everything you came up with." He listened and nodded. "Uh-huh," he said. "Uh-huh . . . uh-huh . . . I see . . . I see . . . uh-huh . . ."

Emma bit her lip.

"Okay. Okay, Meissen, thanks a lot. See you when you get back." C.C. hung up.

"What?" Emma grabbed her father's sleeve. "What did he say, Daddy? What?"

"Now just take it easy and give me back my shirt, Emma. . . . Thank you. I'm afraid there isn't a lot to tell." Emma's body slumped. "Meissen managed to track down Langhorn's daughter—"

"Louis Langhorn had a daughter?" Deirdre asked.

"Yes, and she's a grandmother. He told her exactly what happened, just as you told it to me. She remembered her father's vanishing act—"

"She did?" Emma asked hopefully.

"Yes, she did. But—she doesn't know how he vanished and reappeared in the box."

"*What?*" Emma wailed.

"She said it was the one trick she never figured out from watching backstage. And Langhorn wouldn't tell her how he did it. 'Some things must remain sacred,' he'd say to her. That's what she told Meissen."

"That's *all?*"

"Meissen gave her our number and said if she thought of anything else to please call us, no matter what time of day or night."

"Great," Emma moaned. "Just great!"

●

"Thanks for walking me home, Emma," Jordan said.

Emma didn't answer. She was walking with her head down, staring at her own feet.

"Come on, Emma . . . you haven't spoken to me since Meissen's call. Please . . . don't be upset. This *will* end . . . I *will* be back. I just know it. Emma?"

"It isn't that," Emma said, still looking down. "I mean, it isn't only that."

"Well, what is it then?"

She took a breath, then exhaled. "You know . . . all I wanted was for us to—be away—*together!*"

"I know that."

"But now that *you've* disappeared—alone. I mean, you're—you're *Superman* after all and you're *Mr. Major Magic* and you're—you're *special*. And I'm—I'm—" Now there were tears running down her cheeks as she struggled to find the words to tell him what she was feeling.

Jordan stopped walking, forcing her to stop too. He brushed the tears away with the side of his finger.

"Aw, Emma," he said softly, "I'm sorry. I really am. I got carried away. You know—if that's the way things are, then we might as well make the best of it . . . have some fun. But not if it's at your expense. *You're* the one who's special. To me, anyway. Really."

Emma looked up. "Really?" she asked.

"Really what?" a man said as he passed by. "You talking to me, honey?"

Emma blushed a bright red. "No sir," she said as Jordan grabbed her hand and hurried her down the street, leaving the man gaping after them.

"Yes, really," he whispered. "And I'm going to find some way to prove it to you." He put his arm around her shoulders and hugged her close.

Emma felt warmed in the glow of his affection. She felt herself relax. "Mmmm," she said. "Your sweater smells nice."

•83

"It's the extra-clean-smell fabric softener I use," he answered, smiling.

"Oh, yeah? You do your own laundry?"

"Sure I do. I'm going to make some girl a terrific husband someday. What do you think of that?"

Emma grinned up at him and touched the sweater. "I bet it's your blue one," she said.

"My blue what?" A woman was looking at Emma and pivoting around as she examined her own clothing. "Is there a run in my tights?"

"Uh—" Emma stammered, and Jordan said loudly, "No, but your lipstick's smeared!" Emma grabbed his arm and hurried them down the street.

"That was awful," she giggled. "How could you do that?"

"I'm invisible," he said. "*You're* the one they look at."

"Thanks a lot!"

"By the way, you were right."

"About what?"

"I *am* wearing my blue sweater. And that reminds me . . . You know what's happening Monday?"

"What?"

"They're taking yearbook pictures."

Emma groaned.

"So what should I wear for *that?*"

○

● The next day, Friday, Emma daydreamed through her morning classes. She sighed over some passages in *Romeo and Juliet* in English class, and she sighed when her science teacher asked her the process used to make lipids.

"You're not concentrating today, Emma," her teacher had said in a concerned voice, and Emma nearly answered that that was, in fact, the understatement that ate the Bronx.

"I'm not," she sighed again to herself as she left fifth period. "I'm not concentrating, that's for sure. . . ."

"Talking to yourself, miss?" a voice whispered in her ear, making her jump.

"Jordan?"

"Come here."

"Where?"

"Here! Where we can talk . . ." He pulled her into a recessed area along the corridor.

"I didn't see your mother this morning," Emma said. "I thought maybe you hadn't come. . . ."

"I wouldn't miss important classes," he said. "I can't afford to get behind. Besides, my mother didn't bring me. I told her nobody gets taken to school by his mother when he's sixteen years old. Come on. We're going out."

"Out? Where? Jordan, there are three periods left in school!"

"I know, but you only have lunch, study hall, and gym, and you can afford to miss those."

"How did you know that? Even I wouldn't know that without checking my schedule—"

"Yeah, but *I* know your schedule even if you don't. And that's what's left. So come with me. I thought of the perfect way to prove how special you are to me."

Emma pulled away. "Aw, Jordan," she said, "that's sweet. Really. But I can't cut school."

He pulled her back to him. "Okay, look: Number one, this is a one-time-only. Number two, you won't miss any real work. Neither will I. I set it up that way. Number three . . ."

"Yes? Number three?"

"Number three is . . . look, I know you've been unhappy. I want to show you a special time. Just

the two of us. No one else. And think, Emma
. . . two dates in one week—and both on school
nights!"

"Day."

"What?"

"It's daytime."

"Oh. Yeah. Right. But that's the perfect time
for lunch."

"Lunch?"

"Not just *any* lunch! Not mayo-and-sticks in the
F.P.H. cafeteria! I'm talkin' *lunch!* How does the
Plaza Hotel grab you?"

"The *Plaza?*"

"Please, Emma?" Jordan's voice dropped to a
whisper. "Let's have this special date together
. . . for me and my special girl."

Emma smiled. "Am I dressed for the Plaza?"
she asked.

"Are you kidding? You're dressed for the White
House!"

●

About a half hour after Jordan and Emma had
said good-bye to Franklin Pierce High School,
Olivia Hall was trotting briskly up its front steps.
She was on her lunch hour and decided she had
to talk to the principal one more time. Still a little
upset because Jordan had not allowed her to escort
him that morning, she told herself over and over

again that it was a tribute to her upbringing that the boy was so independent. It didn't help a lot, but it helped some.

"Rest as*sured*, Mrs. Hall, I didn't tell anyone," the principal said patiently. "There isn't a soul who'd believe me, even if the boy demonstrated his—his . . . Anyway, I didn't say anything. Even to my wife."

"I'm worried sick, Mr. Beaumont. You know, Jordan has a wonderful school record—"

"Yes, I know. I've seen it."

"And yesterday he couldn't take his history test. He'd studied hard, Mr. Beaumont! He would have gotten an A, I just know it, but because of this—this—"

"I agree, Mrs. Hall, it's a dilemma! I just don't know how to handle it! I keep thinking, what if this got into the papers? The publicity—I mean, what would you have me *do?*"

"I just want my boy back!" Olivia wailed.

The principal rose from his desk and began to pace. "Mrs. Hall, I'm sorry. I sympathize whole-heartedly! But this is completely beyond me! I must think of the greater good. The entire school population, I mean. Think what would happen if books and papers walked through the halls by themselves. It would be a circus! Mrs. Hall, I promise you, I'll keep thinking, and if there is some way we can work this out . . . well . . ."

Olivia stood up. "I understand," she said. "But don't mark him absent! He's here!"

But he wasn't.

And neither was Emma.

●

"You look great," Jordan whispered.

"I wish I could say the same."

"Well, I look great too. You'll just have to take my word for it. Come on now. Let's go in."

Emma hung back. "Really, Jordan—the Plaza? *The* Plaza Hotel? Do you really think we should?"

"Of *course* I really think we should! Now let's go check out the Palm Court!"

"Jordan, I never even walked in the lobby before!"

"Me neither. Time we did."

"Are we going to have lunch, though?" Emma asked. "I mean, I'll be sitting there . . . all by myself. . . ."

"Just follow me."

●

Emma's eyes took in the soft green, pink, and yellow carpeting, the huge tubs of tall palms framing the room, the arched mirrored and glass windows, the waiters in their white jackets with green

trimming and epaulets and bright green cummer-
bunds, the candelabra illuminating the palms, the
maitre d' in his tuxedo and black bow tie, but
most of all—

"Jordan, *look* at that pastry cart! It's so *tall!* I
never saw so many beautiful cakes and pies and
tarts and—"

"It's gorgeous, all right," Jordan agreed.

There was the usual hum of voices as the lunch-
ers murmured to each other while they unfolded
their yellow linen napkins or studied the large
menus.

"Emma?" Jordan touched her elbow.

"What?" Emma whispered out of the corner of
her mouth.

"See the way the dining room is so open? You
can walk around the outside of it on three sides
and still see almost everything in the whole room?"

"Uh-huh . . ."

"Well, you just hover around here, looking as
if you're waiting for someone. If anyone hassles
you, you can just walk around either side, but I
don't think anyone will bother you. Anyway, hang
out right here . . . and keep an eye on me in
there."

"Easy for you to say . . . Jordan, I'm scared."

"Just—watch," he said.

●

Emma stared into the dining room, but she saw nothing unusual. Her eyes wandered again to the pastry cart. Each delicacy had its name written in swirling script on a tiny card in front of it. Emma began to read the little cards.

Crème caramel, she read. Mmmmmm. Chocolate mousse, English fruitcake, strawberry shortcake . . . ooooh! She stopped at a cake in the corner. A waiter was just bending over it. Its card read "Concord Lake," which made no sense to Emma, but it did make her mouth water. It looked like the entire cake was made of thin sheets of chocolate, rolled into two-inch columns and held together with chocolate custard and whipped cream. Emma licked her lips as the waiter reached for the cake, wondering whom it was for, when suddenly—the cake moved a foot to the left. Emma blinked.

The waiter shook his head a little as if to clear it and reached for the cake again. It moved a foot to the right.

The waiter swallowed. He glanced around, but no one was paying any attention to him except Emma, and she was half hidden by a tub of palms. Again he reached for the Concord Lake. It rose two inches above the cart.

The waiter backed away slowly. He moved across the room to another young man in a waiter's uniform of green and white. He beckoned him with a nod of his head, and the second waiter followed the first back to the pastry cart.

"See the Concord Lake, Jack?" the first waiter said. "Pick it up."

"You crazy?"

"Very possibly. Just pick it up."

The second waiter reached out and hefted the cake plate. "Okay?" he asked his friend.

The first young man nodded. "Thank you," he said.

"You need more sleep, Ron," his friend said, and moved back to his own station.

Ron took a deep breath. He rubbed his hands together. He licked his lips. He closed his eyes, opened them again. He reached for the cake. It rose three feet and began to dance in the air.

This time others saw it, too. Ron, the befuddled waiter, nearly sobbed out loud. He was too relieved to be frightened. There was a hush in the dining room, and then a great gabble of voices. The cake, on its silver plate, danced away from the pastry cart and began to bounce around the room. Some people let out little screams and scooted out of its path as it paid its respects by bowing to each table.

Emma, her fingers clutching the brass rods of the pastry cart, just gaped. Several times she laughed out loud, but no one paid any attention since they saw the same stunts she did. Oh, Jordan, she thought, nibbling a knuckle, what will you do in your *eighteenth* year?

The beautiful cake, the Concord Lake, moved to the entranceway and dipped deeply in a bow to all. Then it hovered over Emma, writhing and dancing until she tentatively extended her arms. Plunk! The Concord Lake sat on her outstretched arms and hands.

Then, almost before she knew it, she was standing on the sidewalk on Fifth Avenue, in front of a hansom cab and its dozing horse.

"Jordan—what—" she managed.

"Wasn't that something?" he chortled. "*Wasn't* it?"

"It was something. . . ."

"Too bad they didn't let you keep the cake. That's what I meant to happen. I think that guy in the suit who took it away from you was from NASA. He said something about studying it. Anyway, wasn't that funny?"

"Well, sure—"

"It was, wasn't it! And no one but *you* knew what was *really* going on! Just you and me! Come on!" He tugged at her arm, and she felt herself being whirled down the street.

"Where are we going?"

"Well, first we'll get a hot dog, since we missed out on that cake. And then—you'll see!"

●

"*Arcade* games?" Emma squeaked.

"Right! Arcade games. Don't you like arcade games?"

"Oh, sure! Only—"

"Come on. Let's go into this one. From a landmark hotel to a Broadway arcade," Jordan crowed. "Now watch." He slipped something heavy into Emma's pocket. "Those are quarters," he whispered, "from my first hard-earned paycheck. You put them in the slots, I'll do the work. Pinball first, okay? Ready?"

"Uh-huh," Emma said hesitantly, as she dropped a coin into the machine.

Suddenly, the pinball machine went into wild action, apparently all by itself. Bells bonged, whistles blew, levers clicked, buzzers buzzed, and the buttons on the front of the machine flew in and out by themselves.

People turned to look at Emma, the person closest to the game. At first she cringed. But as she watched the faces of the people in the crowd and then glanced at the crazy workings of the pinball machine, she began to smile. She raised her hands in the air. "It's not me," she called. "See? I'm not touching a thing." She winked at the air in front of the machine.

The crowd inched closer to the game.

"It's running itself," someone muttered, and someone else said, "Yeah, and look at the score piling up!"

"I've heard of player pianos, but player *pinball?*"

"How does it spring the balls out like that?"

"Now I'm ready for skeeball," Jordan whispered to Emma through the noise, so Emma moved over to the back of the arcade.

"Hey," someone said, "it stopped."

"Yeah. Must've been a fluke. Funny, though, huh?"

Emma plunked a quarter into the skeeball slot.

He's doing this just for me, Emma thought. He's having fun but it's really for me. . . .

A wooden ball rolled up the ramp. By itself.

"Hey! Hey, look! Now the *skeeball's* goin' by itself!"

The crowd moved as one to the back.

"Look at that! Bull's-eye—every time!"

"That's impossible!"

"Yeah! A pinball, that's electric. That could go on the fritz. But wooden balls? Going uphill?"

"Come on," Jordan whispered to Emma. "Let's do the electronic games now."

The manager of the arcade arrived and followed the others as Jordan and Emma moved from game to game.

"I think it's that girl," someone said loudly. "Every game she stands in front of—it goes by itself!"

"Yeah . . . that girl . . . that kid . . ."

But Emma only smiled. "I'm just watching," she said with a shrug. "Like you are. Any machine could go off by itself in this place."

She wanted to touch Jordan—to let him know she shared the joke—but she didn't dare.

And as if on cue, a pinball machine across the room began to ring and buzz.

"See?" Emma said.

"Hey, what's going on?" the manager asked, looking at the group of people clustered together. "What is this, anyway?"

"I think it could be vibrations," Emma offered.

"*What* vibrations?"

Emma sniffed the air. "The vibrations in this particular location," she said. "I think this arcade is built over an ancient Dutch burial ground."

"*Hah?*"

Behind her, Emma heard Jordan's snicker.

"I think it has vibrations," Emma repeated.

"Never had no vibrations before today," the

manager said. The crowd stared at him. "I swear!" he insisted.

The old fortune-teller dummy in her glass cage in a corner began to move from side to side, her mechanical fingers dealing mechanical cards, her eyes scanning the crowd in front of her.

"I'm gettin' outta here!" someone cried. There were some pushes, some grunts, the sound of stamping feet, and in a moment the arcade was empty.

"Hey!" the manager called after the scurrying customers. "Hey!"

"Time to go," Jordan whispered.

"No—wait, Jordan . . . I feel guilty."

They looked around the empty arcade.

"Yeah . . ." Jordan said. "I guess I see what you mean. . . ."

"Let me see if I can fix it a little," Emma whispered and walked up to the bemused manager.

"Listen," she said, "don't worry about this."

"Whaddya mean?"

"Well, from what I've read, this kind of thing only happens once every hundred years."

"Oh, yeah?" The manager hung on her words. "How long does it last?"

Emma looked at her watch. "Probably just a little over an hour," she said.

"Ohhh," the man breathed. "So it oughta be over now."

"Oughta be," Emma agreed and left.

●

"That really was fun," Jordan said, swinging Emma's hand. "Did you like it? You were terrific! That was fast thinking when you said, 'Oh, *any* machine could start up anywhere.' I got over there fast, or they'd be trying you for witchcraft!"

"Mmmm," Emma said, "it was fun."

"And you were great with the manager. You really were. I bet he's got all his customers back already, don't you think?"

"Yeah . . . I hope so." She smiled.

"Listen, Emma, we have some time before we have to be at the store—you know, after school? Hey, maybe we could go out to the aquarium where I used to work! Maybe see if my old friends the fish can sense my presence? What do you think? Emma?" He took her arm as they walked. "What's the matter? You look so far away."

Emma stopped walking. "That's just how I feel, Jordan. Exactly. Far away. I mean, it was fun— playing games all around town—it really was. And I felt in on it . . . but not *part* of it." She pulled him over, next to a wall. "See—you're where you are and I'm here, and no matter what, I can't really share what you're going through and you can't share what *I'm* going through. Do you see what I mean?"

"But—well—" He sighed. "Yeah . . . I guess so. . . ."

She reached up with her fingers to touch his cheek. "But you were so sweet to take me out the way you did. Anyway."

●

That night after dinner, the store extension phone rang in the Majors' loft. Polly picked it up.

"For you, Emma!" she called. "It's guess-who—Jordan!"

"Thanks," Emma said with a sigh and took the receiver from her sister. "Hi," she said.

"Hi. I thought I'd get you on this phone in case everybody was in the kitchen doing dishes or something."

"Oh. Well, they are."

"Did anyone say anything? I mean, about cutting school?"

"Oh no. Everything's okay. We were there for attendance and all."

"Well, look. It's Friday night. I really don't want the day to end. How about going to a movie with me?"

"No . . . thanks anyway, Jordan. I don't feel much like a movie tonight. . . ."

"Well then, how about something else? Where

it won't seem weird that you look like you're alone. Think of something."

"I *am* thinking of something," Emma said. "I'm thinking about that *box*."

"What about it?"

"You got into that box and you disappeared."

"Yeah."

"Well . . . maybe that's what I should do."

"Emma, no! Listen, Emma, that's crazy! Emma, you know that's crazy, don't you?"

Emma sat down on the floor and leaned against the wall. "I just don't know what to do, Jordan!" She was near tears, and he could hear it.

"Aw, Emma, let me come over there. Please? I'd really like to see you."

"No, I'd really like to be by myself tonight, Jordan. I need to think. The truth is, I'd really like to see you, too," she said and hung up.

She sat there on the floor, twisting the phone wire around her ankle until her mother called from the kitchen.

"Em-ma? Where are you?"

"Back here! In the hall!"

"Well, will you take the dishes out of the dishwasher, please? And put them away?"

Emma twisted the phone cord around her other ankle.

"Emma?"

"What?"

"I asked if you'd take these dishes out!"

Suddenly Emma sat up straight. She pulled the cord off her foot. "Mom? Can I do it in a little while? I want to go downstairs for a minute."

"To the *store?*"

"No, the stockroom. There's something I have to check."

"Okay. But don't forget these dishes. Your dad and I are going to bed. We're both bushed. You'll get the lights and locks, won't you?"

But Emma had already gone downstairs.

●

Jordan paced his apartment. He watched "Friday Night Videos" and part of a late movie. He ate two bananas, a ham sandwich, and a stack of Oreo cookies, washed down with two glasses of chocolate milk. He read a chapter of a mystery novel. He folded the laundry. Finally, after Olivia had been asleep for almost an hour, he slipped out of the apartment and took the stairs down instead of the elevator. Once he made up his mind what he was going to do, he fairly sprinted down to West 38th Street!

He didn't have a key to the upstairs apartment, but he could get into the stockroom and try to get Emma's attention somehow! After all, he was invisible—it shouldn't be *that* hard.

Jordan turned the key, let himself in quietly, and felt a sudden chill as he realized that the stock-room light was on.

"Emma!" he whispered. "Oh God, Emma!"

The Langhorn box was in its place against the wall. Its door was shut.

Jordan felt goosebumps all over as he crept softly to the box.

"Emma. . . ." he said again and suddenly swung the door open.

Emma stood inside. Her eyes opened wide and she stared straight ahead. "Jordan?" she asked. "Is that you?"

"Emma," he breathed.

Emma stepped out of the box. She looked tired. Defeated. "You can see me," she said.

"Yeah . . ."

"I've been in here for hours," she said. "I tried everything. I said every incantation I've learned from every magician since Houdini. Since *Langhorn!* I closed the box; I opened the box. I stepped in backwards. I'd even thought maybe it worked, so I went upstairs to put away the dishes, and Polly asked me to set her hair!"

Jordan opened his arms, remembered she couldn't see him, and just pulled her close instead. She began to cry.

"I thought we could at least share it," she said,

sobbing. "At least we could be together, because we're not now . . ."

"Oh, but Emma—" Jordan was nearly crying, too. "This isn't the way to—what did you say this afternoon? 'Be a part of it'? Not like this. Look—" He took her hands and brought her to a large crate where they sat down. "Think about what my mom's going through. Think about *your* parents. Think about school, all the things that I can't do, that *you* won't be able to do. Emma, I thought being invisible would be fun, you know, be a kick. And it would be, too, if I could press a button and be visible again when I wanted to. This is a real drag, Emma. . . ."

"I know." She sniffed.

"Besides, even if you got yourself invisible like me, maybe this invisible boy wouldn't be able to see you as an invisible girl! Maybe we'd both be invisible to each other! Did you think about that?"

"No. . . ." Emma looked horrified. "I *never* thought of that!"

"Well, since there's no precedent for this, we don't know what would happen, do we?"

"No."

"So you see how crazy this was?"

Emma's answer was a deep sigh.

Jordan lifted her chin with his hands and kissed her lips.

•103

And the upstairs phone rang.

"Oh no!" Jordan cried. "My mother discovered I'm gone."

Emma had scooted past him and was on her way up the stairs. "She already discovered that," she said as he hurried. "You *are* gone."

"Well, who could that be?" Jordan said, following her quickly.

Emma pulled open the back door to the apartment. "I don't know, but it's the store extension. I want to get it before my mom and dad get up and—"

But there was Polly, standing in their bedroom doorway, and C.C. and Deirdre wearing matching blue robes. Deirdre had a nightmask pulled up on her forehead.

"What on earth—" she was saying. "Who'd be calling the store at this hour?"

"Don't forget, I'm not here," Jordan whispered to Emma as she picked up the receiver.

"Right," she whispered back. "Hello?"

"What time is it, anyway?" C.C. muttered, but Emma held up a restraining hand as she listened.

"Crank call?" Deirdre asked, but Emma just shook her head impatiently.

"Is it one of your friends?" Polly asked, but Emma ignored her. She was listening. And listening.

Finally she said, "Yes, I understand. At least partly. And thank you, thank you very much for calling. Too late? Oh no, this is the shank of the evening for us. 'Bye!" She hung up.

"The *shank* of the *evening?*" Deirdre asked with raised eyebrow.

"It was Genevre Langhorn Hughes," Emma announced. "Genevre Langhorn. Louis Langhorn's only daughter. Uncle Meissen told her to call us if she remembered anything—any time, night or day."

"Oh," Deirdre said, chewing a long fingernail. "That's right. . . ."

"And she remembered something, so she called."

Behind her, Jordan held his breath.

"Well?" C.C. said.

"She asked if 'the boy' was still gone," Emma said. "And when I said he was, she told me how bad she felt. And then she said she was going over the story she heard from Uncle Meissen, and something struck her. Jordan was chasing a stray cat into the box when he disappeared."

"So?"

"Her father had a cat. It traveled with him everywhere, never left his side. Even at night, it slept on the bed with Langhorn and his wife. Langhorn used to take it on stage with him because he had various ways of disguising it, and the audience never knew it was there. On the road it slept mostly . . . in the box."

"Ooooooh," Polly said. Her mouth formed a small round *o*.

C.C. said, "A cat, huh?"

Emma nodded. "She described it to me. She said it was very unusual-looking because it was all black except for one ear and eye, which were white."

Jordan bit his lips to keep from crying out.

"That's the cat that came through our bedroom window," Polly cried. "The night before school started . . ."

"The day I bought the Langhorn estate," C.C. said.

Deirdre rubbed her arms. "I just got a chill," she said.

"But what happened to it? Mr. Langhorn's cat, I mean?" Polly asked. "Did she say?"

"She was pretty sure it died. On the road, she thought. On tour."

"Well, it couldn't have died," Polly said, "because we saw it."

"Polly, no cat lives for over fifty years," her mother said, "no matter how well you take care of it. And if one ever did, it wouldn't be leaping gracefully into boxes, believe me."

"Of course it died, Polly," her father agreed, "but then how do you explain this carbon copy?"

Emma shook her head.

"Did you ever read the history of Salem, Massachusetts?" Polly asked in a hushed voice. "When they burned witches at the stake? And witches were supposed to have what they called 'familiars'—animals who were really witches, too, and who helped the human witch. Most of the time they were black cats! And I bet *they* could live forever!"

"Oh, Polly, *please*," her mother said.

"I think we ought to nail that box shut," C.C. said. "I really do. It's making me nervous down there. Tomorrow morning. Come on, everybody . . . let's get some sleep now."

"Okay. Let's go, Emma," Polly said and yawned.

"In a minute, Polly. You go ahead. I'm going to have some cocoa or something."

When all of the Majors had returned to their bedrooms, Emma sat at dining room table with Jordan.

"You heard all that?" she asked.

"Sure, I did," he said. "I was right here." There was a grin in his voice.

"Well, sometimes I can't tell. Anyway, what now, Jordan? What do we do?"

"I'll tell you what we do," he said. "We have to make sure your father doesn't nail up that box!"

9

○

● "Why?" Emma wanted to know. "Why shouldn't he nail it up? It's giving all of us the creeping widgies, Jordan! I know you thought you needed it as a gateway of some kind, but—"

"Right. Exactly right, Emma. I still think that box is the only hope I have of getting back. I'm finished with this invisibility stuff. I've got to get back. And that box—and somehow that cat, that magician's cat—they've got to be the key."

"It couldn't possibly be Langhorn's cat," Emma said.

Jordan sighed. "Yeah," he said, "and I couldn't possibly be invisible."

They decided that Jordan should spend the night and sleep on the couch in the living room. He wanted to be there in case Emma's father insisted on sealing the black box. Emma suggested that he call his mother before she even got out of bed the next morning to say that he had gone to Emma's earlier than usual. He just wouldn't say *how* early.

●

Major Magic usually opened at ten, but C.C. came down at nine-thirty and found Emma eating an Egg McMuffin in the stockroom. There were several more McMuffins on the crate in front of her.

"Hm. Early for you, isn't it, Emma?"

"Yes, Daddy."

"You eating all those things by yourself?"

"No, only one. Jordan's eating the rest. He's a growing boy."

"Oh! Jordan's here. I keep forgetting. Hi, kid."

"Hi, Mr. Major."

They both watched as C.C. opened the door of the Langhorn box and peered inside. "You tell Jordan about that call from the Langhorn daughter?" he asked Emma, while still staring at the black emptiness.

"Uh-huh."

"Strange . . . Say, you know, it sure *looks* harmless in there. . . ."

"Morning, Daddy. Hi, Emma," Polly said from the stairs.

"Jordan's here, too," Emma said.

"Hi, Jordan."

"Hi, Polly."

•109

"Anything new? Daddy, what are you doing to the box?"

"I want it sealed," C.C. said, still glowering at the box's dark insides, "Jordan, where are you?"

"Right here, Mr. Major, but—"

"Emma told you. About that cat."

"She did. She told me, but—"

"Well, look in there, will you? See anything? I mean, see anything that I can't see?"

"I don't see a thing, Mr. Major, but—"

"I don't either, Daddy," Polly said. She was on her stomach, gaping into the box from the floor.

"Well, then," C.C. said, "hand me that hammer on the shelf over there and that tin of nails. Let's get this thing—"

"No! *Please*, Mr. Major!" Jordan said, grabbing his arm.

"Jordan's right, Daddy, you *can't* seal the box!" Emma said.

"Why can't I?"

"Jordan feels it's his only chance to become visible again," Emma said. "Daddy—" There was a desperate quality in her voice. "He thinks he'll have to stay this way unless he finds the secret in the box that made him invisible in the first place!"

"It's true, Mr. Major. I can't tell you why I feel so sure—"

There was a sudden sharp knock on the outside door. Everyone jumped.

"Oh," Emma said, "I'll get it. Hold everything, Dad. . . ." She moved to the door and pulled it open.

An older woman, wearing a purple linen suit and very, very red hair piled on top of her head, spread out her arms when she saw Emma.

"Emma, my *dearest* child, you are getting to be absolutely *gorgeous!*" she cried and enveloped Emma in an enormous hug.

"Hello, Artemis," Emma managed, though almost all of the breath had been squeezed out of her.

"Artemis!" C.C. was beaming from ear to ear, and Deirdre called down the stairs: "Did someone say *Artemis* is here? How absolutely *marvelous!*"

Artemis the Artful was wearing enough jewelry to match Deirdre's. When they hugged each other, they clanked.

"I'd like you to meet my boyfriend, Jordan Hall," Emma said without thinking, and then gave her forehead an open-palmed slap. "Sorry," she muttered.

"Boyfriend? Boyfriend? Where is he?" the magician demanded. "I must meet Emma's boyfriend!"

"Uh, he must have just stepped out," Polly said. But she was touching the toe of Jordan's sneaker with her foot. Every time Artemis swept her purple-sleeved arms around, Jordan had to duck.

"A boyfriend of dearest Emma's I must meet," Artemis insisted. "Shall I wait for him before I show you a little trick?"

"No, let's not wait," Polly said. "Show us now, Artemis, please?"

"Well . . . perhaps I will. My favorite people are all here, so-o-o . . ." She twisted her body in an odd way for less than a second, and suddenly a beautiful pale green dove was flying above them.

"Ahhhhhh," Deirdre breathed.

"Ooooooh," Polly said, putting her finger to her mouth.

"Now, that was great, Artemis, really great," C.C. said and shook the magician's hand up and down. "No cape, no lights, no big sleeves, no big tote bag—ver-y nice. I'm not sure how you hid him all this time you've been here. . . ."

"And there was just a slight body twist to release him," Polly finished. "Very neat. And he's beautiful, too."

"Yes, isn't he," Artemis agreed. "I have two of every color of the rainbow. I want scarves to match them all. I have the most marvelous idea! And

C.C., you old devil, you know perfectly well where I hid my baby bird."

C.C. laughed knowingly. "Yes—well, it works very nicely. Artemis, very nicely."

"What is it, C.C.? I don't remember," Deirdre said.

"Sure you do. A kind of pocket made of thin silk mesh—plenty of room, won't hurt a bird's delicate wings and feathers. Then what you do is, you—"

"Daddy!" Polly shrieked, and everyone followed her horrified gaze.

Artemis's green dove was hovering above the Langhorn black box.

"Oh, no—" Deirdre said softly and put her hand to her mouth.

"What is it?" Artemis asked, unconcerned. "What's the matter with all of you?"

"Don't just stand around—someone grab it!" C.C. cried and made a lunge for the bird himself. But it was too late. Startled by C.C.'s quick movements, the dove flew through the open door.

"Sweet," Artemis said. "He thinks it's his cage. They do like to be enclosed." She held out ringed fingers and whistled softly. "Come on, sweetie . . . come to Mama . . . Mama's waiting for her sweet ba-by. . . ." She whistled again. "That's funny," she said. "He always comes when I whistle."

"Not this time," Emma sighed. She was looking at the inside of the box. It was empty.

●

"That's it, that's it!" C.C. bellowed, after they had soothed a tearful Artemis and sent her on her way with a replacement bird—though a white one. "We're sealing that box!"

Jordan sighed and C.C. heard him.

"I'm sorry, Jordan. This time it was a bird. Next time it could be another person, and that's all we need." He slammed the lid of the box closed and grabbed the hammer and box of nails.

Jordan winced as each nail was driven in, as if it were sealing his fate.

●

They all worked hard during the day: Jordan in the stockroom, continuing the sorting and cataloging of the Langhorn estate materials, and the Majors in the store.

As they were closing in the evening, Emma sensed a gloomy cloud around her.

"Jordan?"

"Hm?"

"Now *you're* feeling down."

"How can you tell?"

"The air is thicker. What is it? The sealed box?"

"I'll never get back, Emma, unless I have easy access to it. I just know it. And how long do you think we can keep me a secret, anyway? I have to live my life. What happens when someone outside of us finds out? And they will, you know. Polly—she's a kid, and she thinks this is so great. She's bound to say something."

"Polly?" Emma laughed. "She spilled everything the very next day! She blabbed her head off! She wanted to use you for 'Show and Tell,' no pun intended."

"She *did*? What happened? How come no one mentioned it?"

"There was nothing to mention. Everyone said she had a wonderful imagination but that she's been working in a magic shop too long."

Jordan shook his head. "I'm surprised she didn't bring any of her friends around to catch me in the act."

"Oh, she wanted to, but they're all nine years old and they can't run around the city by themselves. Anyway, some little boy brought a llama to school, so everyone forgot about you."

Jordan laughed feebly.

"Okay, Jordan," Emma said, "now there has to be something we've missed. Sitting around moaning and groaning is getting us rapidly nowhere."

"You're right."

"So let's think. Now. Black box, black cat," she muttered to herself. "One trick that no one can figure out . . . Black box, black box . . . Jordan, you're right—we've got to pry the box open, get the nails out. It's our only hope. Give me a hand."

●

"Are you there?" Emma asked.

"Right here."

"Okay, are you ready?"

"Ready."

Emma had the hammer and a screwdriver, and Jordan had a putty knife, and with these tools they hoped to be able to pry C.C.'s nails from the lid of the box without attracting the family's attention upstairs. Emma began at the bottom and Jordan at the top. The only noise was the occasional clink of a nail as it fell to the floor.

"One more," Jordan said as they met in the center.

"Okay. Here's the last nail." He handed it to Emma, who put it with the others on a crate. The lid of the box creaked and stood open.

"Did you pull it?" Emma whispered.

"No. Look!"

Emma looked. On the floor of the box a bit of white glimmered in the glow of the streetlight.

"Grab that flashlight, Emma, quick," Jordan said

in a hushed voice. "Hurry—I want it *bright* in there!"

Emma clicked the flashlight button and swung the beam around toward the inside of the box. They both gasped. There on the floor of the box was a cat: coal black, with one white eye and ear. Stuck at the edge of his whiskers was a pale green feather.

The cat moved quickly, but not as fast as Jordan, who dove into the box with outstretched arms. All he could grab was the black tail, and though the cat screeched and yowled, Jordan hung on.

Emma, watching the struggle as she chewed her fingernails, barely blinked, but—suddenly—there was Jordan, standing before her, holding a docile black cat curled in his arms.

"I got him!" Jordan cried. "With all that howling, he never even scratched me, can you believe it?"

Emma stared.

"Now we can try it again," Jordan was saying. "The whole experiment! Everything that happened the first time, including the cat! I know the cat had something to do with it all, so maybe if I let him go and chase him into the box again the way I did before, we can—"

"Jordan, stop!"

"Stop?"

"We don't have to recreate it. It worked backwards. I can see you."

"What?"

"You're wearing jeans and a yellow T-shirt."

Jordan looked down at himself. "Yeah. I am. How did you know?"

"I can *see* you! I said I can *see* you!" She rushed to him and hugged him. The black cat dropped silently to the floor as Jordan opened his arms to hug Emma back.

"What's going on down here!" It was C. C.'s usual question.

They looked up. C.C. and Deirdre were on the stairs. Deirdre was holding Polly's lacrosse stick.

"What's with you kids and this stockroom!" C.C. bellowed. "You, Jordan! I thought you knew better than to—" He stopped. "Hey!" he cried. "Hey!"

Deirdre was beaming. "You're back. You're *back!*"

Polly, in Doctor Denton pajamas, peered down the stairs behind her parents.

"How did you do it?" she asked. "Tell me. How?"

"Yes, by all means," Deirdre said, "tell us what you did!"

C.C. looked at the box. "It was nailed shut," he said. "I nailed it myself."

"We opened it, Dad," Emma said. "We just had

to. I told you, both of us had a feeling it was the only chance we had."

"*And?*"

"And when we did, guess what was sitting inside?"

"*What?*"

"The cat," Polly said. "The cat was inside, wasn't it?"

"That's right," Jordan said. "The black cat with—"

"One white eye and ear," Polly said almost reverently. "And then what?"

"Well, it was reflex, really," Jordan told them. "I just knew I'd better grab it, hold it while I could, or it could just disappear on me again. And that's what I did, and I pulled it out of the box. And here I am."

"Well, thank heavens!" Deirdre sank to the steps with relief.

"Where is it?" Polly asked.

"Where's what?"

"Where's the cat now? If you pulled it out of the box, where did it go?"

Deirdre stood up again and everyone looked around. Jordan dragged crates away from the wall, Emma looked behind boxes on shelves, C.C. examined the staircase. Polly ran upstairs to see if the cat had got into their apartment.

"We won't find it," Jordan said after a few minutes. "We'll never see it again."

"He's right," C.C. said. "I'm sure we'll never see it again. Jordan—" He came down the stairs with an outstretched hand. "Good to see you again."

"Thank you, Mr. Major."

"And now there's something we all have to do, and we're going to do it together."

"What, Dad?" Emma asked.

"We're going to go out in back near the dumpster and we're going to take this box apart piece by piece until we have six flat lengths of wood: three sides, one door, one top, one bottom."

"But it's valuable, sir," Jordan said. "What about the sale?"

"I think we'll exclude this piece from the sale. In the interests of—of national security."

"Okay," Jordan said, rubbing his hands together. "Let's get started. . . ." He picked up the screwdriver and hammer from the crate where Emma had put them earlier and began to pry the wooden seams apart.

The upstairs telephone shrilled loudly.

Jordan groaned. "Okay, Mom, I'm coming home," he said.

"Oh, I'll get it," Emma said, hurrying up.

Jordan continued to pull at the door of the box. "I've almost got this door off, Mr. Major—*ow!*"

"What is it?"

Jordan shook his hand, then licked his finger. "I got a bad splinter," he said, frowning at it. ". . . hurts!"

"Hey, everybody!" Emma called down. "It's Harvey! Harvey Leff!"

"Oh! Well!" C.C. smiled.

"Harvey was our shop assistant before you," Deirdre explained to Jordan and then she saw his expression. "Say, how deep is that splinter? Is it bleeding?"

"What does Harvey want?" Polly called.

"He says," Emma yelled back, "that he has really exciting news!"

"What? What is it?" Polly cried.

"He says he just pledged Chi Zeta! It's a fraternity!"

Jordan stared open-mouthed at his hand. The sharp pain was gone. So was the splinter. It had disappeared.

"And he wants to know," Emma said, bending way down over the banister, "what's happening with us?"

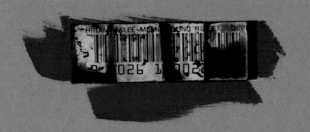